Edited by
Yuan Lin and Xin Chen

Advanced Nano Deposition Methods

Edited by Yuan Lin and Xin Chen

Advanced Nano Deposition Methods

WILEY-VCH
Verlag GmbH & Co. KGaA

Chemical Industry Press

Editors

Prof. Yuan Lin
University of Electronic Science and Technology of China
State Key Laboratory of Electronic Thin films and Integrated Devices
No. 4, Section 2, North Jianshe Road
Chengdu
Sichuan 610054
PR China

Prof. Xin Chen
East China University of Science and Technology
130 Meilong Road
200237 Shanghai
China

Cover
LAGUNA DESIGN @ Getty Images

Library of Congress Card No.: applied for

British Library Cataloguing-in-Publication Data
A catalogue record for this book is available from the British Library.

Bibliographic information published by the Deutsche Nationalbibliothek
The Deutsche Nationalbibliothek lists this publication in the Deutsche Nationalbibliografie; detailed bibliographic data are available on the Internet at <http://dnb.d-nb.de>.

Print ISBN: 978-3-527-34025-5
ePDF ISBN: 978-3-527-69645-1
ePub ISBN: 978-3-527-69647-5
Mobi ISBN: 978-3-527-69646-8
oBook ISBN: 978-3-527-69640-6

Typesetting SPi Global, Chennai, India
Printing and Binding Markono Print Media Pte Ltd., Singapore

Printed on acid-free paper

Contents

List of Contributors

Hongliang Cao
East China University of Science and Technology
School of Materials Science and Engineering
Shanghai Key Laboratory of Advanced Polymeric Materials
Key Laboratory for Ultrafine Materials of Ministry of Education
130 Meilong Road
Shanghai 200237
PR China

Chonglin Chen
University of Texas at San Antonio
Department of Physics and Astronomy
One UTSA Circle
San Antonio, TX 78249
USA

Xin Chen
East China University of Science and Technology
School of Materials Science and Engineering
Shanghai Key Laboratory of Advanced Polymeric Materials
Key Laboratory for Ultrafine Materials of Ministry of Education
130 Meilong Road
Shanghai 200237
PR China

and

Chinese Academy of Sciences
Shanghai Institute of Microsystem and Information Technology
State Key Laboratory of Functional Materials for Informatics
865 Changning Road
Shanghai 200050
PR China

Rabi Ebrahim
University of Houston
Center for Advanced Materials
Houston, TX 77204-5504
USA

Daniel Fisher
University of Houston
Center for Advanced Materials
Houston, TX 77204-5504
USA

Min Gao
University of Electronic Science
and Technology of China
State Key Laboratory of
Electronic Thin Films and
Integrated Devices
No. 4, Section 2, North Jianshe
Road
Chengdu
Sichuan 610054
PR China

Wen Huang
University of Electronic Science
and Technology of China
School of Microelectronics and
Solid-State Electronics
No. 4, Section 2, North Jianshe
Road
Chengdu
Sichuan 610054
PR China

Alex Ignatiev
University of Houston
Center for Advanced Materials
Houston, TX 77204-5004
USA

Yanda Ji
University of Electronic Science
and Technology of China
State Key Laboratory of
Electronic Thin films and
Integrated Devices
No. 4, Section 2, North Jianshe
Road
Chengdu
Sichuan 610054
PR China

Yuan Lin
University of Electronic Science
and Technology of China
State Key Laboratory of
Electronic Thin films and
Integrated Devices
No. 4, Section 2, North Jianshe
Road
Chengdu
Sichuan 610054
PR China

Chunrui Ma
Xi'an Jiaotong University
School of Materials Science and
Engineering
State Key Laboratory for
Mechanical Behavior of
Materials
No. 28, Xianning West Road
Xi'an
Shanxi 710049
PR China

Taisong Pan
University of Electronic Science and Technology of China
State Key Laboratory of Electronic Thin films and Integrated Devices
No. 4, Section 2, North Jianshe Road
Chengdu
Sichuan 610054
PR China

Goran Rasic
North Carolina Central University
Department of Physics
1801 Fayetteville Street
Durham, NC 27707
USA

Zhitang Song
Chinese Academy of Sciences
Shanghai Institute of Microsystem and Information Technology
State Key Laboratory of Functional Materials for Informatics
865 Changning Road
Shanghai
PR China

Serekbol Tokmoldin
Institute of Physics and Technology
11 Ibragimov Street
Almaty 050032
Kazakhstan

Zhongping Wang
University of Science and Technology of China
School of Physical Sciences
Center for Physics Experiments
No. 96 Jinzhai Road
Hefei
Anhui 230026
PR China

Liangcai Wu
Chinese Academy of Sciences
Shanghai Institute of Microsystem and Information Technology
State Key Laboratory of Functional Materials for Informatics
865 Changning Road
Shanghai
PR China

Naijuan Wu
University of Houston
Center for Advanced Materials
Houston, TX 77204-5504
USA

Debiao Xie
East China University of Science and Technology
School of Materials Science and Engineering
Shanghai Key Laboratory of Advanced Polymeric Materials
Key Laboratory for Ultrafine Materials of Ministry of Education
130 Meilong Road
Shanghai 200237
PR China

Weiqing Yang
Southwest Jiaotong University
School of Materials Science and Engineering
Key Laboratory of Advanced Technologies of Materials (Ministry of Education)
Ring North of a Section 111
Chengdu
Sichuan 610031
PR China

Mukhtar Yeleuov
University of Houston
Center for Advanced Materials
Houston, TX 77204-5504
USA

Hulin Zhang
University of Electronic Science and Technology of China
School of Microelectronics and Solid-State Electronics
No. 4, Section 2, North Jianshe Road
Chengdu
Sichuan 610054
PR China

Zengming Zhang
University of Science and Technology of China
School of Physical Sciences
Center for Physics Experiments
No. 96 Jinzhai Road
Hefei
Anhui 230026
PR China

Wangfan Zhou
East China University of Science and Technology
School of Materials Science and Engineering
Shanghai Key Laboratory of Advanced Polymeric Materials
Key Laboratory for Ultrafine Materials of Ministry of Education
130 Meilong Road
Shanghai 200237
PR China

1
Pulsed Laser Deposition for Complex Oxide Thin Film and Nanostructure

Chunrui Ma and Chonglin Chen

1.1
Introduction

Complex oxide thin films and nanostructures are at the heart of new "oxide electronic" applications, such as ultraviolet light-emitting diodes [1–3], resistive switching memories [4, 5], chemical sensor [6, 7], and so on. They are often grown by pulsed laser deposition (PLD) because the technique is believed to be material agnostic. PLD is a thin film deposition technique – a type of physical vapor deposition. A high-power pulsed laser beam is focused on and strikes a target of the material that is to be deposited in a vacuum chamber. This material is vaporized from the target in a plasma plume and deposited as a thin film on a substrate. This process can occur in ultrahigh vacuum or in the presence of a background gas, such as oxygen, which is commonly used when depositing complex oxides.

The synthesis of novel thin films and structures is advancing on two fronts: one is the complexity of materials being deposited; and the other is the reduction in the typical dimensions of the features. As a rule of thumb, any structure that has one or more dimension smaller than about 100 nm is considered to be a nanostructured material. PLD is one of the most promising techniques for the formation of complex oxide heterostructures and nanostructures. The basic setup of PLD is simple relative to many other deposition techniques, and it can stoichiometrically transfer a material from a solid source to a substrate to form its thin film. The first use of PLD to deposit the films of semiconductors and dielectrics by ruby laser is reported in the literature as early as 1965 [8]. PLD for the film growth of $SrTiO_3$ and $BaTiO_3$ was achieved in 1969 [9]. Six years later, stoichiometric intermetallic materials (Ni_3Mn and $ReBe_{22}$) were fabricated by using PLD [10]. In 1987, PLD had a real breakthrough in its successful application to the *in situ* growth of epitaxial high-temperature superconductor films at Bell Communications Research [11]. Since then, PLD has been used extensively in the growth of high-temperature cuprates and numerous other complex oxides, including materials that cannot be obtained by an equilibrium route [12–16].

Advanced Nano Deposition Methods, First Edition. Edited by Yuan Lin and Xin Chen.

This chapter details the PLD setup and focuses primarily on the operating principle, growth mechanism, and parameters of PLD for complex oxide thin film and nanostructure.

1.2
Pulsed Laser Deposition System Setup

The technique of PLD is conceptually simple, as illustrated schematically in Figure 1.1. The system consists of a laser, a vacuum chamber equipped with pumps, a target holder and rotator, and a substrate heater and is typically equipped with various pressure gauges, controllers, and other instruments to control the deposition environment of the system [17]. Film growth can be carried out in reactive environments, such as that for oxides where a partial pressure of oxygen, ozone, or atomic oxygen is carefully controlled. The substrate heater controls the substrate temperature. PLD systems are also often equipped with a set of optics including apertures, attenuators, mirrors, and lenses to focus and direct the laser beam into the target with the right energy density. The bulk material target orients at an angle of 45° toward the incident laser beam. The laser beam is focused onto the target surface by a set of optical components. The target locally absorbs the laser pulse energy and ejects a small amount of target material in the form of a plume containing many energetic species including atoms, molecules, electrons, ions, clusters, particulates, and molten globules. The plasma is then deposited onto a substrate facing the target with a separation distance of 3–5 cm. The substrate temperature can be varied from room temperature to 1000 °C, even higher than 1000 °C, depending on the heater type. The film microstructure depends on various parameters such as substrate

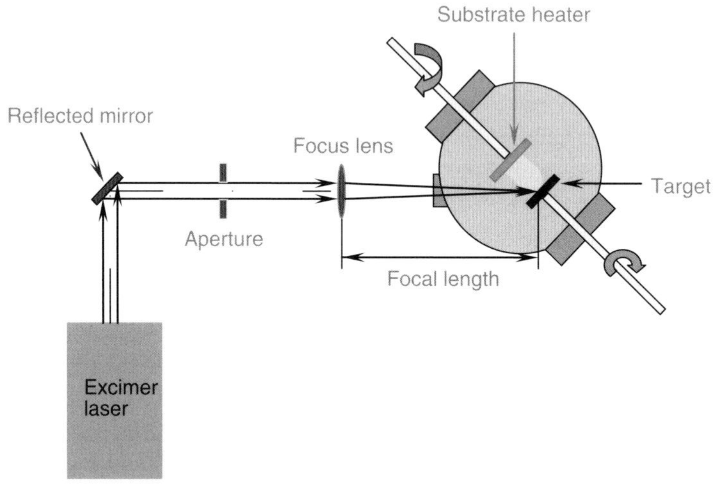

Figure 1.1 Schematic diagram of typical pulsed laser deposition.

temperature; laser energy density and pulse repetition rate; pressure and type of gas inside the chamber; and substrate-to-target distance.

1.3 Advantages and Disadvantages of Pulsed Laser Deposition

PLD exhibits many fascinating properties and practical advantages. Firstly, it has the ability to faithfully keep the stoichiometry of the target material, which is the first aspect that draws the attention of the thin film growth community [18]. Secondly, the energy source for material transport (i.e., the laser) is outside the chamber, minimizing any impurities caused by in-vacuum power components; it is very flexible, cost effective, and fast. Many different materials can be ablated by using the same apparatus, and the different laser wavelengths are available in principle. The isolated local heating by the laser spot means that several different materials can be sequentially ablated in a single vacuum chamber by using a carousel system or a segmented target rod to fabricate heterostructures with little of cross-contamination of the source target material. This avoids the interconnected vacuum transfer and is an important advantage in research environment: one laser can serve many vacuum systems in order to save the laser cost, and high-quality samples can be grown in 10 or 20 min. Finally, it is easy to control film thickness and multilayer film by controlling the pulse repetition rate, growth time, and the use of multiple target holders; it demands a much lower substrate temperature than other film deposition techniques because the high kinetic energy (10–100 eV) of species in the ablation plume promotes surface mobility during film growth.

In spite of the above-mentioned advantages of PLD, there are some drawbacks in using the PLD technique. One of the major problems is limited uniformity because the plasma plume ejected from the target can only provide a narrow forward angular distribution. Another problem is high defect or particulate concentration due to surface boiling. The size of particulates may be as large as a few micrometers, which will greatly affect the growth of the subsequent layers as well as the electrical properties of the films. Therefore, these features limit the large-scale film growth. New techniques, such as rotating both target and substrate and using a shadow mask to block the particulates in order to fabricate a large and uniform film, have been developed to improve the film quality.

1.4 The Thermodynamics and Kinetics of Pulsed Laser Deposition

PLD is a nonequilibrium growth technique owing to the high electronic excitation, degree of ionization, and kinetic energies of flux. There are many distinct stages to film growth: [19] the ablation process of the target material by the laser irradiation; the creation of a plasma plume with high energetic ions; and the crystalline growth of the film itself on the heated substrate. In this section, we will thoroughly describe these processes.

1.4.1 Laser–Material Interactions

After the laser pulse is extinguished, a very hot cloud of vaporized material, typically of 10^4 K or more, has been generated, which is commonly referred to as the *ablation plasma* or *plume*. This process is called *laser ablation*. The mechanisms depend on the laser characteristics as well as on the optical, topological, and thermodynamic properties of the target material. Absorption in a material is defined as

$$I = I_0 \, e^{-\alpha x}$$

where $1/\alpha$ is the absorption length, which is approximately 100 nm for many oxide materials at laser wavelengths commonly used in PLD (<400 nm). In this process, electrons in the target are excited and thermalized within several picoseconds or nanoseconds depending on the energy density, duration, wavelength, and shape of the laser pulse as well as on the material properties (reflectivity, absorption coefficient, heat capacity, thermal conductivity, density, etc.). The next step includes surface melting of the target and conduction of heat into the target. The thermal diffusion length is described as

$$\lambda_{\text{th}} = 2(\alpha_{\text{th}} \, \Delta t)^{1/2}$$

where $\alpha_{\text{th}} = \frac{K}{\rho c}$ is the thermal diffusivity, K is the thermal conductivity, ρ is the mass density, c is the specific heat, and Δt is the pulse duration. During this process, the temperature rises in the surface of the target. The heating rates as high as 10^{11} K/s and instantaneous gas pressures of 10–500 atm are observed at the target surface. Then, the target material will vaporize. During this step, there is multiphoton ionization of the gaseous phase creating the characteristic plasma and the temperature at the surface of the target will exceed the boiling point. The final step of the process is the plasma excitation during which further ionization occurs and free electrons are excited, resulting in Bremsstrahlung absorption in which the hot pulse, at nearly 2000 K, expands in a directed manner.

The ablation threshold of materials, or the minimum energy density required in a material to create a plume, will be discussed. In most oxides, the thermal diffusion length is much longer than the absorption length, especially for UV lasers, because of the fact that oxide materials are often opaque and good thermal conductors. An affected volume is related to the spot size times λ_{th}; thus, a simple estimation of the minimum energy needed to raise this volume to the sublimation point is

$$Q_{\text{heat}} = C_S(T_{\text{melt}} - T) + \Delta H_m + C_m(T_{\text{vap}} - T_{\text{melt}}) + \Delta H_{\text{vap}}$$

where the total energy required (from left to right) is the sum of the energy needed to bring the target material to the melting temperature plus the heat of melting, plus the energy needed to bring the melted material to the vaporization point, plus the heat of vaporization. Although this seems like a large amount of energy, a typical instantaneous power density for a single laser pulse with energy density of approximately 2 J/cm^2 and 20 ns pulse duration is about 10^8 W/cm^2, which is more than enough to ablate nearly all materials.

1.4.2 Dynamics of the Plasma

The material expands in a plasma form parallel to the normal vector of the target surface toward the substrate owing to Coulomb repulsion and recoil from the target surface. The plasma contains many energetic species such as atoms, molecules, electrons, ions, clusters, particulates, and molten globules with the typical plasma temperature around 10 000 K, above the boiling points of most materials. As the plasma expands, it adiabatically cools to 3000–5000 K. The spatial distribution of the plume is strongly dependent on the background pressure inside the PLD chamber. In vacuum, the plume is very narrow and forward directed; at intermediate pressure, it looks like a shock wave due to the splitting of high energetic ions from the less energetic ions; at high pressure, it undergoes a more diffusion-like expansion. The angular distribution of the plume has been fitted to a $\cos^n(x)$ function, with the value of n ranging from 2 to more than 20. In general, n is lower when the plasma is at high pressure because multiple collisions broaden the angular distribution [20]. The propagation of the plasma in the background gas follows a distance–time relation described by a blast wave model:

$$R = \zeta_0 \left(\frac{E_0}{\rho_0}\right)^{1/5} t^{2/5}$$

where ζ_0 is a constant, ρ_0 is the background gas density, and E_0 is the kinetic energy. This model is valid when the mass of the ejected material is smaller than the mass of the background gas. Therefore, the background gas pressure in PLD must be kept at a very low level so that some of the ejected species can avoid being scattered by the background gas and reach the substrate successfully.

1.4.3 Nucleation and Growth of the Film on the Substrate Surface

Nucleation and growth of the film are of importance to determine thin film quality, morphology, and stoichiometry [21]. Growth of thin films resulting from a supersaturated gas condensation is a nonequilibrium phenomenon governed by a competition between kinetics and thermodynamics. Nucleation only occurs when the nucleus size r is equal to or greater than the critical nucleus r^*, as shown in Figure 1.2. It is worth noting that ΔG^* represents a critical energy barrier in the nucleation process. When an aggregate momentarily forms owing to thermodynamic fluctuation but the size of the cluster is smaller than r^*, it is unstable and will disappear by shrinking in size, lowering ΔG in the process. Only clusters with

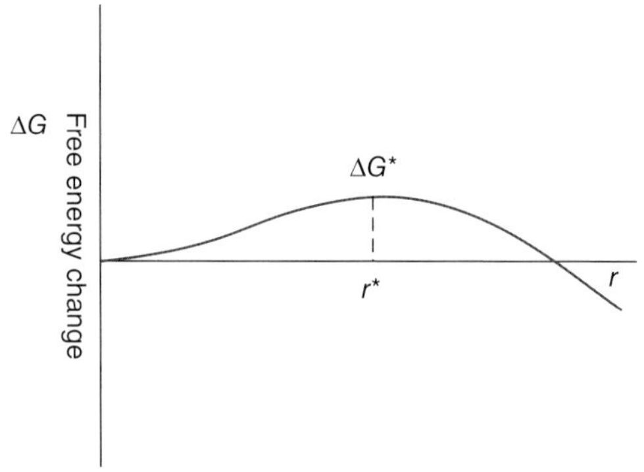

Figure 1.2 Free energy as a function of cluster size r. r^* is the critical nucleus size and ΔG^* is the critical free energy barrier for nucleation.

size larger than r^* can surmount the nucleation energy barrier, and they are stable and tend to grow larger in order to lower the energy of the system.

Kinetic processes in thin film nucleation on an ideal substrate surface [21, 22] are the impinging of the ablated high energetic species from the target onto the substrate surface and the sputtering off of some substrate atoms. These adatoms can subsequently diffuse over the substrate or along the island edge, encounter and bind with each other to form clusters, or attach to an existing island. On the other hand, they can also be re-evaporated from the substrate, from an island, or from a cluster. Meanwhile, it is possible that they are detached from the existing cluster edge but still remain on the substrate surface. In fact, the crystalline surface is not ideally flat but contains lots of defects and step terraces, so other kinetic processes such as interdiffusion and interfacial reactions become significant.

There are several factors that affect the film growth structure. First of all, the nucleation rate is an important parameter: high nucleation rates during deposition will encourage a fine-grained film or even amorphous structure. On the other hand, if nucleation is suppressed, single crystal growth is fostered. Substrate temperature is another factor to influence the film growth. Combining these two parameters, it is concluded that high substrate temperature and low deposition rates favor large crystallites or even monocrystal formation. Alternatively, low substrate temperatures and high deposition rates yield polycrystalline deposits. Basically, there are three thin film growth modes [22, 23]:

1. *Island or Volmer–Weber mode*: As shown in Figure 1.3a, island growth occurs when the smallest stable clusters nucleate on the substrate and grow in three dimensions to form islands. The growth of three-dimensional (3D) islands on the substrate is due to the fact that the bonds between the film atoms are stronger than the binding force of atoms to the substrate.
2. *Layer-by-layer or Frank–van der Merwe mode*: This kind of thin film growth mode (Figure 1.3b) occurs when the atoms are more strongly bound to the

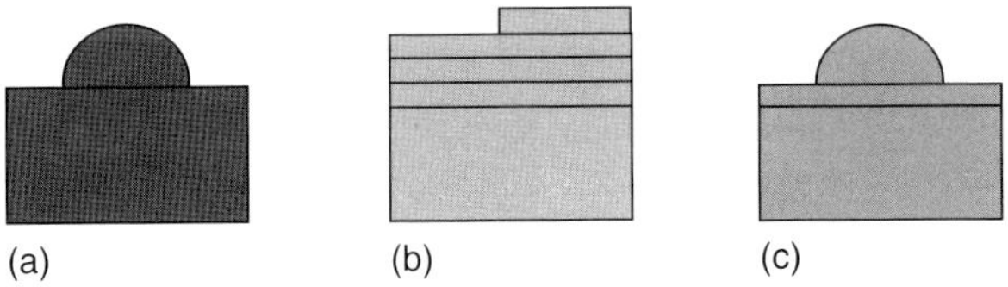

Figure 1.3 Basic modes of thin film growth: (a) island or Volmer–Weber mode, (b) layer-by-layer or Frank–van der Merwe mode, and (c) layer plus island or Stranski–Krastanov mode.

substrate than to each other. Film atoms are completely coalescent to form a monolayer on the surface before they develop into significant clusters on the next layer owing to the no barrier nucleation on this 2D island formation [24, 25]. As long as the decrease in bonding energy is continuous toward the bulk crystal value, the layer growth mode is sustained.

3. *Layer plus island or Stranski–Krastanov mode* (Figure 1.3c): The layer plus island mode follows a two-step process: after forming one or more monolayers, namely, above a critical layer thickness, the layer-by-layer growth mode becomes 3D island forms. The transition mechanism from layer growth mode to island growth mode is unclear until now, but any factor that disturbs the monotonic decrease in binding energy may be the cause, such as the strain from the misfit between the thin film and the substrate, dislocation in the film, chemical potential of the deposited film, and so on.

It is known that the thin film growth mode also depends on the interfacial energies between the three phases present: the substrate, the condensing material, and the vapor [24].

$$\sigma_s < \sigma_{hkl} + \sigma_i - \frac{kT \ln s}{2k_{hkl}\, b^2} \quad \text{(island mode)}$$

$$\sigma_s > \sigma_{hkl} + \sigma_i - \frac{kT \ln s}{2k_{hkl} b^2} \quad \text{(layer mode)}$$

where σ_s and σ_i are the free energy of the substrate surface and interface, respectively, σ_{hkl} is the specific surface energy of the contact plane of the deposit, and b is the contact length.

In PLD systems, the nucleation process and growth kinetics of the thin film depend on several growth parameters: laser parameters (laser energy, pulse repetition rate), ionization degree of the ablated material, substrate temperature, substrate surface, and background pressure. Among them, the substrate temperature T and the supersaturation s are very important, defined by

$$s = \frac{R}{R_e}$$

where R is the actual deposition rate and R_e is the equilibrium deposition value at temperature T. A small supersaturation induces large nuclei, leading to island thin film growth mode. As the supersaturation increases, the critical nucleus keeps shrinking to the atomic level, and then the impinging adatoms will be more likely to add directly to the edges of the existing clusters before they encounter other adatoms. Therefore, the high supersaturation value tends to promote the layer thin film mode [19, 26].

In addition to supersaturation, the substrate temperature also plays an important role in controlling the final microstructure and surface morphology of the as-grown film. At high temperatures, defects at the surface such as void, atomic step intersections, and dislocation intersections can be reduced and the crystallinity of the as-grown films can be improved. However, with the decrease in the substrate temperature, the surface diffusion coefficient decreases and the crystal structure formation slows down, resulting in disordered, or even amorphous, thin film structures.

1.5 Monitoring of Growth Kinetics

1.5.1 Introduction and RHEED Studies

In modern material science, the knowledge and control of the growth processes of oxide materials are important for the fabrication of high-quality epitaxial and heteroepitaxial thin film structures. The quality of a thin film is the key factor to determine its properties and application; thus, monitoring the film growth kinetics becomes the critical issue and has attracted much attention. In this section, we will briefly overview some methods to detect the film growth kinetics.

Reflection high-energy electron diffraction (RHEED) is a technique used to characterize the surface of crystalline materials. A RHEED system basically requires an electron source (gun), photoluminescent detector screen, and a sample with a clean surface. The electron gun generates a beam of electrons, which strikes the sample at a very small angle relative to the sample surface. Incident electrons are diffracted by atoms at the surface of the sample, and then a small fraction of the diffracted electrons interfere constructively at specific angles and form regular patterns on the detector. The diffraction pattern at the detector is a function of the sample surface because the electron interference is related to the position of atoms on the sample surface. A PLD system equipped with a RHEED system can monitor the growth of thin films. The intensities of individual spots on the RHEED pattern fluctuate in a periodic manner resulting from the relative surface coverage of the growing thin film. A simple model similar in shape to those used by film growth experts is shown in Figure 1.4. Each peak represents the formation of a new monolayer since the degree of order is at a maximum once a new monolayer has been formed. The overall intensity of the oscillations drops when more layers are grown because of the fact that the electron beam is focused on the original surface and gets out of focus as more layers are grown. Each full period corresponds to the formation of a single atomic

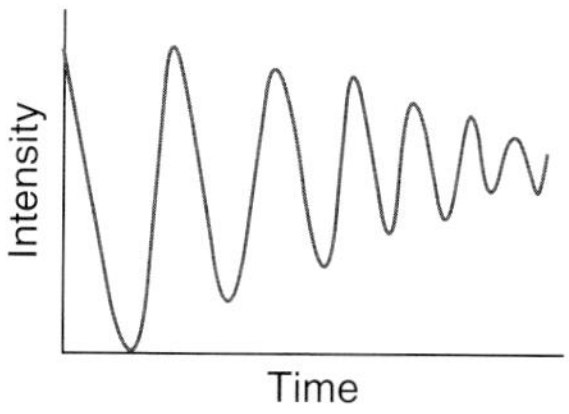

Figure 1.4 The curve is a simple model of the fluctuation of the intensity of a single RHEED point during PLD.

layer thin film. The oscillation period is highly dependent on the material system, electron energy, and incident angle. Thus, RHEED has been widely applied to monitor the growth of complex oxide thin films, especially superlattices such as $SrTiO_3$ [27], $YBa_2Cu_3O_{7-x}$ [28], and superlattice $LaAlO_3/SrTiO_3$ [29].

1.5.2
Growth Kinetics Studies by Surface *X*-ray Diffraction

Although RHEED is a tool for growth kinetics studies, the strong interaction of the electrons with the surface causes multiple scattering and requires a dynamic theory for rigorous interpretation of the intensities. As an alternative technique, surface X-ray diffraction (SXRD) has the unique advantage that kinematic scattering is applicable and the intensity can be interpreted directly by surface coverage [30, 31]. Time-resolved SXRD at crystal truncation rod (CTR) positions allows real-time measurements of interface layer formation; thus the combination of SXRD and PLD is a powerful technique for gathering unique information on interface formation and growth kinetics. In addition, X-rays can avoid surface charging, making it the most suitable technique for studying oxide surfaces and interfaces. Unlike electrons, X-rays are not scattered by the high-pressure background, which is necessary in oxide growth, and do not interact with the surface or alter the growing film. Taking $SrTiO_3$ thin film as an example, the scattered intensity is measured simultaneously at the specular $(00\frac{1}{2})$ and off-specular $(01\frac{1}{2})$ CTRs before, during, and after PLD growth of $SrTiO_3$. Measuring both rods is significant and can get the information about deposition, such as the height distribution of material, since the specular rod has momentum transfer along the surface normal. The lateral ordering on the surface and in-plane registry with the lattice that is synonymous with crystal growth are confirmed by measuring an off-specular rod $(h,k) \neq (0,0)$, which has an in-plane momentum transfer component. Well-developed and persistent RHEED-like SXRD growth oscillations are observed simultaneously at both specular and off-specular CTR positions during homoepitaxial growth of $SrTiO_3$ at temperatures from 310 to 780 °C. The time resolution of 10 μs in SXRD has the ability to measure the crystalline layer formation on the same timescale as the plume arrival time and reveals new details, which advance our understanding and change the transitional view of how PLD works.

1.6 Fundamental Parameters in Thin Film Growth

It is well known that many parameters, such as substrate temperature, background gas, pre- and post-treatment, as well as the lattice misfit between the film and the substrate, can significantly influence the microstructure, defect density, epitaxial quality, and interface/surface morphology of the as-grown films. However, the optimum deposition conditions for excellent highly epitaxial thin films and the desired physical properties are still semi-empirical and require intense investigations. Therefore, in this section, we will mainly discuss the influence of substrate temperature, background gas, laser energy, target–substrate distance, post-annealing, and lattice misfit on thin-film quality.

1.6.1 Substrate Temperature

Substrate temperature greatly determines the crystal structure, orientation, stoichiometric composition, and morphology of the as-grown film [32]. For a highly epitaxial single-crystalline phase, the deposition temperature should be in epitaxial temperature zone III, which is usually when the ratio of the deposition temperature to the melting point of the deposited material is above 0.5. High deposition temperatures can increase the surface mobility of the absorbed atoms and the crystallization rate, thus benefiting the epitaxial film growth. However, if substrate temperature is too high, the interdiffusion reaction between film and substrate and the substrate reconstruction may induce more distortion, thus reducing single crystallinity of the film. Sometimes, even within the optimal growth temperature range, different substrate temperatures can result in different crystal structures and film orientations. Therefore, suitable substrate temperature is an extremely important factor to determine the quality of the film. In addition, Sambri *et al.* reported that substrate temperature can significantly influence the plume dynamics of $La_{0.7}Sr_{0.3}MnO_3$ and $SrTiO_3$ and observed that a considerable reduction of the background gas resistance to plume propagation as the substrate temperature was increased, leading to a remarkable change in the velocity of species impacting the substrate during film growth and in their stopping distance. At the same time, they argued that the effect can be safely expected to be universal [33].

1.6.2 Background Gas Pressure

Energetic species can be deactivated or quenched by collisional processes with a background gas [34]. The ideal range of energies of depositing species for promoting surface diffusion on the growing film is between about 10 and 40 eV. Although PLD produce a deposition flux just in this energy region, it tends to have high-energy tails that can result in implantation and produce significant disruption of film crystallinity. Therefore, introducing a background gas (especially oxygen gas for oxide thin film) in PLD makes the entire kinetic energy distribution shift to lower values, which is a critical factor in controlling the stoichiometry and stabilizing the desired crystal phase in thin film growth [34]. Firstly, the background gas affects all the plume species and reduces their effectiveness in promoting surface diffusion. Secondly, the background gas pressure increases the deposition rate, especially for species such as Ba, Sr, Pb, and Ag, which show large re-sputtering yields, but decreases other species' deposition rate because the plume is strongly scattered by the gas and becomes less directed [35, 36]. Thirdly, too low oxygen pressures can lead to stress, usually compressive, in the film [37], caused by the bombardment of the film during the deposition by highly energetic species originating from the plume. High oxygen pressures, however, need higher deposition temperatures and, therefore, improved crystallinity is expected. Finally, an oxygen background is needed to ensure stoichiometric transfer from the target to the film. If the partial pressure of oxygen is too low, the film will grow off stoichiometry, which will affect the nucleation density and film quality. On the other hand, if the applied oxygen pressure is too high, the velocities of various species in the plume will be slowed down due to collisions with the oxygen molecules. Therefore, optimal oxygen pressure is necessary for getting a uniform and stoichiometric thin film.

1.6.3 Laser Fluence and Ablation Area

Laser fluence and ablation area are critical factors to control the stoichiometry. Inadequacy of either the laser fluence or the ablation area (the beam spot area) results in nonstoichiometric transfer of material from the ablation target [38]. Therefore, proper simultaneous choice of ablation laser fluence and ablation area is essential and often more important than the growth temperature and oxygen pressure for obtaining bulk-like properties in oxide heterostructures [39]. Ngom *et al.* reported the photographic imaging of plasma generated by KrF excimer laser ablation of $Sm_{0.55}Nd_{0.45}NiO_3$ and investigated the effect of plasma dynamics on the growth of $Sm_{0.55}Nd_{0.45}NiO_3$ solid solution during PLD [40]. The increase in the laser fluence can enhance the total optical emission of ZnO thin film [41]. Also, the ablation laser fluence has a very strong effect on the lattice constant and defect structure of the films, and 0.3 J/cm^2 is the critical laser fluence, not threshold value, for stoichiometric $SrTiO_3$ film growth [38].

1.6.4 Target–Substrate Distance

For most materials, the most important deposition parameters are the ambient O_2 pressure (P) and the substrate temperature (Ts). It has been shown that the equilibrium P–Ts phase diagram is a good guide for fabricating the *in situ* structural phase of the deposited oxides [42]. One of the major problems in PLD technology is spatial uniformity of the deposited films, which is related to a peaked angular distribution of plasma arriving on the substrate. At a characteristic target–substrate distance (D), the plasma loses its unidirectional velocity, scatters, and thermalizes, and this distance provides the optimum conditions for film growth. Several works have investigated experimentally and theoretically the relationship between P and D [43], which show that the P–D correlation is more stringent than the P–Ts requirement, since a slight deviation can cause a significant change in the film quality. A scaling law $PD^3 = \text{constant}$ from thermodynamic data and the experimental volumetric erosion rate of the metallic elements ejected from the target was proposed in 2005 and successfully applied to the growth of $CdTeO_3$, $AlVO_4$, and $PbFe_{12}O_{19}$ [44].

1.6.5 Post-Annealing

Post-annealing has a great effect on the microstructure and local structure of films and is critical in order to obtain high-quality films [45]. ZnO thin films were post-growth annealed in air at various annealing temperatures for 30 min. Increasing the annealing temperatures favors the diffusion of atoms absorbed on the substrate and accelerates the migration of atoms to an energy-favorable position, resulting in enhancement of the crystallinity and c-axis orientation of ZnO films [46].

It is commonly observed that annealing in vacuum or a reducing atmosphere results in oxygen vacancies in complex oxide thin films and enlarges the lattice constant [47]. On the other hand, annealing in oxygen atmosphere will insert the oxygen into the film and reduce the lattice constant [48]. Hundley *et al.* reported that after the $La_{0.7}Ca_{0.3}MnO_3$ thin film was fabricated, it was annealed in flowing oxygen at 950 °C for 10 h, which raised the Curie temperature, dropped resistivity by two or three orders of magnitude, and sharpened the magnetic signature of the ferromagnetic transition [49]. However, transition metal-doped ZnO thin film annealed in vacuum or a reducing atmosphere results in enhanced ferromagnetism of the sample due to the introduction of oxygen vacancies, with interstitial Zn critical for room temperature ferromagnetism in transition metal-doped ZnO [45].

1.6.6 Lattice Misfit

When thin films are grown on a single crystal substrate, two types of epitaxy can be distinguished and each has important scientific and technological implications [32]. Homoepitaxy refers to the case where the film and substrate are of the same material, such as epitaxial Si deposited on Si wafers. Heteroepitaxy is the second type and refers to the case where films and substrates are of different materials, which have a different crystal structure and lattice constant. An important quantity to characterize heteroepitaxy is the lattice misfit f defined as

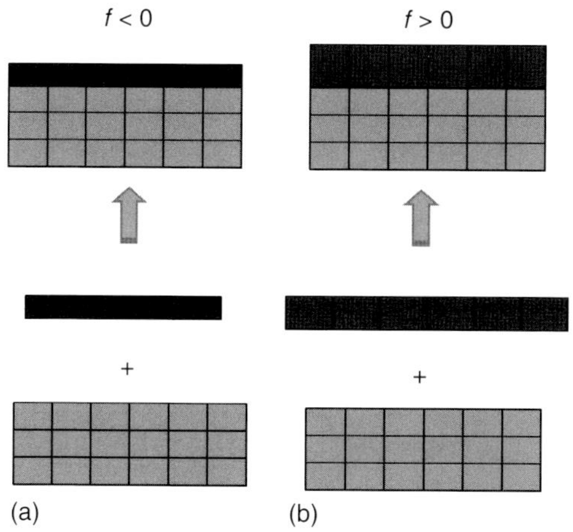

Figure 1.5 Schematic representation: (a) the mismatch between the film and the substrate is negative; (b) the mismatch is positive.

$$f = \frac{a_f - a_s}{a_s}$$

a_f and a_s are the unstrained lattice constants of the film and the substrate, respectively. A negative f (lattice constant of the film smaller than that of the substrate) implies that the initial layers of the epitaxial film will be stretched in tension (Figure 1.5a), and a positive f (lattice constant of the film larger than that of the substrate) means the film is under compression (Figure 1.5b).

Differences in lattice constant and thermal expansion behavior between the film and the underlying substrate will be imparted into strain in thin films [50], which will influence the physical properties and quality of thin films. A good lattice match between the substrate and the film can minimize the interface energy, especially the electrostatic repulsion potential, thus making epitaxial film growth occur easily. Usually, a small lattice mismatch value of less than 5% is required for epitaxial film growth. The small lattice mismatch correspondingly ensures small strain energy at the interface and leaves the film homogeneously strained and commensurate with the substrate.

1.7 Pulsed Laser Deposition for Complex Oxide Thin Film Growth

PLD has been widely used to fabricate complex oxide thin films because of the above-mentioned advantages. In this section, we will focus on the PLD for complex oxide thin film growth, such as ferroelectric, ferromagnetic, and other functional oxide thin films. Also, we will discuss the film thickness, lattice mismatch, and post-annealing effect on the properties of complex oxide thin films.

1.7.1 Pulsed Laser Deposition for Superconductor Thin Film

Since 1987, PLD has been extensively applied to the *in situ* growth of epitaxial high-temperature superconductor films [11, 51–54]. In 1996, Norton *et al.* reported that *c*-axis-oriented $YBa_2Cu_3O_7$ films with superconducting critical current density as high as 700 000 A/cm^2 at 77 K were fabricated on thermomechanically rolled textured nickel (001) tapes by PLD, which is comparable to those grown on single crystal oxide substrates. This result represents a viable approach for the production of long superconducting tapes for high-current, high-field application at 77 K by PLD [55]. Later on, PLD is widely used to fabricate various superconductor thin films and improve the superconductor properties by interface strain, such as copper oxides, $La_{2-x}Sr_xCuO_4$ [56], $SmBa_2Cu_3O_y$ [57], and so on.

1.7.2 Pulsed Laser Deposition for Ferroelectric Thin Films

Ferroelectric oxide thin films, especially those grown by PLD, have fascinated scientists and engineers because of their interesting physical properties and important application in various areas, such as ferroelectric memories, ferroelectric field effect transistors, optical, microwave devices, and so on [58–60]. Lead zirconate titanate (PZT) has long been the leading material considered for ferroelectric memories, though strontium bismuth tantalite (SBT) is also a popular choice because of its superior fatigue resistance and the fact that it is lead free [61].

Ferroelectric thin film as the gate of ferroelectric field effect transistors has been widely investigated and reviewed [58, 62, 63]. A ferroelectric transparent thin film transistor was fabricated by PLD, consisting of a high-mobility, Sb-doped, n-type SnO_2 semiconductor layer: $PbZr_{0.2}Ti_{0.8}O_3$ as a ferroelectric insulator and $SrRuO_3$ as a gate electrode [62]. It is found that a pulse of small voltage (3 V) can induce a change in the remnant conductance by a factor of 2. Also, epitaxial cuprate superconductor ferroelectric heterostructures were investigated by PLD [64].

In the past two decades, ferroelectric barium strontium titanate ($Ba_xSr_{1-x}TiO_3$) thin films have been considered to be one of the most important candidates for microwave elements because the dielectric constant can be tuned by an external field and the Curie temperature can be adjusted by varying the solution concentration between barium titanate and strontium titanate. Chen *et al.* recently systematically investigated the microwave properties of epitaxial BST or doped BST by fabricating single-layered or multilayered ferroelectric thin films. The tenability can reach 62% at 2G frequency by interface strain from the substrate, which is the maximum value reported in the literature [65–70].

It is known that the interface strain can significantly influence the properties of thin films, especially the perovskite structure's ferroelectric thin film. Predictions and measurements of the effect of biaxial strain on the properties of epitaxial ferroelectric thin films and superlattices were reviewed in 2007 [71].

1.7.3 Pulsed Laser Deposition for Ferromagnetic Thin Film

Ferromagnetic oxide thin films fabricated by PLD, such as perovskite manganites, cobaltates, and ion oxides, have been intensively investigated worldwide over the past several decades because they span a wide range of crystal structure and exhibit an incredible variety of physical phenomena and spectacular functionality, for example, magnetic storage, sensors and activators, spintronic and colossal magnetoresistance (CMR), and so on [72, 73].

Perovskite cobaltates with a chemical formula (Ln, A)$CoO_{3-\delta}$ (Ln = lanthanide, A = alkaline earth metal: Ca, Sr, or Ba) have recently attracted significant interest owing to important magnetic and electronic transport properties such as CMR, magnetic ordering, electronic phase separation, insulator–metal transition, and large thermoelectric power at low temperatures. These properties offer various application prospects, that is, magnetic data storage, oxidation catalyst, fuel cell cathode, gas sensor, and so on [74–76]. Chen *et al.* recently systemically investigated an epitaxial $LaBaCo_2O_{5+\delta}$ (LBCO) thin film, fabricated by PLD, from high-temperature fuel cell properties to low-temperature magnetic properties by interface strain; it is found that the magnetoresistance of epitaxial LBCO thin film is almost three times that of the bulk materials and increases the application temperature from 10 to 40 K [77–79]. Also, a series of transparent ferromagnet thin films, that is, $Sn_{0.95}Fe_{0.05}O_2$, were fabricated by PLD [80].

1.7.4 Pulsed Laser Deposition for Multiferroics Thin Film

Multiferroics is magnetism and ferroelectricity coexisting in materials and the quest for multiferroic materials is of great technological and fundamental importance since magnetism and ferroelectricity are essential to many forms of current technology [81, 82]. PLD system has been widely applied to fabricate not only the single-phase multiferroic thin film but also the horizontal and vertical heterostructures. To our knowledge, only two classes of single-phase multiferroic hexagonal manganites and the Bi-based perovskites have been prepared as single-phase films. The first multiferroic to be investigated in thin film form was the hexagonal manganite $YMnO_3$ [83, 84]. Currently, perovskite structure bismuth ferrite is the most studied single-component multiferroic because of its high Curie temperature (820 °C) and large polarization, making it appealing for application in ferroelectric nonvolatile memories and high-temperature electronics [85, 86].

Horizontal heterostructures consisting of alternating layers of conventional ferromagnets and ferroelectrics [87], such as $Pr_{0.85}Ca_{0.15}MnO_3$ and $Ba_{0.6}Sr_{0.4}TiO_3$ superlattice, have been reviewed elsewhere [88]. Vertical heterostructures such as the nanopillar geometry, recently, have attracted much attention owing to the reduced substrate-imposed mechanical clamping and larger interfacial surface area in three dimensions, which allows for stronger coupling between ferroelectric and magnetic components [89]. Although many different combinations of perovskite ($BaTiO_3$, $BiFeO_3$, $PbTiO_3$) and spinels ($CoFe_2O_4$, $NiFe_2O_4$, and Fe_3O_4) have been grown, the design and control of such heterostructures remain a challenge.

1.7.5 Interface Strain Engineering the Complex Oxide Thin Film

It is well known that strain engineering in the complex oxide thin film has received considerable interest owing to its significant impact on the microstructure and physical properties of epitaxial thin film for several decades [90–96]. Interesting and anomalous new electrical and magnetic properties can be achieved by strain modification, such as with superconducting thin films grown on different substrates. The influence of lattice distortions on the magnetic properties of perovskite-related manganite films has been established both theoretically and experimentally [91, 97, 98].

1.7.5.1 Thickness Effect

The thickness of a thin film is among the first quoted attributes of its nature because physical properties and behaviors of the thin film, as well as crystal structures, are strongly dependent on its thickness. In general, for very thin films, the material will adapt itself to the structure imposed by the substrate. At the same time, the thin film stores the biaxial strain energy below a critical thickness because of the lattice mismatch between the film and the substrate. With thin film growth, the strain energy will be rapidly accumulated and stored in the film. When the thickness of the thin film is above the critical thickness, edge dislocations are usually formed to fully or partially release the strain, so the influence of the substrate diminishes and the material adopts the structure of the bulk, with the lattice constant of the thin film being equal or close to its bulk lattice constant under similar growth conditions. It is evident that a thin film with thickness less than thecritical thickness will have obviously different physical properties from those with a larger thickness. Highly epitaxial LBCO thin films on (001) MgO are systematically investigated as a function of the thin film thickness and a 7-nm strained layer is observed, which is marked in Figure 1.6(e) below the black line [77].

As shown in Figure 1.7a, the resistivity of the films and the metal–insulation transition temperature increase rapidly with decreased thicknessbecause of the interface strain from the lattice misfit. In Figure 1.7b, the variable range hopping model fits better than the other two models in the $40\,K \leq T \leq 146\,K$ range, suggesting that the LBCO is dominantly a disordered system in which the charge carriers move by hopping between various localized electronic states [99].

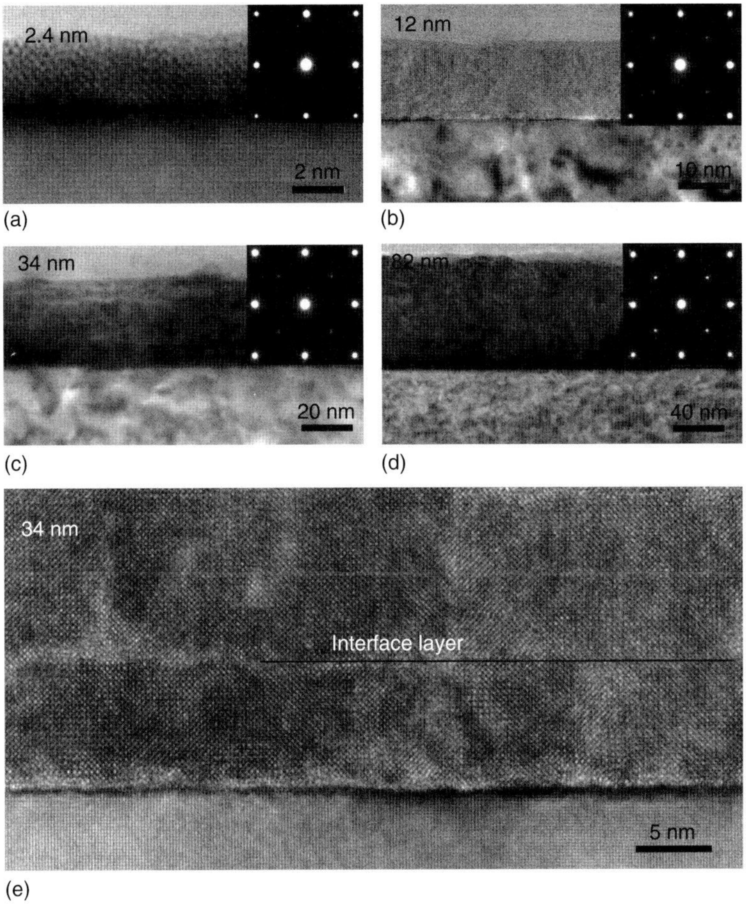

Figure 1.6 (a–d) Transmission electron microscopic images for four different thicknesses of LBCO thin films on the MgO substrate and (e) cross-sectional high-resolution transmission electron microscopy for 34-nm thin film. (Ma *et al.* [77]. Reproduced with permission of American Institute of Physics.)

1.7.5.2 **Substrate Effect**

Substrate is another very important factor engineering thin film structure and physical properties such as the Curie temperature shift, interface dislocation, and formation of nanocolumns. Selecting different substrates will produce either tensile strain or compressive strain in the thin film, which will influence the physical properties performance of a material. Consider a highly epitaxial LBCO thin film on four different substrates, (001) $LaAlO_3$ (LAO), (001) $SrTiO_3$ (STO), (110) $NdGaO_3$ (NGO), and (001) MgO, as example; the LBCO film is under tensile strain on MgO and STO substrates while LAO and NGO impose compressive strain on it [79]. As shown in Figure 1.8, the film/substrate lattice mismatch can significantlyaffect the electrical conductivity and magnetoresistance of the

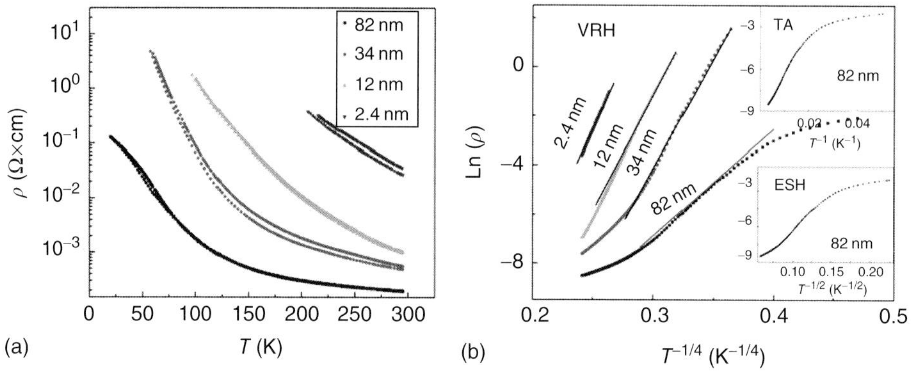

Figure 1.7 (a) Temperature dependence of resistivity with different thickness and (b) logarithm of theresistivity versus $T^{-1/4}$ plot. Insets in (b): logarithm of the resistivity versus $T^{-1/n}$ plots (where $n = 1, 2$) for 82-nm LBCO thin film on MgO substrate. (Ma *et al.* [77]. Reproduced with permission of American Institute of Physics.)

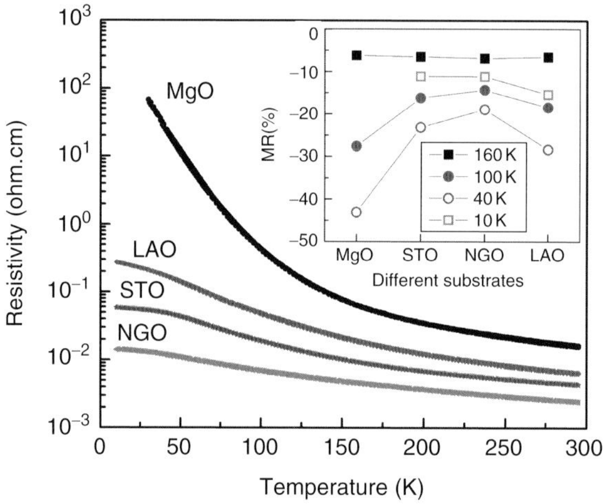

Figure 1.8 Temperature dependence of the resistivity for LBCO thin filmson various substrates. (Ma *et al.* [78]. Reproduced with permission of American Chemical Society.)

LBCO films. The resistivity for the LBCO thin film system is found to have the highest resistivity value on the MgO substrate, followed by the LAO, STO, and NGO in accordance with the sequence of the interface strain values. The large interface strain also induces the maximum magnetoresistance value of 43% for the LBCO films on the MgO substrates, which is almost three times that of the bulk materials, and increases the application temperature from 10 to 40 K.

In addition to the strain generated from the lattice mismatch between the substrate and the LBCO film, substrate surface structures such as surface step, terrace, and kink sites as well as other defect distribution are also important factors that affect the microstructures and physical properties of the films. Only by carefully choosing the substrate's miscut angle (vicinality) can the step-flow growth be promoted, resulting in atomically smooth film surfaces. Chen *et al.* have systematically reported the magnetic and electrical transport properties of highly epitaxial LBCO thin films on vicinal (001) $SrTiO_3$ substrates with different miscut angles (0.5°, 3.0°, and 5.0°) and observed various interesting electronic transport properties and anomalous magnetic phenomena [78].

As shown in Figure 1.9, the electrical resistance of the LBCO films and the metal–insulation transition temperature decrease rapidly with increase in the miscut angles of vicinal (001) $SrTiO_3$ substrates. Magnetoresistance measurements were performed for the as-grown LBCO films as a function of the magnetic field ranging from 0 to 7 T using isothermal MR measurements. The MR is defined as $MR(\%) = (\rho(H) - \rho(0))/\rho(0) \times 100\%$, where $\rho(H)$ and $\rho(0)$ are the resistivities with applied magnetic field and zero magnetic field, respectively. Obviously, the MR value of the LBCO on 0.5° miscut (001) $SrTiO_3$ is much larger than those on the other two vicinal substrates from the inset of Figure 1.9. The maximum MR value of approximately 34% is achieved at 60 K, which is much larger than the maximum value of 15% from its bulk material at 10 K [100], or the maximum value of 24% for the films grown on the regular (001) $SrTiO_3$ [101], which is represented as 0° in Figure 1.9.

They proposed a model to explain these phenomena as shown in Figure 1.10. When a number of film unit cells fill up along the terrace, a mismatch gap can be generated at the end of the step terrace (i.e., the "residual mismatching gap"), as seen in Figure 1.10a. Practically, this residual mismatch gap will not exist at

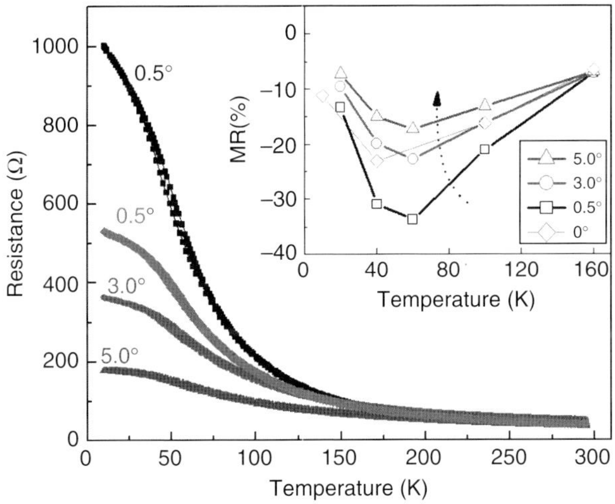

Figure 1.9 Temperature dependence of resistivity from 300 to 20 K for the as-grown LBCO thin films on different miscut angle (001) $SrTiO_3$ substrates. The inset is the MR value of as-grown LBCO at different temperatures under 7 T magnetic field. (Ma *et al.* [78]. Reproduced with permission of American Chemical Society.)

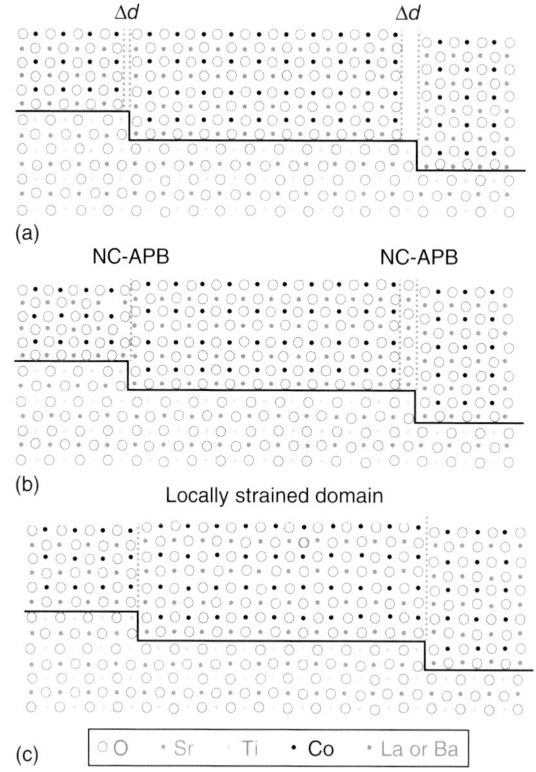

Figure 1.10 The thin film/substrate mismatch in the epitaxial LBCO film grown on the vicinal (001) $SrTiO_3$ substrates with low miscut angle: (a) the generation of the "residual mismatching gap" at the end of the step terrace; (b) the formation of nonconservative antiphase domain boundary; and (c) the formation of locally strained domains induced by the "residual mismatching gap."

the end of the step terrace in an epitaxial film. The last atomic plane of the film will always occupy the atomic position of the terrace end by rearranging the local atomic structure. Two possible atomic rearrangements for the highly epitaxial oxide thin films can be considered: the formation of nonconservative antiphase domain boundary [102] and that of the locally strained domain induced from the "residual mismatching gap," as seen in Figure 1.10b,c. For the half unit cellresidual mismatching gap, it is highly improbable to form a nonconservative antiphase domain boundary at the end of the substrate terrace since a huge amount of interface strain energy will be generated because of the same charge repulsion at the domain boundary and interface. Therefore, the residual matching gap will be released via the formation of a locally strained domain, resulting in tensile strain over the interface and compressive strain along the out-of-plane direction. The local strain can significantly alter the epitaxial quality and the resulting physical properties [103].

As seen in Figure 1.9, the resistance behavior for the film on the 0.5° miscut surface is more like semiconductors rather than the semimetallic behavior found for the films on the 3.0° and 5.0° vicinal surfaces, resulting from the larger tensile strain in 0.5° miscut surface, which induces distortion of the CoO_6 octahedral structure and change of the Co–O bond length and Co–O–Co bond angles, reducing the hopping of the charge carriers hopping.

1.7.5.3 **Post-Annealing**

Complex oxide thin films are very sensitive to oxygen stoichiometry, since different oxygen content will induce various crystal structures and directly influence the properties of complex oxide thin films [46, 48]. Similar to cobalt oxides, cobalt has three valencies ($Co^{2+}/Co^{3+}/Co^{4+}$) and various spin states, which are related to the content of oxygen inside. Oxygen stoichiometry, thus, is an important factor in determining Co ion coordination (octahedral or pyramidal), the Co spin state, the metal–insulator transition, as well as magnetic and electronic transport properties of the perovskite-based cobalt oxides. For higher cobalt valency, the material easily exhibits metallic ferromagnetic behavior and a metal–metal transition, whereas for lower cobalt valency one obtains weakly ferromagnetic semiconductors [104].

Highly epitaxial LBCO thin films were annealed at 465 °C in a tube furnace under pure oxygen (1 atm) for 5 h to examine the post-annealing effect, especially oxygen effect. From Figure 1.11, both the as-grown and oxygen-annealed thin films show excellent single crystal quality. The only difference is that the peak positions are shifted slightly to a higher angle by annealing the thin film in oxygen flow (see the inset (a) of Figure 1.11 for the (200) diffraction peak), suggesting that

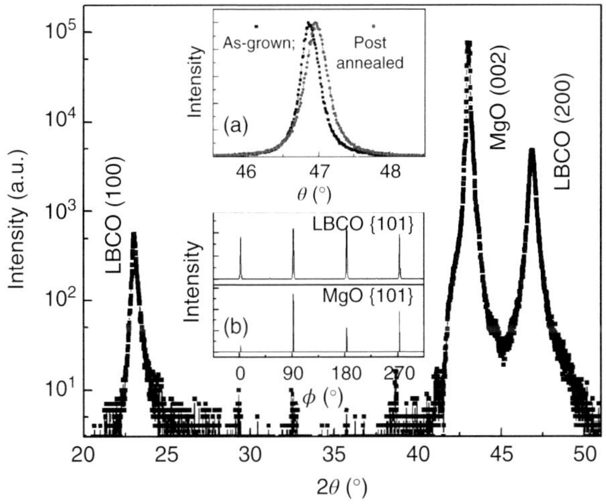

Figure 1.11 A typical XRD pattern of the as-grown LBCO thin films on (001) MgO substrates. The inset (a) shows the (200) diffraction peaks of the as-grown and the oxygen-annealed thin films. The inset (b) shows the ϕ scans taken around the (101) diffraction of the LBCO thin films and MgO substrates.

the lattice parameter along the a-axis becomes smaller in the oxygen-annealed thin films because of the increase in the oxygen content, the decrease in the oxygen vacancies, and the increase in the attractive force between the anion and the cation.

Figure 1.12a shows that both oxygen-annealed and as-grown LBCO thin films exhibit spontaneous magnetizations M below approximately 170 K, but the magnetic moment of the oxygen-annealed sample is much higher than that of the as-grown sample, resulting from the increase in Co^{4+}, followed by the increase in double exchange of Co^{4+}–O–Co^{3+}. The resistivities of the as-grown and

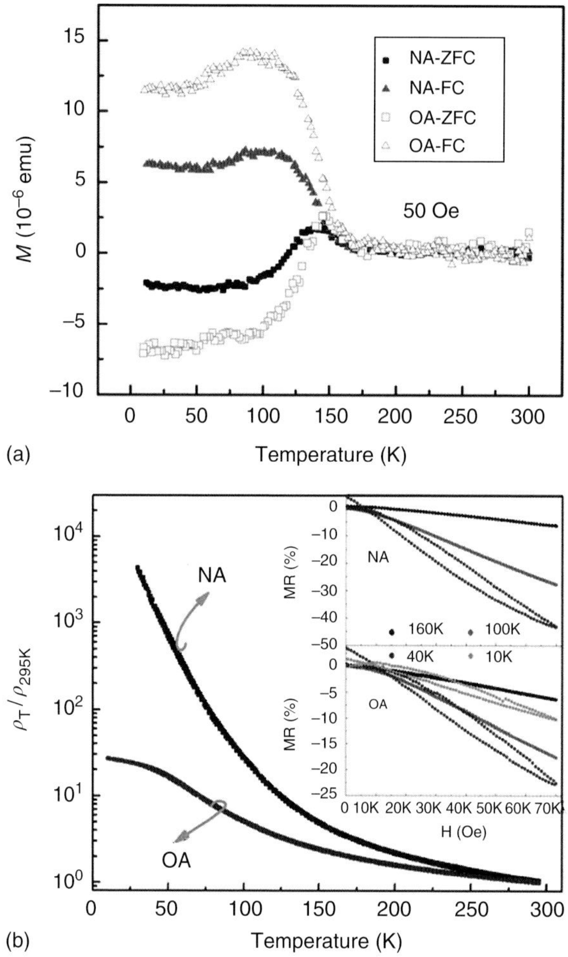

Figure 1.12 (a) ZFC and FC magnetization of no annealing and oxygen annealing of LBCO thin film on (001) MgO substrate. (b) Resistivity change for the two samples with temperature. Inset is the applied magnetic field dependence of the MR effects at isothermal magnetoresistance measurements.

oxygen-annealed samples increase exponentially with decreasing temperature, indicating both of them to be semiconductors. However, the resistivity for the oxygen-annealed thin film exhibits a semimetallic behavior below 50 K (Figure 1.12b). These phenomena indicate that post-annealing is a critical factor to influence the crystalline structure and physical properties of complex oxide thin films.

1.8 Pulsed Laser Deposition for Nanostructure Growth

Miniaturization of electronic devices, such as the nanoscale design of memories, sensors, and actuators, is prompting the development of surface patterning techniques. Although traditional techniques such as focused ion-beam patterning, electron-beam etching, or nanoimprint lithography have been widely used, their common drawback is the need for a resist or polymer process and, hence, numerous chemical, thermal, and etching associated steps, which will influence the properties of devices. The only two ways to avoid the expensive and contaminated top-down processes are self-assembly [105, 106] and stenciling [107–111].

1.8.1 Self-Assembled Nanoscale Structures

Self-assembly process allows to pattern large-area arrays of high quality, possible epitaxial nanostructure with lateral dimensions at the nanoscale, in the sub-100 nm range. The self-assembled nanoscale structure is a good platform for in-depth investigation of the effects induced by defects and size on structural quality and properties. Ordered self-assembly complex oxide Ce:$BaTiO_3$ quantum dots shaped as a pyramid with a square base were fabricated by PLD, and the nonlinear effect is improved by the low-dimensional structures [112].

1.8.2 Geometrically Ordered Arrays

Well-ordered arrays of nanoscale complex oxide structures without using the classic top-down process is technically very demanding. A stencil mask, which is a pattern of apertures in a thin free-standing membrane, is developed to fabricate geometrically ordered arrays [13, 113]. A nanoscale-ordered arrays of $BaTiO_3$ thin films in a single deposition step by PLD is achieved through nanostencil.

The nanostencil method is quite easy and elegant, but it suffers from a few disadvantages, such as the lack of long-range order and the individual final size. Lee *et al.* have developed an alternative mask. They used porous anodic alumina oxide (AAO) as template and electrodeposition to make gold nanotube membranes, which is used as a stencil mask for the deposition of complex oxides nanostructure.

Perfectly, two-dimensional (2D) arrays of ferroelectric PZT nanodots (50 nm in diameter) were fabricated using PLD by ultrathin AAO membrane stencil masks to investigate vortex polarization states in nanoscale ferroelectric arrays [114].

1.9 Variation of Pulsed Laser Deposition

As with most techniques, PLD has several variants. Most variations of PLD have been summarized by Gupta in 1994 [115]. Since then, several further adaptations have been presented. Combinatorial PLD extends the basic PLD setup to include multiple lasers and targets and currently shows great potential for precision materials engineering [116]. The combined setup can be used to grow designer materials or alloys with deliberately engineered physical properties [117] and enables greater control over film stoichiometry. Also, it can facilitate deposition of material that cannot currently be fabricated by conventional, single-laser, single-target PLD [118 120]. In addition, continuous compositional spread technique is based on PLD and applied to the growth of epitaxial films [121].

1.10 Conclusion

PLD system is a very useful tool to fabricate high-quality complex oxide thin films and nanostructures because of the technical simplicity and the ease with which deposition can be switched from one material to another. The most important is that it can maintain the stoichiometric target material in the thin film.

Although the setup of PLD is simple, the operation principles are complex. Various factors can significantly influence the crystalline structure and physical properties of the complex oxide thin film and nanostructure, such as the substrate temperature, background gas, laser energy density, the target–substrate distance, post-annealing, and so on.

PLD can be used to fabricate various complex oxide thin films, for instance, high-temperature superconductor epitaxial thin film, high-quality ferroelectric and ferromagnetic thin film, and multiferroic single-phase thin film or horizontal/vertical heterostructures.

With the development of PLD system, there are a lot of variants of PLD. For example, a multiple lasers and target PLD system can fabricate designer materials or alloys with deliberately engineered physical properties and can even realize the deposition of material that cannot currently be fabricated by conventional, single-laser, single-target PLD.

Like any deposition technique, PLD has some weaknesses, but it is a promising research tool for investigating new growth phenomena and systems, offering some new and unique features.

References

1. Ozgur, U., Alivov, Y.I., Liu, C., Teke, A., Reshchikov, M.A., Dogan, S., Avrutin, V., Cho, S.J., and Morkoc, H. (2005) A comprehensive review of ZnO materials and devices. *J. Appl. Phys.*, **98**, 041301.
2. Kim, H., Gilmore, C.M., Pique, A., Horwitz, J.S., Mattoussi, H., Murata, H., Kafafi, Z.H., and Chrisey, D.B. (1999) Electrical, optical, and structural properties of indium-tin-oxide thin films for organic light-emitting devices. *J. Appl. Phys.*, **86**, 6451–6461.
3. Janotti, A. and Van de Walle, C.G. (2009) Fundamentals of zinc oxide as a semiconductor. *Rep. Prog. Phys.*, **72**, 126501.
4. Choi, D.H., Lee, D., Sim, H., Chang, M., and Hwang, H.S. (2006) Reversible resistive switching of $SrTiO_x$ thin films for nonvolatile memory applications. *Appl. Phys. Lett.*, **88**, 082904.
5. Panda, D. and Tseng, T.Y. (2013) Growth, dielectric properties, and memory device applications of ZrO_2 thin films. *Thin Solid Films*, **531**, 1–20.
6. Liu, J., Liu, M., Collins, G., Chen, C.L., Jiang, X.N., Gong, W.Q., Jacobson, A.J., He, J., Jiang, J.C., and Meletis, E.I. (2010) Epitaxial nature and transport properties in $LaBaCo_2O_{5+\delta}$ thin films. *Chem. Mater.*, **22**, 799–802.
7. Liu, M., Ren, S.P., Zhang, R.Y., Xue, Z.Y., Ma, C.R., Yin, M.L., Xu, X., Bao, S.Y., and Chen, C.L. (2015) Gas sensing properties of epitaxial $LaBaCo_2O_{5.5+\delta}$ thin films. *Sci. Rep.-UK*, **5**, 10784.
8. Smith, H.M. and Turner, A.F. (1965) Vacuum Deposited Thin Films Using a Ruby Laser. *Appl. Opt*, **4**, 147–148.
9. Schwarz, H. and Tourtell, H. (1969) Vacuum deposition by high-energy laser with emphasis on barium titanate films. *J. Vac. Sci. Technol*, **6**, 373–378.
10. Desserre, J. and Eloy, J.F. (1975) Interaction of pulsed coherent light-beam with complex target – Application to production of thin-film compounds. *Thin Solid Films*, **29**, 29–41.
11. Dijkkamp, D., Venkatesan, T., Wu, X.D., Shaheen, S.A., Jisrawi, N., Minlee, Y.H., Mclean, W.L., and Croft, M. (1987) Preparation of Y-Ba-Cu oxide superconductor thin-films using pulsed laser evaporation from high-Tc bulk material. *Appl. Phys. Lett.*, **51**, 619–621.
12. Zheng, H., Wang, J., Mohaddes-Ardabili, L., Wuttig, M., Salamanca-Riba, L., Schlom, D.G., and Ramesh, R. (2004) Three-dimensional heteroepitaxy in self-assembled $BaTiO_3$-$CoFe_2O_4$ nanostructures. *Appl. Phys. Lett.*, **85**, 2035–2037.
13. Cojocaru, C.V., Harnagea, C., Rosei, F., Pignolet, A., van den Boogaart, M.A.F., and Brugger, J. (2005) Complex oxide nanostructures by pulsed laser deposition through nanostencils. *Appl. Phys. Lett.*, **86**, 183107.
14. Ji, Y.D., Zhang, Y., Gao, M., Yuan, Z., Xia, Y.D., Jin, C.Q., Tao, B.W., Chen, C.L., Jia, Q.X., and Lin, Y. (2014) Role of microstructures on the M1-M2 phase transition in epitaxial VO_2 thin films. *Sci. Rep.-UK*, **4**, 4854.
15. Izyumskaya, N., Alivov, Y., Cho, S.J., Morkoc, H., Lee, H., and Kang, Y.S. (2007) Processing, structure, properties, and applications of PZT thin films. *Crit. Rev. Solid State Mater. Sci.*, **32**, 111–202.
16. Zhou, Y.C., Cao, W.J., Liu, L.C., Agaian, S., and Chen, C.L.P. (2015) Fast Fourier transform using matrix decomposition. *Inf. Sci.*, **291**, 172–183.
17. Martin, L.W., Chu, Y.H., and Ramesh, R. (2010) Advances in the growth and characterization of magnetic, ferroelectric, and multiferroic oxide thin films. *Mater. Sci. Eng., R*, **68**, 89–133.
18. Ohnishi, T., Yamamoto, T., Meguro, S., Koinuma, H., and Lippmaa, M. (2007) Pulsed laser ablation and deposition of complex oxides. *J. Phys. Conf. Ser.*, **59**, 514–519.
19. Singh, R.K. and Narayan, J. (1990) Pulsed-laser evaporation technique for deposition of thin films: physics and theoretical model. *Phys. Rev. B*, **41**, 8843–8859.
20. Willmott, P.R. and Huber, J.R. (2000) Pulsed laser vaporization and deposition. *Rev. Mod. Phys.*, **72**, 315–328.

21. Ratsch, C. and Venables, J.A. (2003) Nucleation theory and the early stages of thin film growth. *J. Vac. Sci. Technol., A*, **21**, S96–S109.
22. Venables, J.A., Spiller, G.D.T., and Hanbucken, M. (1984) Nucleation and growth of thin-films. *Rep. Prog. Phys.*, **47**, 399–459.
23. Vook, R.W. (1984) Nucleation and growth of thin-films. *Opt. Eng.*, **23**, 343–348.
24. Markov, I. and Kaischew, R. (1976) Influence of supersaturation on mode of crystallization on crystalline substrates. *Thin Solid Films*, **32**, 163–167.
25. Reichelt, K. (1988) Nucleation and growth of thin-films. *Vacuum*, **38**, 1083–1099.
26. Metev, S. and Meteva, K. (1989) Nucleation and growth of laser-plasma deposited thin-films. *Appl. Surf. Sci.*, **43**, 402–408.
27. Klein, J., Hofener, C., Alff, L., and Gross, R. (1999) *In situ* monitoring of the growth of oxide thin films at high oxygen pressure. *Supercond. Sci. Technol.*, **12**, 1023–1026.
28. Rijnders, G., Koster, G., Blank, D.H.A., and Rogalla, H. (1997) *In situ* monitoring during pulsed laser deposition of complex oxides using reflection high energy electron diffraction under high oxygen pressure. *Appl. Phys. Lett.*, **70**, 1888–1890.
29. Huijben, M., Brinkman, A., Koster, G., Rijnders, G., Hilgenkamp, H., and Blank, D.H.A. (2009) Structure-property relation of $SrTiO_3/LaAlO_3$ interfaces. *Adv. Mater.*, **21**, 1665–1677.
30. Tischler, J., Eres, G., Larson, B., Rouleau, C.M., Zschack, P., and Lowndes, D.H. (2006) Nonequilibrium interlayer transport in pulsed laser deposition. *Phys. Rev. Lett.*, **96**, 226104.
31. Braun, W., Jenichen, B., Kaganer, V.M., Shtukenberg, A.G., Daweritz, L., and Ploog, K.H. (2003) Layer-by-layer growth of GaAs(001) studied by *in situ* synchrotron X-ray diffraction. *Surf. Sci.*, **525**, 126–136.
32. Ohring, M. (1991) *The Materials Science of Thin Films*, Academic Press, Inc.
33. Sambri, A., Amoruso, S., Wang, X., Radovic, M., Granozio, F.M., and Bruzzese, R. (2007) Substrate heating influence on plume propagation during pulsed laser deposition of complex oxides. *Appl. Phys. Lett.*, **91**, 151501.
34. Willmott, P.R. (2004) Deposition of complex multielemental thin films. *Prog. Surf. Sci.*, **76**, 163–217.
35. Sturm, K., Fahler, S., and Krebs, H.U. (2000) Pulsed laser deposition of metals in low pressure inert gas. *Appl. Surf. Sci.*, **154**, 462–466.
36. Scharf, T. and Krebs, H.U. (2002) Influence of inert gas pressure on deposition rate during pulsed laser deposition. *Appl. Phys. A-Mater. Sci. Process.*, **75**, 551–554.
37. Yang, W.Q., Wei, Z.R., Gao, M., Chen, Y., Xu, J., Chen, C.L., and Lin, Y. (2013) Fabrication and field emission properties of needle-shaped MoO_3 nanobelts. *J. Alloys Compd.*, **576**, 332–335.
38. Ohnishi, T., Lippmaa, M., Yamamoto, T., Meguro, S., and Koinuma, H. (2005) Improved stoichiometry and misfit control in perovskite thin film formation at a critical fluence by pulsed laser deposition. *Appl. Phys. Lett.*, **87**, 241919.
39. Ohnishi, T., Shibuya, K., Yamamoto, T., and Lippmaa, M. (2008) Defects and transport in complex oxide thin films. *J. Appl. Phys.*, **103**, 103703.
40. Ngom, B.D., Lafane, S., Dioum, A., Manyala, N., Abdelli-Messaci, S., Kerdja, R.T., Madjoe, R., Nemutudi, R., Maaza, M., and Beye, A.C. (2011) The influence of plasma dynamics on the growth of $Sm_{0.55}Nd_{0.45}NiO_3$ solid solution during pulsed laser deposition. *J. Phys. Chem. Solids*, **72**, 1218–1224.
41. Claeyssens, F., Cheesman, A., Henley, S.J., and Ashfold, M.N.R. (2002) Studies of the plume accompanying pulsed ultraviolet laser ablation of zinc oxide. *J. Appl. Phys.*, **92**, 6886–6894.
42. Hammond, R.H. and Bormann, R. (1989) Correlation between the in situ growth-conditions of YBCO thin-films and the thermodynamic stability-criteria. *Physica C*, **162**, 703–704.
43. Kim, H.S. and Kwok, H.S. (1992) Correlation between target-substrate distance

and oxygen-pressure in pulsed laser deposition of $YBa_2Cu_3O_7$. *Appl. Phys. Lett.*, **61**, 2234–2236.

44. Castro-Rodriguez, R., Coronado, D.R., Iribarren, A., Watts, B.E., Leccabue, F., and Pena, J.L. (2005) Correlation between target-substrate distance and oxygen pressure in pulsed laser deposition of complex oxide thin films. *Appl. Phys. A*, **81**, 1503–1507.
45. Pan, F., Song, C., Liu, X.J., Yang, Y.C., and Zeng, F. (2008) Ferromagnetism and possible application in spintronics of transition-metal-doped ZnO films. *Mater. Sci. Eng., R*, **62**, 1–35.
46. Rusop, M., Uma, K., Soga, T., and Jimbo, T. (2006) Post-growth annealing of zinc oxide thin films pulsed laser deposited under enhanced oxygen pressure on quartz and silicon substrates. *Mater. Sci. Eng., B*, **127**, 150–153.
47. Liu, M., Ma, C.R., Enriquez, E., Xu, X., Bao, S.Y., and Chen, C.L. (2014) Effects of annealing ambient on electrical properties of $LaBaCo_2O_{5+\delta}$ thin films. *J. Nano Res.-Sw*, **27**, 25–30.
48. Liu, M., Ma, C.R., Liu, J., Collins, G., Chen, C.L., He, J., Jiang, J.C., Meletis, E.I., Sun, L., Jacobson, A.J., and Whangbo, M.H. (2012) Giant magnetoresistance and anomalous magnetic properties of highly epitaxial ferromagnetic $LaBaCo_2O_{5.5+\delta}$ thin films on (001) MgO. *ACS Appl. Mater. Interfaces*, **4**, 5524–5528.
49. Hundley, M.F., Hawley, M., Heffner, R.H., Jia, Q.X., Neumeier, J.J., Tesmer, J., Thompson, J.D., and Wu, X.D. (1995) Transport-magnetism correlations in the ferromagnetic oxide $La_{0.7}Ca_{0.3}MnO_3$. *Appl. Phys. Lett.*, **67**, 860–862.
50. Opel, M. (2012) Spintronic oxides grown by laser-MBE. *J. Phys. D: Appl. Phys.*, **45**, 033001.
51. Roas, B., Schultz, L., and Endres, G. (1988) Epitaxial-growth of $YBa_2Cu_3O_{7-X}$ thin-films by a laser evaporation process. *Appl. Phys. Lett.*, **53**, 1557–1559.
52. Buzea, C. and Yamashita, T. (2001) Review of the superconducting properties of MgB_2. *Supercond. Sci. Technol*, **14**, R115–R146.
53. Yamada, Y., Takahashi, K., Kobayashi, H., Konishi, M., Watanabe, T., Ibi, A., Muroga, T., Miyata, S., Kato, T., Hirayama, T., and Shiohara, Y. (2005) Epitaxial nanostructure and defects effective for pinning in $Y(RE)Ba_2Cu_3O_{7-x}$ coated conductors. *Appl. Phys. Lett.*, **87**, 132502.
54. Lee, P.A., Nagaosa, N., and Wen, X.G. (2006) Doping a Mott insulator: physics of high-temperature superconductivity. *Rev. Mod. Phys.*, **78**, 17–85.
55. Norton, D.P., Goyal, A., Budai, J.D., Christen, D.K., Kroeger, D.M., Specht, E.D., He, Q., Saffian, B., Paranthaman, M., Klabunde, C.E., Lee, D.F., Sales, B.C., and List, F.A. (1996) Epitaxial $YBa_2Cu_3O_7$ on biaxially textured nickel (001): an approach to superconducting tapes with high critical current density. *Science*, **274**, 755–757.
56. Locquet, J.P., Perret, J., Fompeyrine, J., Machler, E., Seo, J.W., and Van Tendeloo, G. (1998) Doubling the critical temperature of $La_{1.9}Sr_{0.1}CuO_4$ using epitaxial strain. *Nature*, **394**, 453–456.
57. Miura, S., Yoshida, Y., Ichino, Y., Matsumoto, K., Ichinose, A., and Awaji, S. (2015) Characteristics of high-performance $BaHfO_3$-doped $SmBa_2Cu_3O_y$ superconducting films fabricated with a seed layer and low-temperature growth. *Supercond. Sci. Technol.*, **28**, 065013.
58. Dawber, M., Rabe, K.M., and Scott, J.F. (2005) Physics of thin-film ferroelectric oxides. *Rev. Mod. Phys.*, **77**, 1083–1130.
59. Setter, N., Damjanovic, D., Eng, L., Fox, G., Gevorgian, S., Hong, S., Kingon, A., Kohlstedt, H., Park, N.Y., Stephenson, G.B., Stolitchnov, I., Tagantsev, A.K., Taylor, D.V., Yamada, T., and Streiffer, S. (2006) Ferroelectric thin films: review of materials, properties, and applications. *J. Appl. Phys.*, **100**, 051606.
60. Lee, H.N., Hesse, D., Zakharov, N., and Gosele, U. (2002) Ferroelectric $Bi_{3.25}La_{0.75}Ti_3O_{12}$ films of uniform *a*-axis orientation on silicon substrates. *Science*, **296**, 2006–2009.
61. Scott, J.F. (1998) High-dielectric constant thin films for dynamic random

access memories (DRAM). *Annu. Rev. Mater. Sci.*, **28**, 79–100.

62. Prins, M.W.J., GrosseHolz, K.O., Muller, G., Cillessen, J.F.M., Giesbers, J.B., Weening, R.P., and Wolf, R.M. (1996) A ferroelectric transparent thin-film transistor. *Appl. Phys. Lett.*, **68**, 3650–3652.
63. Watanabe, Y., Tanamura, M., Matsumoto, Y., Asami, H., and Kato, A. (1995) Ferroelectric (La,Sr)(2)Cuo_4 epitaxial heterostructure with high thermal-stability. *Appl. Phys. Lett.*, **66**, 299–301.
64. Ramesh, R., Inam, A., Chan, W.K., Wilkens, B., Myers, K., Remschnig, K., Hart, D.L., and Tarascon, J.M. (1991) Epitaxial cuprate superconductor ferroelectric heterostructures. *Science*, **252**, 944–946.
65. Liu, M., Collins, G., Silva, E., Ma, C.R., Liu, J., Chen, C.L., He, J., Jiang, J.C., Meletis, E.I., Qu, S.W., Zhang, Q.Y., and Bhalla, A. (2011) Interface engineered ferroelectric $BaTiO_3//SrTiO_3$ heterostructures with anomalous clamped polarization on Si (100). *Integr. Ferroelectr.*, **131**, 89–94.
66. Liu, M., Liu, J., Ma, C.R., Collins, G., Chen, C.L., Alemayehu, A.D., Subramanyam, G., He, J., Jiang, J.C., Meletis, E.I., and Bhalla, A. (2013) Enhanced dielectric properties of (Ba, Sr)TiO_3//Ba(Zr, Ti)O_3 heterostructures with optimized structure design. *CrystEngComm*, **15**, 6641–6644.
67. Liu, M., Ma, C.R., Collins, G., Liu, J., Chen, C.L., Alemayehu, A.D., Subramanyam, G., Ding, Y., Chen, J.H., Dai, C., Lin, Y., and Cole, M.W. (2013) Ferroelectric $BaTiO_3/SrTiO_3$ multilayered thin films for room-temperature tunable microwave elements. *Nanoscale Res. Lett.*, **8**, 338.
68. Liu, M., Ma, C.R., Collins, G., Liu, J., Chen, C.L., Dai, C., Lin, Y., Shui, L., Xiang, F., Wang, H., He, J., Jiang, J.C., Meletis, E.I., and Cole, M.W. (2012) Interface engineered $BaTiO_3/SrTiO_3$ heterostructures with optimized high-frequency dielectric properties. *ACS Appl. Mater. Interfaces*, **4**, 5761–5765.
69. Liu, M., Ma, C.R., Collins, G., Liu, J.A., Chen, C.L., Shui, L., Wang, H., Dai, C., Lin, Y.A., He, J., Jiang, J.C., Meletis, E.I., and Zhang, Q.Y. (2010) Microwave dielectric properties with optimized Mn-doped $Ba_{0.6}Sr_{0.4}TiO_3$ highly epitaxial thin films. *Cryst. Growth Des.*, **10**, 4221–4223.
70. Liu, M., Ma, C.R., Liu, J., Collins, G., Chen, C.L., Xiang, F., Wang, H., He, J., Jiang, J.C., Meletis, E.I., and Bhalla, A. (2014) Microwave dielectric properties of Mn-doped (Ba, Sr)TiO_3//Ba(Zr, Ti)O_3 multilayered thin films: optimization of designed structure. *Integr. Ferroelectr.*, **150**, 116–122.
71. Schlom, D.G., Chen, L.Q., Eom, C.B., Rabe, K.M., Streiffer, S.K., and Triscone, J.M. (2007) Strain tuning of ferroelectric thin films. *Annu. Rev. Mater. Res.*, **37**, 589–626.
72. Ueda, K., Tabata, H., and Kawai, T. (2001) Magnetic and electric properties of transition-metal-doped ZnO films. *Appl. Phys. Lett.*, **79**, 988–990.
73. Haghiri-Gosnet, A.M. and Renard, J.P. (2003) CMR manganites: physics, thin films and devices. *J. Phys. D: Appl. Phys.*, **36**, R127–R150.
74. Akahoshi, D. and Ueda, Y. (2001) Oxygen nonstoichiometry, structures, and physical properties of $YBaCo_2O_5+x$ ($0.00 \leq x \leq 0.52$). *J. Solid State Chem.*, **156**, 355–363.
75. Podlesnyak, A., Streule, S., Mesot, J., Medarde, M., Pomjakushina, E., Conder, K., Tanaka, A., Haverkort, M.W., and Khomskii, D.I. (2006) Spin-state transition in $LaCoO_3$: direct neutron spectroscopic evidence of excited magnetic states. *Phys. Rev. Lett.*, **97**, 247208.
76. Seikh, M.M., Simon, C., Caignaert, V., Pralong, V., Lepetit, M.B., Boudin, S., and Raveau, B. (2008) New magnetic transitions in the ordered oxygen-deficient perovskite $LnBaCo_2O_{5.5+\delta}$. *Chem. Mater.*, **20**, 231–238.
77. Ma, C.R., Liu, M., Collins, G., Liu, J., Zhang, Y.M., Chen, C.L., He, J., Jiang, J.C., and Meletis, E.I. (2012) Thickness effects on the magnetic and electrical transport properties of highly epitaxial $LaBaCo_2O_{5.5+\delta}$ thin films on MgO substrates. *Appl. Phys. Lett.*, **101**, 021602.

78. Ma, C.R., Liu, M., Collins, G., Wang, H.B., Bao, S.Y., Xu, X., Enriquez, E., Chen, C.L., Lin, Y., and Whangbo, M.H. (2013) Magnetic and electrical transport properties of $LaBaCo_2O_{5.5+\delta}$ thin films on vicinal (001) $SrTiO_3$ surfaces. *ACS Appl. Mater. Interfaces*, **5**, 451–455.
79. Ma, C.R., Liu, M., Liu, J., Collins, G., Zhang, Y.M., Wang, H.B., Chen, C.L., Lin, Y., He, J., Jiang, J.C., Meletis, E.I., and Jacobson, A.J. (2014) Interface effects on the electronic transport properties in highly epitaxial $LaBaCo_2O_{5.5+\delta}$ films. *ACS Appl. Mater. Interfaces*, **6**, 2540–2545.
80. Coey, J.M.D., Douvalis, A.P., Fitzgerald, C.B., and Venkatesan, M. (2004) Ferromagnetism in Fe-doped SnO_2 thin films. *Appl. Phys. Lett.*, **84**, 1332–1334.
81. Cheong, S.W. and Mostovoy, M. (2007) Multiferroics: a magnetic twist for ferroelectricity. *Nat. Mater.*, **6**, 13–20.
82. Ramesh, R. and Spaldin, N.A. (2007) Multiferroics: progress and prospects in thin films. *Nat. Mater.*, **6**, 21–29.
83. Fujimura, N., Azuma, S., Aoki, N., Yoshimura, T., and Ito, T. (1996) Growth mechanism of $YMnO_3$ film as a new candidate for nonvolatile memory devices. *J. Appl. Phys.*, **80**, 7084–7088.
84. Salvador, P.A., Doan, T.D., Mercey, B., and Raveau, B. (1998) Stabilization of $YMnO_3$ in a perovskite structure as a thin film. *Chem. Mater.*, **10**, 2592–2595.
85. Ueda, K., Tabata, H., and Kawai, T. (1999) Coexistence of ferroelectricity and ferromagnetism in $BiFeO_3$-$BaTiO_3$ thin films at room temperature. *Appl. Phys. Lett.*, **75**, 555–557.
86. Palkar, V.R., John, J., and Pinto, R. (2002) Observation of saturated polarization and dielectric anomaly in magnetoelectric $BiFeO_3$ thin films. *Appl. Phys. Lett.*, **80**, 1628–1630.
87. Murugavel, P., Saurel, D., Prellier, W., Simon, C., and Raveau, B. (2004) Tailoring of ferromagnetic $Pr_{0.85}Ca_{0.15}MnO_3$/ferroelectric $Ba_{0.6}Sr_{0.4}TiO_3$ superlattices for multiferroic properties. *Appl. Phys. Lett.*, **85**, 4424–4426.
88. Singh, M.P., Prellier, W., Simon, C., and Raveau, B. (2005) Magnetocapacitance effect in perovskite-superlattice based multiferroics. *Appl. Phys. Lett.*, **87**, 022505.
89. Zheng, H., Wang, J., Lofland, S.E., Ma, Z., Mohaddes-Ardabili, L., Zhao, T., Salamanca-Riba, L., Shinde, S.R., Ogale, S.B., Bai, F., Viehland, D., Jia, Y., Schlom, D.G., Wuttig, M., Roytburd, A., and Ramesh, R. (2004) Multiferroic $BaTiO_3$-$CoFe_2O_4$ nanostructures. *Science*, **303**, 661–663.
90. Yang, F., Kemik, N., Biegalski, M.D., Christen, H.M., Arenholz, E., and Takamura, Y. (2010) Strain engineering to control the magnetic and magnetotransport properties of $La_{0.67}Sr_{0.33}MnO_3$ thin films. *Appl. Phys. Lett.*, **97**, 092503.
91. Xiong, Y.M., Wang, G.Y., Luo, X.G., Wang, C.H., Chen, X.H., Chen, X., and Chen, C.L. (2005) Magnetotransport properties in $La_{1-x}Ca_xMnO_3$ (x=0.33, 0.5) thin films deposited on different substrates. *J. Appl. Phys.*, **97**, 083909.
92. Xiong, Y.M., Chen, T., Wang, G.Y., Chen, X.H., Chen, X., and Chen, C.L. (2004) Raman spectra in epitaxial thin films of $La_{1-x}Ca_xMnO_3$ (x=0.33, 0.5) grown on different substrates. *Phys. Rev. B*, **70**, 094407.
93. Park, S., Ryan, P., Karapetrova, E., Kim, J.W., Ma, J.X., Shi, J., Freeland, J.W., and Wu, W.D. (2009) Microscopic evidence of a strain-enhanced ferromagnetic state in $LaCoO_3$ thin films. *Appl. Phys. Lett.*, **95**, 072508.
94. Lin, Y., Dai, C., Li, Y.R., Chen, X., Chen, C.L., Bhalla, A., and Jia, Q.X. (2010) Strain relaxation in epitaxial (Pb,Sr)TiO_3 thin films on $NdGaO_3$ substrates. *Appl. Phys. Lett.*, **96**, 102901.
95. Grutter, A., Wong, F., Arenholz, E., Liberati, M., Vailionis, A., and Suzuki, Y. (2010) Enhanced magnetism in epitaxial $SrRuO_3$ thin films. *Appl. Phys. Lett.*, **96**, 082509.
96. Grutter, A., Wong, F., Arenholz, E., Liberati, M., and Suzuki, Y. (2010) Enhanced magnetization in epitaxial $SrRuO_3$ thin films via substrate-induced strain. *J. Appl. Phys.*, **107**, 09e138.

97. Millis, A.J., Darling, T., and Migliori, A. (1998) Quantifying strain dependence in "colossal" magnetoresistance manganites. *J. Appl. Phys.*, **83**, 1588–1591.

98. Radaelli, P.G., Iannone, G., Marezio, M., Hwang, H.Y., Cheong, S.W., Jorgensen, J.D., and Argyriou, D.N. (1997) Structural effects on the magnetic and transport properties of perovskite $A_{1-x}A_x'MnO_3$ ($x = 0.25, 0.30$). *Phys. Rev. B*, **56**, 8265–8276.

99. Kundu, A.K., Raveau, B., Caignaert, V., Rautama, E.L., and Pralong, V. (2009) Electron transport and thermoelectric properties of layered perovskite $LaBaCo_2O_{5.5}$. *J. Phys.: Condens. Matter*, **21**, 056007.

100. Rautama, E.L., Boullay, P., Kundu, A.K., Caignaert, V., Pralong, V., Karppinen, M., and Raveau, B. (2008) Cationic ordering and microstructural effects in the ferromagnetic perovskite $La_{0.5}Ba_{0.5}CoO_3$: impact upon magnetotransport properties. *Chem. Mater.*, **20**, 2742–2750.

101. Liu, M., Liu, J., Collins, G., Ma, C.R., Chen, C.L., He, J., Jiang, J.C., Meletis, E.I., Jacobson, A.J., and Zhang, Q.Y. (2010) Magnetic and transport properties of epitaxial $LaBaCo_2O_{5.5+\delta}$ thin films on (001) $SrTiO_3$. *Appl. Phys. Lett.*, **96**, 132106.

102. Jiang, J.C., Lin, Y., Chen, C.L., Chu, C.W., and Meletis, E.I. (2002) Microstructures and surface step-induced antiphase boundaries in epitaxial ferroelectric $Ba_{0.6}Sr_{0.4}TiO_3$ thin film on MgO. *J. Appl. Phys.*, **91**, 3188–3192.

103. Lu, H.L., Zhang, C.D., Guo, H.M., Gao, H.J., Liu, M., Liu, J.A., Collins, G., and Chen, C.L. (2010) Surface-step-terrace-induced anomalous transport properties in highly epitaxial $La_{0.67}Ca_{0.33}MnO_3$ thin films. *ACS Appl. Mater. Interfaces*, **2**, 2496–2499.

104. Raveau, B., Pralong, V., Caignaert, V., Hervieu, M., and Maignan, A. (2005) Primordial role of cobalt valence in the magnetotransport properties of oxygen deficient perovskites $Sr_{1-x}Ln_xCoO_{3-\delta}$. *J. Phys.: Condens. Matter*, **17**, 7371–7382.

105. Szafraniak, I., Harnagea, C., Scholz, R., Bhattacharyya, S., Hesse, D., and Alexe, M. (2003) Ferroelectric epitaxial nanocrystals obtained by a self-patterning method. *Appl. Phys. Lett.*, **83**, 2211–2213.

106. Roelofs, A., Schneller, I., Szot, K., and Waser, R. (2002) Piezoresponse force microscopy of lead-titanate nanograins possibly reaching the limit of ferroelectricity. *Appl. Phys. Lett.*, **81**, 5231–5233.

107. Lee, W., Han, H., Lotnyk, A., Schubert, M.A., Senz, S., Alexe, M., Hesse, D., Baik, S., and Gosele, U. (2008) Individually addressable epitaxial ferroelectric nanocapacitor arrays with near Tb inch(-2) density. *Nat. Nanotechnol.*, **3**, 402–407.

108. Lee, W., Schwirn, K., Steinhart, M., Pippel, E., Scholz, R., and Gosele, U. (2008) Structural engineering of nanoporous anodic aluminium oxide by pulse anodization of aluminium. *Nat. Nanotechnol.*, **3**, 234–239.

109. Lee, S.K., Lee, W., Alexe, M., Nielsch, K., Hesse, D., and Gosele, U. (2005) Well-ordered large-area arrays of epitaxial ferroelectric $(Bi,La)_4Ti_3O_{12}$ nanostructures fabricated by gold nanotube-membrane lithography. *Appl. Phys. Lett.*, **86**, 152906.

110. Shin, H.J., Choi, J.H., Yang, H.J., Park, Y.D., Kuk, Y., and Kang, C.J. (2005) Patterning of ferroelectric nanodot arrays using a silicon nitride shadow mask. *Appl. Phys. Lett.*, **87**, 113114.

111. Vrejoiu, I., Alexe, M., Hesse, D., and Goesele, U. (2008) Functional perovskites – From epitaxial films to nanostructured arrays. *Adv. Funct. Mater.*, **18**, 3892–3906.

112. Shi, W.S., Chen, Z.H., Liu, N.N., Lu, H.B., Zhou, Y.L., Cui, D.F., and Yang, G.Z. (1999) Nonlinear optical properties of self-organized complex oxide Ce: $BaTiO_3$ quantum dots grown by pulsed laser deposition. *Appl. Phys. Lett.*, **75**, 1547–1549.

113. Cojocaru, C.-V., Nechache, R., Harnagea, C., Pignolet, A., and Rosei, F. (2010) Nanoscale patterning of functional perovskite-type complex oxides by pulsed laser deposition through a nanostencil. *Appl. Surf. Sci.*, **256**, 4777–4783.

114. Rodriguez, B.J., Gao, X.S., Liu, L.F., Lee, W., Naumov, I.I., Bratkovsky, A.M., Hesse, D., and Alexe, M. (2009) Vortex polarization states in nanoscale ferroelectric arrays. *Nano Lett.*, **9**, 1127–1131.
115. Gupta, A., Chrisey, D.B., and Hubler, G.K. (1994) *Pulsed Laser Deposition of Thin Films*, Wiley-Interscience, New York.
116. Sloyan, K.A., May-Smith, T.C., Eason, R.W., and Lunney, J.G. (2009) The effect of relative plasma plume delay on the properties of complex oxide films grown by multi-laser, multi-target combinatorial pulsed laser deposition. *Appl. Surf. Sci.*, **255**, 9066–9070.
117. Gazia, R., May-Smith, T.C., and Eason, R.W. (2008) Growth of a hybrid garnet crystal multilayer structure by combinatorial pulsed laser deposition. *J. Cryst. Growth*, **310**, 3848–3853.
118. Darby, M.S.B., May-Smith, T.C., and Eason, R.W. (2008) Deposition and stoichiometry control of Nd-doped gadolinium gallium garnet thin films by combinatorial pulsed laser deposition using two targets of Nd:$Gd_3Ga_5O_{12}$ and Ga_2O_3. *Appl. Phys. A*, **93**, 477–481.
119. Holesinger, T.G., Jia, Q., Maiorov, B., Civale, L., Dowden, P.C., and Gibbons, B.J. (2007) Ultrafine multilayers of complex metal oxide films. *Adv. Mater.*, **19**, 1917–1920.
120. Willmott, P.R., Herger, R., and Schleputz, C.M. (2004) Multilayers, alloys, and complex profiles by pulsed laser deposition using a novel target geometry. *Thin Solid Films*, **453**, 436–439.
121. Christen, H.M., Silliman, S.D., and Harshavardhan, K.S. (2001) Continuous compositional-spread technique based on pulsed-laser deposition and applied to the growth of epitaxial films. *Rev. Sci. Instrum.*, **72**, 2673–2678.

2
Electron Beam Evaporation Deposition

Zhongping Wang and Zengming Zhang

2.1
Introduction

A thin film is defined as a low-dimensional material fabricated on a substrate by condensing, one-by-one, atomic/molecular/ionic species of matter [1]. The thickness of the thin film ranges from a few nanometers to several tens of micrometers. The controllable growth and properties of thin films have greatly benefited from the significant advances in vacuum technology, development of surface-sensitive probes, and electron diffraction techniques [2, 3]. High degrees of vacuum help reduce surface contamination prior to the growth of a thin film, avoiding contamination of the growing thin film and preparing thin films with reproducible properties. With greatly advanced techniques, investigators have developed many dozens of deposition technologies to grow thin films used as optical components, electronic devices, magnetic sensors, and so on [4–13]. Thin films can be deposited on a substrate by various techniques such as physical vapor deposition (PVD), chemical vapor deposition (CVD), liquid-phase processes, and writing/printing methods, which are described in more detail in other chapters.

From the viewpoints of manufacturing and applications, it is very important to comprehensively consider factors such as substrate (metal, semiconductor, glass, ceramics, plastics, etc.), growth rate, uniformity of deposition, wide window of processing, adaptability to different conditions of operation, cost of instruments, reproducibility, throughput, yield, and reliability of the product [8–12]. Every deposition method has its own advantages and disadvantages for thin film growth. For example, the advantage of resistively heated evaporation is the ability to employ most of the source materials, but the disadvantages are the easy contamination by crucibles, heaters, and support materials and the limitation of relatively low input power levels. This makes it difficult to deposit highly pure films or evaporate high melting point materials at appreciable rates.

Electron beam deposition uses electrons to evaporate source materials or stimulate dissociation of molecular precursors under vacuum for fabricating thin films. The method eliminates the disadvantages of resistively heated evaporation

Advanced Nano Deposition Methods, First Edition. Edited by Yuan Lin and Xin Chen.

sources. With the commercial development of electron beam systems, electron beam deposition becomes a popular method for fabricating metal films because of its highly controllable rate and the ability to easily evaporate high melting point metals.

Electron beam deposition includes two modes: one is the electron beam evaporation deposition method, mainly for growing thin films with high melting point materials or highly pure metals. The source material is evaporated with gas release and then transferred to the substrate for deposition to form a thin film or coating. Electron beam deposition systems have gained popularity because a large variety of materials can be rapidly and cleanly evaporated and deposited with minimum consumption of energy. The highest temperature of the evaporating system appears at the area of contact of the electron beam with the evaporant surface. Different kinds of thin films of metals or nonmetals (such as Au, Pt, Ta, Pd, α-Fe, Al, Hf, Cu, Si, B, Be, Sb [14–37]), metal oxides (such as Ta_2O_5, TiO_2, CeO_2, HfO_2, In_2O_3, Fe_2O_3, SiO_2, ZnO, ZrO_2 [38–73]), metal alloys (such as Fe/Pt, Mg/Al, Co/Pt, Al/Ni, Sm/Co [74–85]), other compounds (such as BC, CdS, ZnS, CdSe, CdTe, ZnTe, MgF_2, TaN, SiN, Eu^{3+}:Y_2O_3 [86–110]), organic materials (such as InBrPcs (InBr-phthalocyanines), polyethene, polyaniline, graphene, fullerene [111–119]), and also multilayers (such as TiO_2/Cu/TiO_2, CdTe/Co/CdTe, CdTe/Cr/CdTe, Pt/Ga/C [120–126]) have been successively fabricated on various substrates by many research groups using the electron beam evaporation technique. The optical, electrical, magnetic, and other properties and also surface morphology of these films have been investigated. The quality of the thin film becomes better, which is reflected in the uniform surface morphology and precise thickness.

The other mode, electron-beam-induced deposition, which is a kind of focused ion-beam-induced deposition, is a direct-writing technique with nanometer resolution that allows the design and fabrication of special nanostructures at any position on a substrate. The basic principle of beam-induced deposition is described as follows and can also be found in the literature [127–131]. Gas molecules locally introduced into the sample chamber as a precursor via a small dosing nozzle are adsorbed on a substrate. The absorbed molecular precursor is dissociated into volatile and nonvolatile materials, and then the non-volatile materials deposit on the substrate with high purity. This beam-induced deposition occurs only around the beam irradiation positions. Therefore, a three-dimensional nanostructure can be fabricated at any position by controlling the beam scanning patterns and beam dwell times at each position [130, 131]. Furthermore, various materials (e.g., Au [15], Pt [14], Fe [18, 19], or SiO_x [56]) can be deposited depending on the precursor gas species (such as $Pt(PF_3)_4$, CH_3-Au-acetylacetone, $Fe(CO)_5$, $Si(OCH_3)_4$, $Co_2(CO)_8$). Chris et al. have grown Pt nanostructures with 94% purity using $Pt(PF_3)_4$ as the precursor by electron-beam-induced deposition. Electron-beam-induced deposition has now reached a state of maturity considering the advantages and disadvantages of each type of primary beam; beam-induced deposition techniques are available for the fabrication of field-emitter arrays with a needle shape with high aspect ratio, which is hard to fabricate by semiconductor nanofabrication processes.

The main contents in this chapter will focus on the introduction of electron beam evaporation deposition, including the principle, equipments for electron beam evaporation, application, and so on. We have also introduced some techniques for the characterization of thin films and potential applications of the electron beam technique.

2.2 Electron Beam Evaporation System

From the time Marcello van Pirani applied for patent on the focused electron beam melting of high melting point materials in vacuum in 1907, many kinds of electron beam heating sources have been designed and used in thin film deposition; the most common ones are the e-type electron gun, hollow-cathode electron gun, and high-voltage electron gun, of which the hollow-cathode electron gun and high-voltage electron gun can generate higher power and are mainly used in ion plating equipment, scanning electron microscopy (SEM), and transmission electron microscopy [1, 127]. The e-type electron gun, which is mainly used for thin film deposition, will be introduced in this chapter.

Electron beam evaporation deposition is an important coating technique using electron beams directly for heating target materials placed in a water-cooling crucible and then forming thin films on substrates. Electron beam evaporation can heat target materials up to 6000 °C; almost all of the common materials can be melted and can be deposited on substrates of metal, alloy, oxide, semiconductor, and plastic at a high rate. The deposition rate for a compound is about 100 nm/s, but the deposition rate can reach 1 μm/s for metals, which is two to three orders of magnitude more than the deposition rate achieved with sputtering methods.

2.2.1 Heating Principle and Characters of Electron Beams

2.2.1.1 Heating Principle of Electron Beams

A wire filament is passed through a high current to heat it to a high temperature. A stream of electrons is emitted from the filament surface at this high temperature. These electrons with kinetic energies are accelerated by the acceleration of the anode, directed by magnetic lens, and focused to bombard the source material for melting. This high-energy bombardment heats up the source to the point where it melts and evaporates (or sublimes).

If the initial velocity is not considered, the kinetic energy of an electron after acceleration is $\frac{1}{2}mv^2$ and it should be equal to the electric potential energy of the electron at the initial position. The relationship is as follows:

$$\frac{1}{2}mv^2 = eU \tag{2.1}$$

where m is the mass of electron (9.1×10^{-31} kg), e is the electron charge (1.6×10^{-19} C), and U is the voltage of acceleration. The velocity of electrons hitting the source materials should be reduced:

$$v = 5.93 \times 10^5 \sqrt{U}\,(\mathrm{m/s}) \tag{2.2}$$

If $U = 10$ kV, the electron velocity should reach 6×10^4 m/s. These electrons flowing with high velocity are focused into a beam by an electromagnetic field and they bombard the surface of the evaporant, and the kinetic energy is transformed into heat energy. The power of the electron is

$$W = neU = IU \tag{2.3}$$

where n is the electron flow (s^{-1}) and I is the electron beam current (A). If t is the dwell time of the electron beam, the heat Q(J) should be

$$Q = 0.24Wt \tag{2.4}$$

When a high accelerating voltage is used, the heat generated as in the above-mentioned formula should be adequate to make the target material evaporate; thus electron beam heating becomes a good heating source in the vacuum evaporation technique.

The target material can be held in a crucible that may or may not be cooled. Some materials operate using a water-cooled metallic crucible and rely on the thermal gradient between the crucible and the surface to enable a molten pool to be formed at the surface for evaporation but for the material coming in contact with the crucible to remain a solid. In this way, the material stays pure and is much less likely to react with the crucible than in the molten form and to come in contact with the crucible where it might be possible to form an alloy or to have some crucible material diffusing into the molten pool.

2.2.1.2 Characters of Electron Beams

The advantages of electron beam evaporation are as follows:

1. The energy density obtained with an electron beam bombarded heat source with high-beam current density is far more than that obtained with a source of resistance heating. Power densities of 10^4-10^9 W/cm^2 can be reached in a smaller area so that it can make high melting point material evaporation easy, and further get higher evaporation rates. For example, Au, Pt, Mo, Ge, Al, Hf, Fe, CeO_2, TiO_2, SiO_2, and Al_2O_3 can be evaporated by an electron beam [15–73].
2. The source material is placed in the water-cooling crucible, which can avoid contamination from the evaporation of the material of the container, and the reaction between the material of the container and the coating material, which is crucial for improving the purity of the thin film that is coating on the substrate.
3. The heat quantity of the electron beam can directly act on the surface of the evaporation material, so that the heat efficiency is higher than that with

resistance heating; loss due to heat conduction and the loss due to heat radiation are less.

4. The structure of the electron gun is complex, and the cost of fabrication of the equipment system of electron beam deposition is high. The instruments need to be in high voltage when the equipments are working.

2.2.2 Equipments of Electron Beam Source

The instrumental system for electron beam evaporation contains mainly the vacuum chamber and the electron beam source (including electron beam control, power supply, crucible, and feed system) and is also assisted with other deposition accessories such as effusion cell, ion-beam gun, target materials container, sample manipulator for substrate, shutter and other accessories for *in situ* characterization, and the conjoint vacuum chamber for the measurements of the thin films in the same condition. The whole process of electron beam evaporation is performed in vacuum, which requires a vacuum less than 10^{-2} Pa for its operation while the vacuum in the electron beam generating chamber of the gun is usually 10^{-4} Pa or less. The vacuum available in the instrumental system has been enhanced to 10^{-10} Pa now [2, 3, 130, 131].

2.2.2.1 Filament and Electron Emission

Firstly, the core component for electron beam evaporation in vacuum is the electron beam source. Figure 2.1 shows the schematic design. A bright white hot filament can produce electrons by using a power supply with low voltage and high current. The electrons are emitted from the filament surface with random directions. By surrounding the filament with a negatively charged cathode surface except for a slit, the electrons through the slit can move in a specific direction. To avoid electron loss due to gas collisions at low pressure, the electron beam gun with high power has its own pumping system. When AC or DC is used as power supply to the filament, the high current flow may generate a magnetic field and affect electron emission. DC power supply is the best choice as it tends to make the electron beams tighter, and AC power supply can induce a broader electron beam [132–136].

The quantity of emitted electrons depends on the current passed through the filament, the achieved temperature, and the work function along with other factors. The three factors are relative to the material chosen for the filament. The materials used as filament to emit electrons are usually tungsten (W), tungsten–rhenium alloys (W/Re), tantalum (Ta), and lanthanum hexaboride (LaB_6). Most electron guns use tungsten as a filament [5–9].

Now many designs for high-power electron beam systems have a separate vacuum compartment for the electron beam filament, which has its own pumping system. This chamber can be isolated from the rest of the system, thus allowing fast access for filament changes.

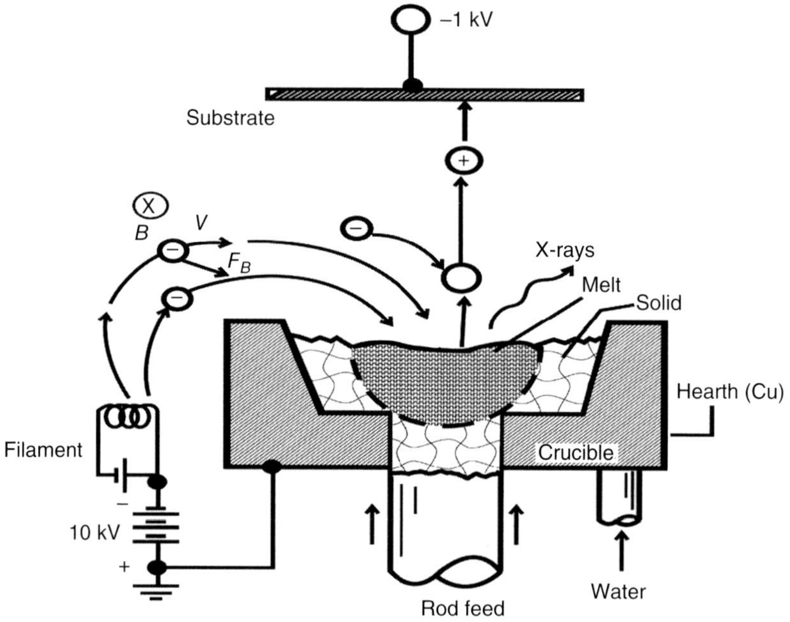

Figure 2.1 The deposition process of electron-beam-induced evaporation by e-type electron gun. (Harsha [5]. Reproduced with permission of Elsevier.)

When the electrons hit the evaporant material, many interactions occur between the incident electrons and the atoms of the evaporant material. Backscattered electrons, secondary electrons, thermionic electrons, and X-rays have been produced by bombardment and detected at the upper surface of the evaporant material. However, a large portion of the incident kinetic energy of the electron beam is converted into heat [132–137]. Any of the lost electrons coming from the electron beam and all of secondary electrons produced from the evaporant material need to be captured by an electron trap to avoid generating electrets in some cases that remain in the substrate, and arcing, which is often caused by oxide buildup [132].

2.2.2.2 **Electron Beam Control**

The electron beam can be accelerated by the anode. The acceleration voltage can get as high as 100 kV, at a power of 1.2 MW now [11]. Once this beam has been generated, it can be focused and manipulated in the same way using electromagnetic sweep coils. There are several sources designed with different electron paths of electron beam guns as seen in Figure 2.2. When using in-focused electron beam deposition, the beam is pointed straight at the source material and the sweep coils are used to move the beam across the whole surface to provide a large area rather than a point source, as in the *straight gun mode*. In others, the electron beam is bent through as much as 270° into the crucible by a magnetic lens, as in the *e-type gun mode*, which was utilized in electron beam evaporation deposition.

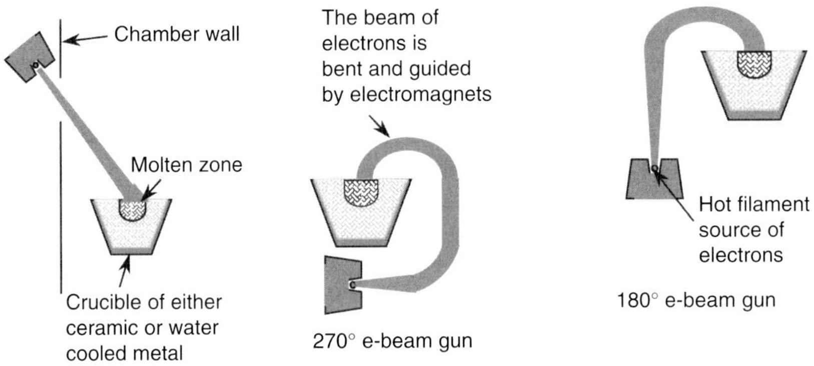

Figure 2.2 Schematic drawings showing several different sources with different electron paths. (McGuire [6]. Reproduced with permission of Elsevier.)

Most e-beam evaporators are designed with an e-type gun for the controlled evaporation of materials under ultrahigh vacuum conditions. Electron beam systems become miniature with more than one pocket and can be arranged in small miniarrays; it is possible to dispense with any scanning system on each of the electron beam guns. This can make the uniformity of heating and stability of deposition rate better than that for a single scanned source [6].

The electron beam can be controlled by the power, the beam shape (diameter), the raster pattern, and the residence time. The electron beam is extremely powerful and can damage the target material if used wrongly. The source material can be in a variety of forms from powder to solid lumps that may be anything from a good to a poor thermal conductor. When a high-power electron beam hits the source material, it blows all the powder out of the crucible, warms up the pockets of gas trapped in the block, and blows the block apart. The heating rate of the source materials is determined by the thermal conductivity and power supply. Materials with a poor thermal conductivity may be heated up slowly by building up the power supply slowly. Furthermore, the beam can be defocused and moved across the whole surface for preheating, thus bringing the whole surface up to the required temperature at the same time. To evaporate a high melting point metal, it may be necessary to keep the beam more tightly focused to attain high temperatures and to limit the size of the molten pool by limiting the raster pattern. If there are greater heat losses at the ends of the crucible, it is common to compensate for this by increasing the beam residence time at each end. This also results in a small increase in the evaporation rate, enabling an increase in the deposition that not only compensates for the end losses but also improves the deposition profile. For mixed materials, the electron beam can be scanned in predetermined patterns with slow or stationary points such as at the ends of the crucible and quicker points across the center of the crucibles.

2.2.2.3 Power Supply, Crucibles, and Feed Systems

The power supply must do more than supply a highly stabilized voltage and current for electron beam guns. The power supply has to be well protected from arcing. The emission circuit is dependent on a source with very high stability and a current-sensing resistor, which ensure a constant electron emission current flow. The voltage for acceleration also utilizes a constant voltage to control the circuit, thus keeping everything as stable as possible. The raster coils have sinusoidal or saw tooth waveforms available as well as the basic beam direction controls.

Crucibles of various high melting point materials, such as Ti, W, Mo, Cu, WC, BN, Al_2O_3, and graphite, are employed for containing source materials according to the requirements of the experimental condition. Water-cooled copper crucibles as the common choice are adequate for the evaporation of most target materials. Containing the evaporant in water-cooled crucibles permits the evaporation of high melting and reactive materials. It is crucial to ascertain that oxide layers are absent to ensure efficient heat transfer between the evaporant and the copper crucible. If high heat losses cannot be tolerated because the thermal conductivity of the evaporant material is large, then one would require higher powers of operation to melt the evaporation materials. Crucible inserts of insulating materials (such as ceramics) are often utilized to act as heat barriers. The heat losses to the crucible can thus be reduced so that extremely high evaporation rates are achievable as a consequence of enhanced utilization of energy.

The material to be evaporated, the desired evaporation rate, energy requirements, and the nature of vapor distribution dominate the dimensions and kinds of crucible. The crucible type and the electron beam control should be optimized to suit the different materials according to all the requirements. There are many variations to the design of crucibles to allow for compensation of the source material. There have been water-cooled crucibles where the source material has been fed as a solid rod through the base of the crucible and into the molten source. The feed of the rod is aimed at keeping the liquid level constant. Others have used a specially designed annulus crucible with the electron beam directed into the crucible at one point; fresh material is fed into the annulus directly opposite to that point and the annulus crucible is made to rotate slowly. Another alternative is to add pellets to the molten pool periodically. One option is to have two or more parallel crucibles and a single electron beam gun that moves between sources, keeping both evaporating. The composition of the alloy or the multilayer can be varied by controlling the residence time on each crucible and hence the rates of individual deposition.

2.2.2.4 **Source Materials**

The attraction of electron beam systems derives from the ability to evaporate not only elements with low melting points (e.g., Zn, Cd, As, Al, Ag) and high melting points (e.g., Mo, Pt, W, C) [14–37] but also elements with highly reactivity (e.g., Ti, Ta, Nb) by making use of the skull that forms inside a water-cooled crucible. Oxides, and other compounds and alloys (e.g., SiO_2, Al_2O_3, HfO_2, ZnS, ZnSe, Ru-β-FeSi, TaN, InAlN, FePt, SmCo) can also be deposited on the substrate, forming a thin film by electron beam evaporation [38–82]. Now, some molecular precursors, such as $Co(CO)_8$, $Fe(CO)_5$, $W(CO)_6$, $Si(OCH_3)_4$, and $Pt(PF_3)_4$, can also be used as evaporants to fabricate thin films or nanostructures with high purity by focused electron beam deposition [15–19]. Table 2.1 shows the evaporation characteristics of source materials used in recent research.

When extremely high power densities for pure metal evaporation are accommodated by splashes of molten liquid to be ejected from the metal surface, the volatile contaminants of thin films may come from gaseous carbon in the vacuum system, or a slag accumulating on the surface of the molten material by reaction with background gases with the metal or by precipitation of bulk contaminates on metal melting, or a gas pocket trapped within the sintered material. Consequently, in order to minimize the contamination as far as possible, source materials should be of high purity and be melted under vacuum. It is also possible to make crucible-free evaporation by using a large ingot of material in which crucibles have to be continuously fed during evaporation. The material can be supplied either in the form of a rod or wire or slice or supplied through a storage hopper.

In the process of depositing alloys, it is very difficult to achieve uniform composition and thickness over the whole substrate area. Many alloys, such as Fe/Pt, Mg/Al, Au/Pd, Ti/Pt, Pt/C, Co/Pt, Pt/Si, and Pt/Ga/C, with wide vapor pressures between the elements and alloys with lower concentration of solutes have been evaporated with the given stoichiometric composition successfully using the technique of multiple crucibles [74–82]. Since the evaporation rate depends exponentially on temperature, even a small fluctuation in source temperature results in a large variation in the composition of the alloy thin film. The vapor stream of the composition is regulated by controlling the temperature of the individual sources. The ratio of the separation distance between the crucibles and the distance of the substrate from the sources gives the range over which adequate mixing of the vapor occurs because of the directional nature of the vapor stream. Controlling the distribution of electron beam power to the crucibles helps adjust the evaporation rates by controlling the dwell time and location of the electron beam. Techniques such as the use of separate sources from adjacent crucibles and continuous feeding of the evaporant source with excess of high-pressure evaporating material are necessary to keep a steady-state composition for the depositing alloys. If the evaporant has impurities, such as oxides and carbides, they rise to the molten surface because of their low density and cover the liquid surface partially. The impurities result in the evaporation rates being reduced considerably.

Oxides and other compounds, such as SiO_2, Ta_2O_5, In_2O_3, HfO_2, InAlN, Ge-β-FeSi, and Ru-β-FeSi, can be deposited when they evaporate suitably [38–73, 83–109]. Problems encountered in charging the evaporant as well as in cracking the evaporant material are due to the lack of electrical conductivity in a compound phase. These are solved by controlling power densities, scanning an electron beam in a specific pattern, and controlling the density of the target used as the evaporant source (e.g., using a powder compact). In order for the compounds to dissociate on evaporation, reactive gases are added and transmitted into the working vacuum chamber. However, these gases increase the pressure, which leads to the formation of porous thin films due to the thermalization of vapors by intermolecular collisions. Accumulation of local surface charge can occur while vaporizing solid surfaces of electrically insulating materials. This eventually leads to surface arcing,

Table 2.1 Evaporation characteristics of target materials.

Materials		Minimum evaporation temperature[a)]	State of evaporation	Recommended crucible materials	Deposition rate (nm/s)	Power (kW) of e-beam[b]
Metals or nonmetals	Be	1000	Melts	BeO, graphite	10	1.5
	B	1800	Melts	WC, graphite	1	1.5
	Al	1010	Melts	BN	2	5
	Si	1337	Melts	BeO	1.5	0.15
	Ti	1453	Melts	Graphite	2	1.5
	Fe	1180	Melts	BeO, Al_2O_3	5	2.5
	Co	1200	Melts	BeO, Al_2O_3	2	2.0
	Ni	1262	Melts	Al_2O_3	2.5	3.0
	Cu	1017	Melts	Al_2O_3, graphite	5	0.2
	Ge	1167	Melts	Graphite	2.5	3.0
	Zr	1987	Melts	W	2	5.0
	Mo	2117	Melts	Graphite	4	4.0
	Cd	180	Melts	Quartz, Al_2O_3	3	0.3
	In	742	Melts	Al_2O_3	10	0.1
	Sn	997	Melts	Graphite, Al_2O_3	1	2.0
	Sb	425	Melts	Al_2O_3, BN	5	0.5
	Ta	2590	Semimelts	Mo	10	1.5
	W	2757	Melts	Mo	2	5.5
	Pt	1747	Melts	Graphite	2	4.0
	Au	1132	Melts	BN, Al_2O_3	3	6.0
	Pb	497	Melts	Al_2O_3	3	0.1
	C	2140	Sublimes	Mo	3	1.0
	As	210	Sublimes	Al_2O_3	10	0.1
	Cr	1157	Sublimes	W	1.5	0.3
	Zn	250	Sublimes	Al_2O_3	5	0.25
	Mg	327	Sublimes	Graphite	10	0.04
Alloys	FePt	1573	Melts	Mo	0.03	9
	SmCo	1818	Melts	Mo	0.5	10
	MgAl	736	Melts	Mo	0.2	10
Oxides	BeO	2650	Melts	Graphite	0.4	1.0
	Al_2O_3	1325	Semimelts	Copper	1	0.5
	SiO_2	850	Semimelts	Ta	2	0.7
	CeO_2	2670	Melts	Ti	0.1	26
	MnO_2	808	Melts	Graphite	0.35	0.42
	TiO_2	1300	Melts	W	1	1.0
Others compounds	BC	2623	Semimelts	Mo	3.5	1.0
	CdS	250	Sublimes	Graphite	0.2	0.2
	CaF_2	1696	Semimelts	Mo, graphite	3	0.05
	LiF	1180	Melts	W, Mo	1	0.15
	MgF_2	1540	Semimelts	Al_2O_3	3	0.01
	ZnS	300	Sublimes	Mo	0.12	1.6
	ZnSe	660	Sublimes	Quartz	0.5	6

a) Temperature (°C) at which vapor pressure is 10^{-2} Pa.
b) For 10 kV, copper hearth, source-substrate distance of 400 mm.
Adapted from Refs. [15–126].

which can cause particulate contamination in the deposition system. The properties of the compound thin film are sensitively dependent on the stoichiometry of the thin film.

Electron beam evaporation makes vapor streams travel in straight lines until they hit the substrate. It is difficult to deposit atoms in selected areas such as deep trenches and vias in semiconductor technology or to coat fibers fully in composites because of the line of sight path of the vapor atoms. Most of the evaporant deposits in undesired locations contribute to poor material utilization efficiency. A new technique of focused electron-beam-induced deposition has helped realize the deposition of atoms in selected areas, which has received increasing attention in the last decade [127–131].

2.2.3 Application of Electron Beam Evaporation

2.2.3.1 Cooling of Electron Beam Gun

An electron beam gun needs high voltage for acceleration, and beam energy is therefore very high; but the whole of the gun head device has poor heat dissipation in the vacuum chamber, so cooling of the gun head is necessary. Cooling includes water cooling of the crucible, the pole of electron collector, the magnetic pole, and the high-voltage electrode. When the electron energy is less than 5 keV, water flow rate is about 5–6 L/min for the cooling of the copper crucible and the inner diameter of the external cooling tube is about 5–6 mm. The power of the electron gun promotes 1 kW and the flow rate of the cooling water increases by about 0.8–1 L. When the output power of the electron gun reaches 10 kW, the water flow rate should reach 10–12 L/min and the inner diameter of the external water-cooling pipe correspondingly increased by 8–9 mm. The cooling of the high-voltage electrode should account for the magnitude of the resistance of water; the cooling water pipe should have sufficient length to ensure that the outlet potential tends to zero. Furthermore, pipes for cooling water cannot come in contact with any metal parts at ground potential, and the outer wall of the water pipe should be kept dry. Cooling can be achieved with deionized water or high-quality tap water with resistivity of 10^4–10^5 Ω cm. In order to spare the water resource, a low-temperature circulation bath is utilized for cooling the electron beam system; the water temperature in the bath can be adjusted to remain constant, which can maintain the consistency of the temperature of cooling water and ensure normal operation of the electron beam instrument system.

2.2.3.2 Evaporation of Source Materials by Electron Beam

The electron beam cannot be focused directly on source materials; when it is in a defocused state, it can preheat the entire area of the source material in the crucible at the same time. During deposition, the electron beam is focused on a spot and made to impact on the source material, which accepts the beam energy, and the temperature of the site of impact rises rapidly to above 2000 °C.

If the focused spot of the electron beam does not move, then only the source material at the spot melts partially and evaporates, especially for oxides, because of the poor heat transfer of oxides. It is possible to penetrate the crucible containing the source material. Source materials have smooth surfaces before heating; if the electron beam does not move, then only electron beam hits parts of the source material ablation, elsewhere in the evaporation rarely, only in the local place caused ablation becomes a deep pit, finally bad result is the breakdown of the crucible. Thin films are related to the evaporation of source materials contained in the crucible, and crucibles are of many kinds, commonly made of molybdenum; not only the molybdenum material is expensive but processing costs are also high. So it is forbidden to heat the focused spot to hit one point, in order to avoid damage to the crucible. The solution lies in controlling the electron beam to scan with a speed along two directions so that the source material can be uniformly heated and evaporated.

So when the electron gun is used, the electron beam in the defocused state can be used for the initial preheating, the focusing electron beam used for coating thin films, and the electron beam scanned in the X and Y directions at a certain speed to make the heating of the source materials uniform.

2.2.3.3 Vacuum Deposition Process of Electron Beam Evaporation

The following is an example of the process of electron beam evaporation of Ag, Cu, and other metals or Ta_2O_5, SiO_2 and other oxide thin film coating [15–82].

1. Transfer the clean substrate into the vacuum chamber and fix it. Put an Ag rod or Ta_2O_5 (WO_2, SiO_2) pellets into the water-cooled crucible, rotate the baffle to the above position of the evaporation source, and finally close the vacuum chamber door.
2. The vacuum in the electron gun chamber is below 5×10^{-8} Pa and that in the sample chamber is below 5×10^{-4} Pa; the substrate should be heated and baked to a certain temperature firstly and then for the preheating and melting of source materials.
3. When the pressure of the vacuum chamber is decreased to 5×10^{-3} Pa, the heating current is increased to evaporate the source material. When the evaporation rate is stable, the substrate holder is rotated and the baffle is opened above the evaporation source, and the atoms of source material can be deposited onto the substrate at a stable rate.
4. When the thickness of the thin film reaches the required level, monitored by a quartz crystal resonator, the evaporation source stops heating, and the baffle is rotated above the evaporation source and the substrate rotation is stopped. The vacuum should be pumped for more than 20 min and then the pumping system closed.
5. After the source is cooled to a safe temperature, the vacuum chamber is filled with high-purity nitrogen gas to get the pressure balance; it is then opened, the sample is transferred out, and the vacuum chamber is closed.

If you need to deposit multilayer thin films, then you need to choose more than two crucibles placed with different source materials in the vacuum chamber; when one layer is deposited, rotate the crucible and align the next material to be evaporated with the electron beam, and then evaporate for other layers. The operation is simple and convenient.

2.2.3.4 Attention and Warning for Electron Beam Evaporation

The following aspects need to be paid attention to when electron beam evaporation coating is in process:

1. Firstly, preheating the source materials is done using the defocused electron beam.
2. When to evaporate insulated powder of source materials such as CeO_2, due to the accumulation of electron in the surface of source material, which make the powder particles carrying negative charges, and charged particles are mutually exclusive, leading to source material in the vacuum chamber splash. In order to avoid this phenomenon, at initial heating the heating current needs to be smaller so that the melting of the source material is partial, and then the electron beam flux can be increased so that no phenomenon of source material splashing occurs.
3. When multicrucible systems are used for electron beam evaporation, the evaporated source material may deposit at the apertures, resulting in making the lid of crucible stuck or polluting other evaporation sources.

The following warnings need to be borne in mind during operation of the evaporation source:

1. The source can deliver lethal currents at voltages reaching 2000 V or more, so extreme caution should be used at all times. Ensure that the power supply is properly grounded. Never, under any circumstances, open any power supplies. Never run the source without water cooling.
2. Never attach the power cable to the source unless the source is in an evacuated, grounded vacuum chamber. Do not remove the electron beam source from the chamber unless all cables, gas lines, and water lines have been removed.
3. The molybdenum crucibles that hold the evaporation sources are electrically isolated from the cooling plate by a beryllium oxide spacer. Beryllium oxide is very toxic if inhaled. Never perform any kind of abrasive or other process that will result in generation of dust. Consult a Material Safety Data Sheet for beryllium oxide before handling it in any way.

2.3 Characterization of Thin Film

The application of thin films is guided by their physical and chemical properties, which need to be closely controlled. The structural properties of thin films are decided by the thickness, chemical composition, crystal phase and crystallinity, microstructure and the surface and interface structures of morphology, stresses in the films, and so on. Various measurements have been developed for research on thin film properties such as the crystal structure; these can be performed by X-ray diffraction (XRD), energy dispersive X-ray spectroscopy (EDXS), and reflected low-energy electron diffraction (RLEED); morphology can be characterized by atomic force microscopy (AFM), SEM, and scanning tunneling microscopy (STM) [138–142]. The thickness of thin films can be monitored by a quartz crystal resonator and verified by interference meters and spectroscopic ellipsometry [143, 144].

Now a new method for the determination of thickness of thin films by using ellipsometry iterated with transmittance methods has confirmed the effective optical constants of Ag thin film determined with effective thickness simultaneously [145].

2.3.1 Surface Morphology by AFM

In contrast to many other surface analytical techniques such as STM, AFM does not require vacuum. AFM is used to study the morphology of the sample surface, which can measure almost any of the sample surfaces, conduct or not. Atomic resolution can be obtained and imaging can also be performed on areas larger than 100 μm^2. A further important advantage of AFM is the fact that images contain direct depth information, which makes the technique a valuable metrological tool. AFM can test and verify the thickness reciprocal with the result measured by spectroscopic ellipsometry [142].

The properties of thin films are related to the thickness and surface morphology. Figure 2.3 displays AFM images of the copper thin films for three different effective thickness values: 5.3, 9.1, and 13.1 nm, which are prepared by electron

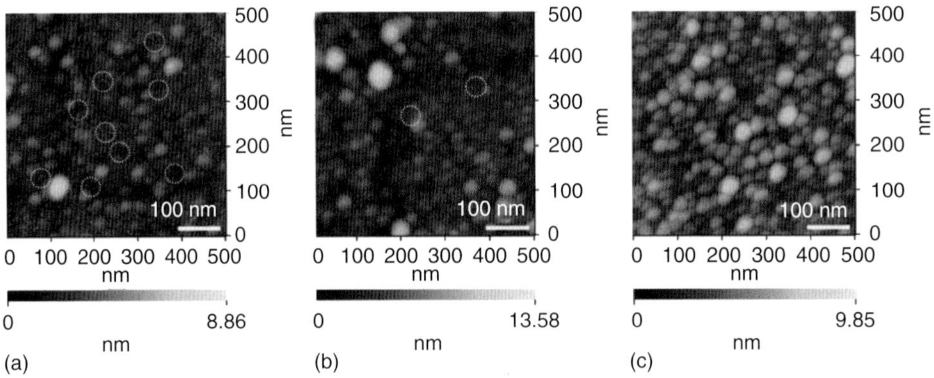

Figure 2.3 AFM images (500 nm × 500 nm) of Cu films with effective thickness values of (a) 4.3 nm, (b) 9.1 nm, and (c) 13.1 nm. Some interspaces between islands are marked by white circles. Copper thin films with nanometer-range thickness are prepared by electron beam evaporation of QUAD-EV-C HP mini e-beam evaporator (Mantis Deposition Ltd). (Gong *et al.* [24]. Reproduced with permission of Institute of Physics.)

beam evaporation [24]. The images show that these ultrathin films do not form a smooth uniform layer. Instead, they all consist of nanoparticles grown from nanoislands. For a film of thickness 5.3 nm, the islands are individually separate from each other and there exist some obvious interspaces between them on the substrate. Some interspaces are marked by white circles as in Figure 2.3. When the thickness increases to 9.1 nm, the islands are connected to form nanoparticles that cover the entire substrate surface with smaller interspaces. However, the metallic behavior of bulk copper does not appear because of the presence of insufficient copper atoms on the substrate. At a thickness of 13.1 nm, the substrate is completely covered; the metallic behavior starts to appear above the percolation threshold at 10 nm. The AFM images indicate that the percolation threshold occurs somewhere between 9.1 and 13.1 nm, which can be verified with the value determined from the optical constants measured by ellipsometry.

2.3.2 Thickness Measurement by Spectroscopic Ellipsometry

Ellipsometry is an optical technique that characterizes the change in the polarization state through light reflection or transmission to analyze the film thickness, the optical constant of the sample, material gradients and mixtures of different materials, layer stacks, and substrate materials with very high sensitivity [143, 144]. Film thicknesses from 0.1 nm to 100 μm can be measured; depending on the spectral range used for the analysis and the homogeneity of thicker films, thicknesses less than 1 μm can be determined with sensitivity better than 0.01 nm; thickness in the microrange can be analyzed with sensitivity typically better than 1 nm. In general, the measurement using spectroscopic ellipsometry is carried out in the ultraviolet/visible region, but measurements in the infrared region have also been performed widely.

The key feature of ellipsometry is that it measures the change in polarized light upon light reflection or transmission on a sample. Figure 2.4 presents the measurement principle of ellipsometry [143]. The polarization state is expressed by superimposing waves propagating along two orthogonal axes. The polarization

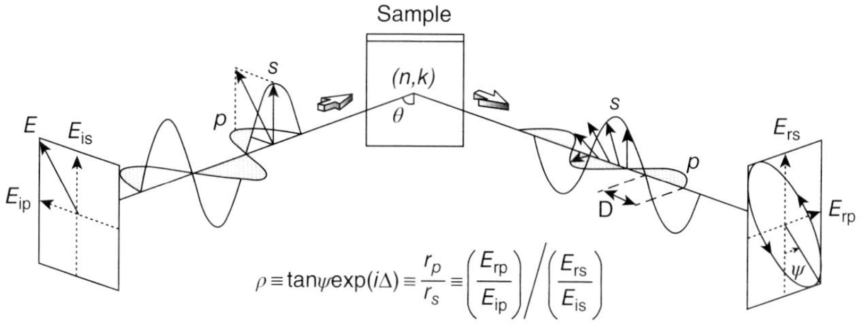

Figure 2.4 Measurement principle of ellipsometry. (Fujiwara [143]. Reproduced with permission of Wiley.)

states of incident and reflected light waves are described by the coordinates of p- and s-polarizations. It can be seen from the incident vectors $\boldsymbol{E}_{ip}$ and $\boldsymbol{E}_{is}$ that the directions of the electric field vectors for p- and s-polarizations are reversed on both incident and reflection sides as in Figure 2.4 and that the incident light is linear polarization oriented at +45° relative to the E_{ip} axis. In particular, the expression $E_{ip} = E_{is}$ holds for this polarization since the amplitudes of p- and s-polarizations are the same and the phase difference between the polarizations is zero. The amplitude reflection coefficients for p- and s-polarizations differ significantly because of the difference in electric dipole radiation. Thus, upon light reflection on a sample, p- and s-polarizations show different changes in amplitude and phase. As shown in Figure 2.4, ellipsometry measures the two values (Ψ, Δ) that represent the amplitude ratio Ψ and phase difference Δ between p- and s-polarized light waves. In spectroscopic ellipsometry, (Ψ, Δ) spectra are measured by changing the wavelength of light. In particular, when a sample structure is simple, the amplitude ratio Ψ is characterized by the refractive index n, while Δ represents light absorption described by the extinction coefficient k. In this case, the two values (n, k) can be determined directly from the two ellipsometry parameters (Ψ, Δ) obtained from a measurement by applying the Fresnel equations [143].

The information of the properties of the sample is contained in the complex ratio, ρ, which describes the connection between the measured quantities Ψ and Δ and the sample properties included in the coefficients of reflection and hence in ρ. This complex equation can be solved analytically for pure substrates only. Each complex structure requires optical modeling of the sample and fitting of the calculated ratio ρ to the measured quantities Ψ and Δ. Film thickness, refractive index, extinction coefficient, and other properties are calculated on the basis of a model. The choice of the correct optical model for the sample is a fundamental assumption for correct values of all the properties calculated.

The thickness of the film influences the interference of light waves reflected from the front and back of the film, and hence the reflectance. Therefore, the thickness of an absorbing film can be measured only as long as there is still contribution from the back of the film to the reflectance of the sample. Ellipsometric measurements rely on the incident angle, which is close to the Brewster angle of the substrate, and is highly sensitive for very thin layers on top of the substrate. Multiple angle measurements are suitable for confirming optical models of sample structure.

Two measured ellipsometric angles Ψ and Δ at a fixed wavelength and a fixed angle of incidence enable the calculation of a maximum of two other properties, for example, the film thickness and refractive index of a transparent layer. Multiple angle measurements increase the number of measured quantities and hence the number of properties that can be determined for a specific sample. Spectroscopic ellipsometry further increases the number of measurable properties, and hence the complexity of the analyzable sample structure.

The relationship between the wavelength and Ψ or Δ angle is presented in Figures 2.5 and 2.6 using data measured by a spectroscopic ellipsometer with a Woollam M-2000U system. Figures 2.5 and 2.6 show that the measured angles Ψ

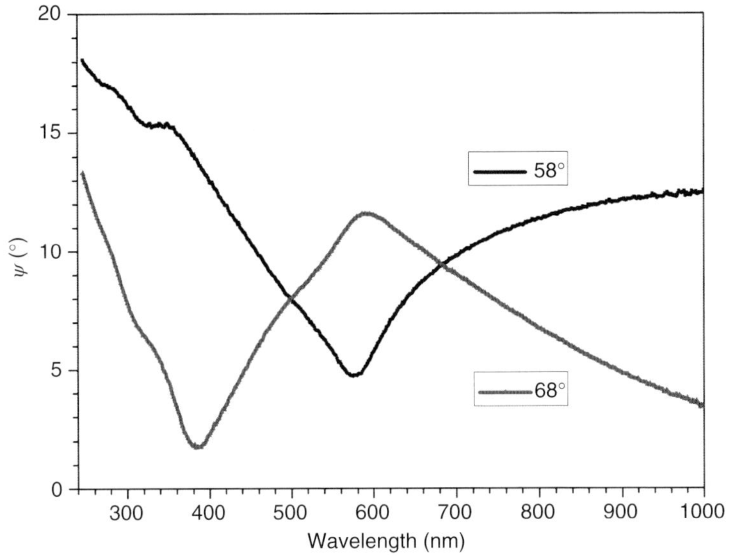

Figure 2.5 The measured angles Ψ change with wavelength from 240 to 1000 nm at fixed angles 58° and 68°. Copper thin films with nanometer-range thickness are prepared by electron beam evaporation of QUAD-EV-C HP mini e-beam evaporator (Mantis Deposition Ltd) [24].

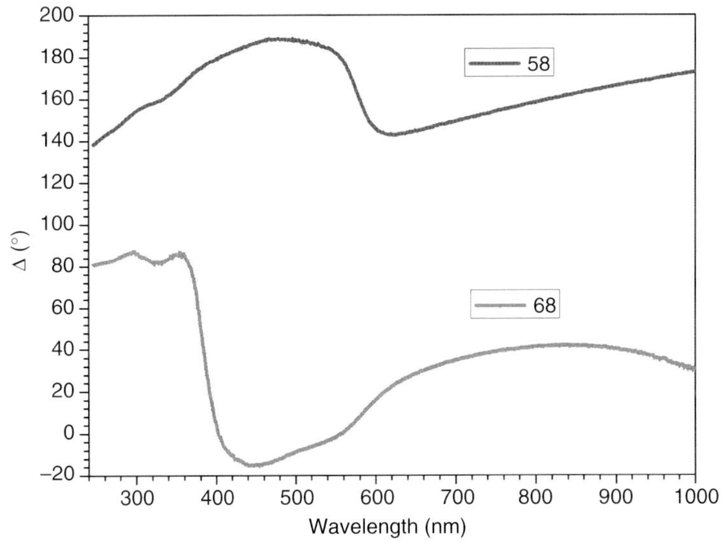

Figure 2.6 The measured angles Δ change with wavelength from 240 to 1000 nm at fixed angles 58° and 68°. Copper thin films with nanometer-range thickness are prepared by electron beam evaporation of QUAD-EV-C HP mini e-beam evaporator (Mantis Deposition Ltd) [24].

and Δ change with wavelength from 240 to 1000 nm at incident angles of 58° and 68°, respectively [24].

According to the measured data in Figures 2.5 and 2.6, thin film thickness, optical constants, and other properties are the quantities calculated on the basis of a model of ellipsometry equation and Hadley equation [143, 144]. This effective thickness of thin film compares with mass thickness measured by quartz crystal resonator and physical thickness tested by direct height measurement by AFM. The results of thickness measured by the present spectroscopic ellipsometric method can effectively characterize the thickness values of ultrathin films. This also lays a solid foundation for studying the thickness dependence of the effective optical constants and oxidation of ultrathin copper films.

The optical properties of ultrathin films are significantly different from those of bulk materials; the optical database that is commonly in use is actually not applicable to ultrathin films of thickness under several tens of nanometers considering that it has been proved that the optical constants of a film are dependent on thickness when the film is very thin. Some thin films become more and more transparent with thickness reduction to several nanometers, accompanied by change in the ratio of reflectance and absorbance. Furthermore, the optical constants of ultrathin films need to be determined precisely. Ellipsometry, as the most effective optical method to obtain optical constants, also faces a challenge when applied to ultrathin absorbing films. To solve this problem, an ellipsometry iterated with transmittance method has been proposed and used to determine effective optical constants of nanothin films of silver. The flowchart of this method is shown in Figure 2.7. First, with an assumed initial thickness value and the measured ellipsometric parameters, Ψ and Δ, the initial optical constants are obtained on the basis of the ellipsometry equation. Then a transmittance spectrum is computed

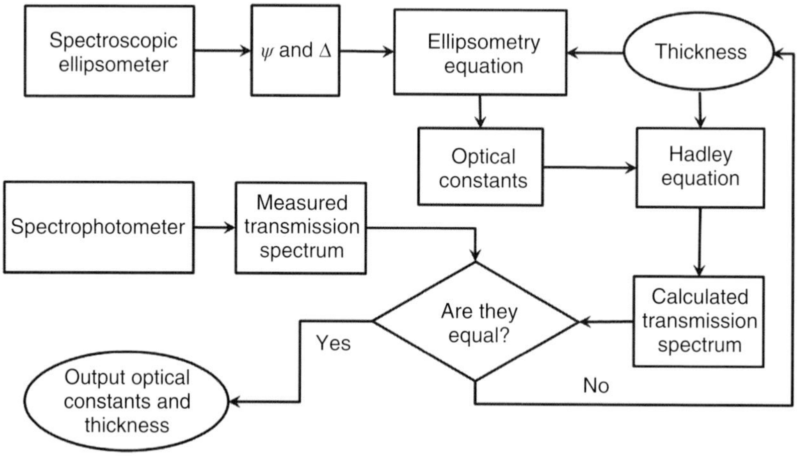

Figure 2.7 Schematic ellipsometry iterated with transmittance method. (Gong *et al.* [145], http://www.nature.com/articles/srep09279. Used under CC BY 4.0 http://creativecommons.org/licenses/by/4.0/)

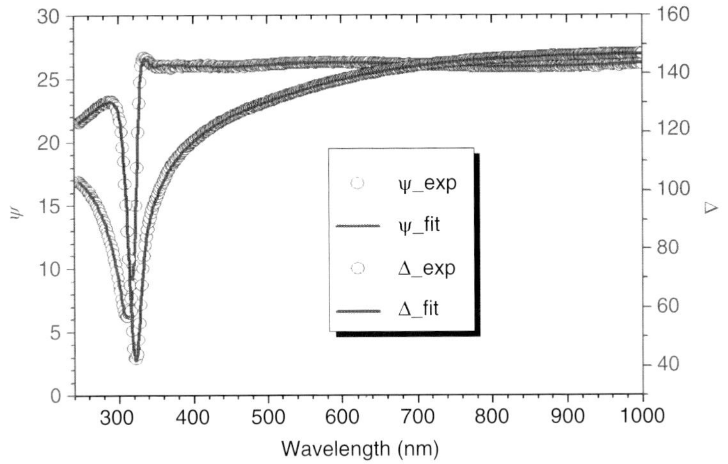

Figure 2.8 Ellipsometric parameter-fitting spectra for 9.6-nm silver film at 58° incident angle. (Gong *et al.* [145], http://www.nature.com/articles/srep09279. Used under CC BY 4.0 http://creativecommons.org/licenses/by/4.0/)

by using this initial thickness and the optical constants based on Hadley equation [24, 145]. The calculated transmittance spectrum $T_{cal}(\lambda)$ is then compared with the measured spectrum $T_{exp}(\lambda)$.

Mean square error (MSE) is used to quantify the difference, and then the thickness value is altered to minimize MSE. The iteration procedure is performed until a minimum MSE value is achieved, indicating that the final thickness value derived and the optical constants altogether satisfy the ellipsometry equation and the Hadley equation. Figure 2.8 presents ellipsometric parameter-fitting spectra for a 9.6 nm silver thin film at 58° incident angle. The typical fitted ellipsometry spectra for ellipsometric parameters and the calculated parameters match very well with the measured data [24].

The measured transmittance spectra and spectroscopic ellipsometric spectra are combined into a new method to define an effective thickness and effective optical constants, which can satisfy the ellipsometry equation and the Hadley equation simultaneously for ultrathin metal films in the form of islands. In addition, considering that the size of the optical beam in millimeters is much larger than that of islands, the film can be regarded as homogeneous on a macroscale, and thus the film properties can be described by the effective optical constants of the homogeneous layer. The same values of effective optical constants and of effective thickness measured at different positions of the sample also confirm that this assumption is suitable. Figure 2.9 shows the dependence of the refractive index n (a) and extinction coefficient k (b) on the wavelength for silver thin films at different thicknesses. The inset in Figure 2.9b shows the optical constants n and k of thin films with 4.7 and 6.0 nm effective thicknesses [145].

The film with thickness below 8 nm displays a nonmetallic behavior considering that the value of refractive index is larger than that of the extinction coefficient

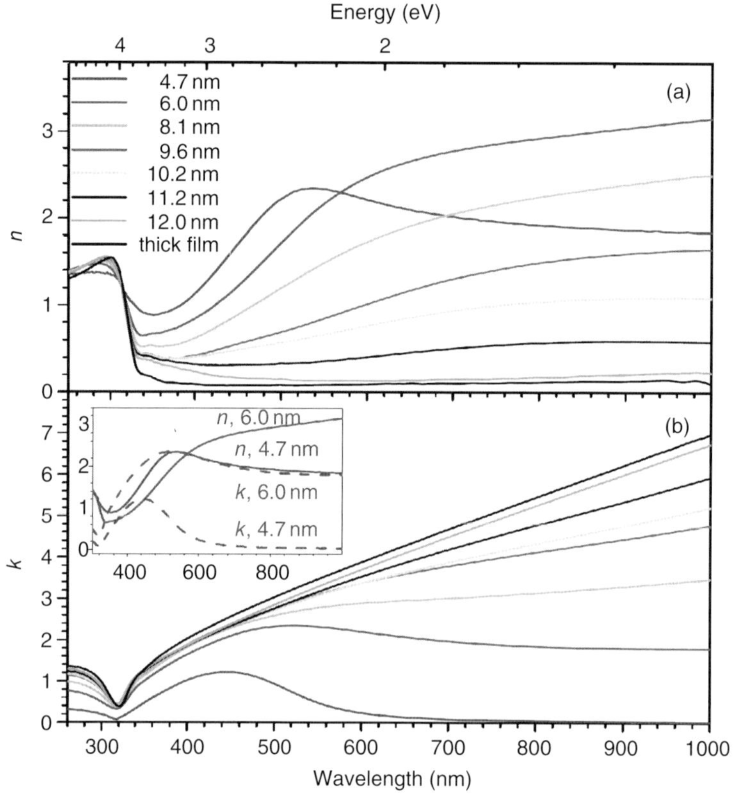

Figure 2.9 Dependence of the refractive index n (a) and extinction coefficient k (b) on the wavelength for silver films at different thicknesses. Inset in (b) shows the optical constants n and k of films with 4.7 and 6.0 nm effective thicknesses. (Gong *et al.* [145], http://www.nature.com/articles/srep09279. Used under CC BY 4.0 http://creativecommons.org/licenses/by/4.0/)

in the wavelength range of 600–1000 nm (inset of Figure 2.9). With increasing thickness the film changes to metallic behavior, the refractive index in the range of 600–1000 nm gets larger at first from 4.7 nm thickness, goes down from 6 nm thickness, and finally tends to be stable. Meanwhile, the extinction coefficient monotonously increases with thickness while the increment gradually gets smaller. When the thickness is larger than 12 nm, both n and k tend to be stable and become closer to those of bulk Ag.

A number of experiments have proved that spectroscopic ellipsometry is a powerful tool for analyzing optical constants and thickness of thin films with high sensitivity. Furthermore, employing the ellipsometry iterated with transmittance method has helped to simultaneously and correctly obtain the effective thickness and effective optical constants of thin Ag films deposited on quartz glass. This optical measurement method based on ellipsometry is also helpful in studying effective energy loss function and plasmon properties of ultrathin films.

2.4 Summary

In this chapter, a general description and applications of electron-beam-induced deposition, especially electron beam evaporation deposition, for thin film coating have been described. There are a number of process parameters (electron beam power, time and position programming of electron beams, evaporant materials, vacuum, substrate heating, carrier gas flux, etc.) that influence the efficiency of deposition. The greatest advantage of the electron beam evaporation technique is the direct heating of the evaporant material in which the highest temperature of the evaporating system appears at the area of contact of the electron beam with the evaporant surface. Electron beam evaporation can be applicable to all of the metals and alloys and can also be used for compounds such as oxides and ZnS (especially high melting point materials) for thin film fabrication. The characterization of thin films by AFM and ellipsometry, including improved ellipsometric method, has been mentioned. Accompanied by the development of advanced techniques, electron beam sources will continue to be incorporated into deposition arrangements and applied to other materials to provide high-quality thin films with improved properties and special nanostructures.

Acknowledgments

The authors acknowledge the financial support from the National Natural Science Foundation of China (Grant Nos 11074232 and 21002097) and "973" project (No. 2011CB932801).

References

1. Seshan, K. (ed.) (2002) *Handbook of Thin-Film Deposition Processes and Techniques: Principles, Methods, Equipment and Applications*, 2nd edn, William Andrew Inc., New York.
2. Betts, D.S. (ed.) (2005) *Modern Vacuum Physics*, CRC Press, New York.
3. Hoffman, D.M., Thomas, J.H., and Singh, B. (1997) *Handbook of Vacuum Science and Technology*, Elsevier Science & Technology, Park Ridge, NJ.
4. Wasa, K., Kitabatake, M., and Adachi, H. (2004) *Thin Film Materials Technology, Sputtering of Compound Materials*, William Andrew Inc., New York.
5. Harsha, S. (ed.) (2006) *Principles of Physical Vapor Deposition of Thin Films*, Elsevier Science & Technology, Amsterdam, Boston, MA.
6. McGuire, G.E. (2007) *Vacuum Deposition onto Webs, Films and Foils*, William Andrew Inc., Leicestershire.
7. Tian, M.B. (ed.) (2006) *Thin Film Technologies and Materials*, Qinghua University Press, Beijing.
8. Decher, G. and Schlenoff, J.B. (2002) *Multilayer Thin Films: Sequential Assembly of Nanocomposite Materials*, Wiley-VCH Verlag GmbH, Weinheim.
9. Tracton, A.A. (ed.) (2007) *Coatings Technology Fundamentals Testing and Processing Techniques*, CRC Press, Boca Raton, FL.
10. Ohring, M. (ed.) (2002) *Materials Science of Thin Films: Deposition and Structure*, 2nd edn, Academic Press.
11. Ramanathan, S. (ed.) (2010) *Thin Film Metal-Oxides: Foundamentals and*

Application in Electronics and Energy, Springer, New York.

12. Freund, L.B. and Suresh, S. (2003) *Thin Film Materials: Stress, Defect Formation and Surface Evolution*, Cambridge University Press, New York.
13. Venables, J.A. (2003) *Introduction to Surface and Thin Film Processes*, Cambridge University Press, New York.
14. Zou, P., Chen, S., Gao, J., and Geng, T. (2013) *Res. Prog. Solid State Electron.*, **33** (5), 462–465.
15. Belic, D., Shawrav, M.M., Gavagnin, M., Pollach, M.S., Wanzenboeck, H.D., and Bertagnolli, E. (2015) *ACS Appl. Mater. Interfaces*, **7**, 2467–2479.
16. Elbadawi, C., Tilos, M., and Lobo, C.J. (2013) *ACS Appl. Mater. Interfaces*, **5**, 9372–9376.
17. Guisbiers, G., Strehle, S., Overschelde, O.V., and Wautelet, M. (2006) *AIP Conf. Proc.*, **817**, 317–321.
18. Zhang, W., Shimojo, M., Takeguchi, M., Che, R.C., and Furuya, K. (2006) *Adv. Eng. Mater.*, **8** (8), 711–714.
19. Lukasczyk, T., Schirmer, M., Steinruck, H.P., and Marbach, H. (2008) *Small*, **4** (6), 841–846.
20. Jiang, C., Liu, H., and Liu, D. (2013) *Chin. Opt. Lett.*, **11** (Suppl. 1), S10304.
21. He, L.J., Li, C., and Liu, X.Z. (2013) *Mater. Lett.*, **101**, 1–4.
22. Dong, H., Santosh, K., Azcatl, A., Cabrera, W., Qin, X., Brennan, B., Zhernokletov, D., Cho, K., and Wallace, R.M. (2013) *J. Appl. Phys.*, **114**, 203505–203511.
23. Stolz, C.J., Sheehan, L.M., von Gunten, M.K., Ronald, P., and Douglas, J. (1999) *Proc. SPIE Int. Soc. Opt. Eng.*, **3738**, 318–324.
24. Gong, J.B., Dong, W.L., Dai, R.C., Wang, Z.P., Zhang, Z.M., and Ding, Z.J. (2014) *Chin. Phys. B*, **23** (8), 087802–087806.
25. Netrvalová, M., Vavruňková, V., Müllerová, J., and Šutta, P. (2009) *J. Electr. Eng.*, **60** (5), 279–282.
26. Pan, Y. and Huang, G. (2011) *Infrared Laser Eng.*, **40** (11), 2233–2237.
27. Yun, J., Varlamov, S., Huang, J., Kim, K., and Green, M.A. (2014) *Appl. Phys. Lett.*, **104**, 242102–242107.
28. Zhao, M., Tan, M.Q., Zhou, D.B., Wu, X.M., and Wang, X.D. (2007) *J. Optoelectron. Laser*, **18** (1), 40–42.
29. Zhao, M., Zhou, D., Tan, M., Wang, X., and Wu, X. (2006) *Chin. J. Semicond.*, **27** (9), 1586–1589.
30. Alzamil, M.A. (2013) *Indian J. Phys.*, **87** (10), 1005–1010.
31. Golzan, M.M., Hassanzadeh, A., Maleki, M.H., and Kiani-Dehkiani, A. (2010) *Superlattices Microstruct.*, **48** (3), 321–329.
32. Eygia, Z.D., Kulakci, M., and Turand, R. (2014) *Appl. Surf. Sci.*, **318**, 116–120.
33. Kumar, R.R., Yuvaraj, D., and Rao, K.N. (2010) *Mater. Lett.*, **64** (16), 1766–1768.
34. Rebholz, C., Ziegele, H., Leyland, A., and Matthews, A. (1998) *J. Vac. Sci. Technol., A*, **16**, 2851–2858.
35. Burdovitsyn, V.A., Medovnik, A.V., and Oks, E.M. (2011) *Nanotechnol. Russ.*, **6** (3–4), 265–267.
36. Hart, A.J., Slocum, A.H., and Royer, L. (2006) *Carbon*, **44** (2), 348–359.
37. Grechanyuk, N.I., Minakova, R.V., and Kopylova, G.E. (2013) *Powder Metall. Met. Ceram.*, **52** (3-4), 228–236.
38. Todorova, Z., Donkov, N., Ristic, Z., Bundaleski, N., Petrovic, S., and Petkov, M. (2006) *Plasma Processes Polym.*, **3**, 174–178.
39. Pyun, M.W., Kim, E.J., Yoo, D.H., and Hahn, S.H. (2010) *Appl. Surf. Sci.*, **257** (4), 1149–1153.
40. Jang, H.K., Whangbo, S.W., Choi, Y.K., Chung, Y.D., Jeong, K., Whang, C.N., Lee, Y.S., Lee, H.-S., Choi, J.Y., Kim, G.H., and Kim, T.K. (2000) *J. Vac. Sci. Technol., A*, **18** (6), 2932–2937.
41. Mohanty, P., Kabiraj, D., Mandal, R.K., Kulriya, P.K., Sinha, A.S.K., and Rath, C. (2014) *J. Magn. Magn. Mater.*, **355**, 240–245.
42. Chen, T., Wang, D.S., and Xiong, Y.Q. (2013) *Adv. Mater. Res.*, **189-193**, 1233–1237.
43. Jin, B., Li, W., Zheng, Z.H., and Wang, X. (2004) *J. Funct. Mater. Devices*, **10** (1), 24–29.
44. Sakamoto, N., Inoue, T., and Kato, K. (2003) *Cryst. Growth Des.*, **3** (2), 115–116.

45. Yamada, N., Oyama, Y., Higuchi, T., and Yamaguchi, S. (2004) *Solid State Ionics*, **172** (1–4), 293–297.
46. Vasil'ev, A.G., Zakharov, R.A., Orlikovskii, A.A., Rogozhin, A.E., Sonin, M.S., and Khorin, I.A. (2009) *Russ. Microlectron.*, **38** (5), 327–333.
47. Shen, Y.M., He, H.B., Shao, S.Y., and Fan, Z.X. (2006) *Chin. J. Lasers*, **33** (6), 827–831.
48. Xing, S., Zhang, N., Song, Z., Shen, Q., and Lin, C. (2003) *Microelectron. Eng.*, **66** (1-4), 451–456.
49. Wang, Y.J., Lin, Z.L., Cheng, X.L., Xiao, H.B., Zhang, F., and Zou, S.C. (2004) *Appl. Surf. Sci.*, **228** (1-4), 93–99.
50. Keshavarzi, R., Mirkhani, V., Moghadam, M., Dastjerdi, M.J.V., and Modayemzadeh, H.R. (2013) *Mater. Res. Bull.*, **46** (4), 615–620.
51. Chiu, P.K., Cho, W.H., Chen, H.P., Hsiao, C.N., and Yang, J.R. (2011), the 4th International Nanoelectronics Conference (INEC 2011) Proceedings of the International Nanoelectronics Conference.
52. Yu, H.K. and Lee, J.L. (2014) *Sci. Rep.*, **4**, 6589–6596.
53. Kumar, R.R., Gaddam, V., Rao, K.N., and Rajanna, K. (2014) *Mater. Res. Express*, **1**, 035008–035015.
54. Basnet, P., Larsen, G.K., Jadeja, R.P., Hung, Y.C., and Zhao, Y.P. (2013) *ACS Appl. Mater. Interfaces*, **5**, 2085–2095.
55. Ohya, T. (2000) the 43rd SVC Annual Technical Conference, Proceedings of the 43rd Annual Technology Conference SVC, pp. 368–372.
56. Perentes, A. and Hoffmann, P. (2007) *Chem. Vap. Deposition*, **13**, 176–184.
57. Wu, X.M. (2001) *J. Optoelectron. Laser*, **12** (6), 569–571.
58. Wei, Z.-R., Yang, D.-Y., Zhu, X.H., and Yang, W.Q. (2011) *J. Univ. Electron. Sci. Technol. China*, **40** (6), 933–936.
59. Tesfamichael, T. (2010) *IEEE Sens. J.*, **10** (11), 1796–1802.
60. Qiu, S.C., Chen, T., Wu, Z.H., and Huang, H.Y. (1998) *Microfab. Technol.*, **1**, 49–54.
61. Sahu, D.R., Lin, S.Y., and Huang, J.L. (2008) *Thin Solid Films*, **516** (15), 4728–4732.
62. Asmar, R., Atanas, J.P., Zaatar, Y., Podlecki, J., and Foucaran, A. (2011) *Microelectron. J.*, **37** (10), 1080–1085.
63. Gruzintsev, A.N., Volkov, V.T., and Matveeva, L.N. (2002) *Russ. Microlectron.*, **31** (3), 193–199.
64. Aghamalyan, N.R., Gambaryan, I.A., Goulanian, E.K., Hovsepyan, R.K., Kostanyan, R.B., Petrosyan, S.I., Vardanyan, E.S., and Zerrouk, A.F. (2003) *Semicond. Sci. Technol.*, **18** (6), 525–529.
65. Shen, H., Xu, L., Zheng, G., Su, J., and Zhu, R. (2013) *J. Optoelectron. Adv. Mater.*, **15** (3-4), 244–248.
66. Wu, S.G. (2008) *High Power Laser Part. Beams*, **20** (10), 1724–1726.
67. Zhang, D.P., Shao, S.Y., Huang, J.B., Fan, Z.X., and Shao, J.D. (2006) *Opto-Electron. Eng.*, **33** (6), 37–40.
68. Hao, D., Wu, F., Ma, L., Yan, B., and Zhang, X. (2005) *Acta Photonica Sin.*, **35** (2), 224–227.
69. Zhan, M.Q., Zhang, D.P., He, H.B., Shao, J.D., and Fan, Z.X. (2004) *Chin. J. Lasers*, **31** (11), 1356–1360.
70. Liu, G.D., Wu, L.Y., and Zhang, F.P. (2015) *Mater. Sci. Semicond. Process.*, **31**, 14–18.
71. Cheng, Y., Yang, K., Peng, Y., Yin, Y., Chen, J.X., Jing, B., Liang, H.W., and Du, G.T. (2013) *J. Mater. Sci.: Mater. Electron.*, **24**, 5122–5126.
72. Xiong, S.M., Zhang, Y.D., and Tang, J.F. (2001) *Opto-Electron. Eng.*, **28** (1), 13–15.
73. Schiller, S., Neumann, M., Morgner, H., and Schiller, N. (1993) the 36th SVC Annual Technical Conference, Society of Vacuum Coaters, pp. 278–292.
74. Castaldi, L., Giannakopoulos, K., Travlos, A., Boukos, N., Niarchos, D., Boukari, S., and Beaurepaire, E. (2008) *Nanotechnology*, **19**, 135702–135707.
75. Garcia, G., Domenech-Ferrer, R., Pi, F., Santiso, J., and Viejo, J.R. (2007) *J. Comb. Chem.*, **9**, 230–236.
76. Castaldi, L., Giannakopoulos, K., Travlos, A., Boukari, S., and Beaurepaire, E. (2005) *J. Magn. Magn. Mater.*, **290-291**, 544–546.
77. Kita, K., Sasaki, H., Nagahora, J., and Inoue, A. (2000) *J. Jpn. Soc. Powder Powder Metall.*, **47** (4), 406–411.

78. Saravanan, P., Vinod, V.T.P., Černík, M., Vishnuraj, R., Chelvane, J.A., Kamat, S.V., and Hsu, J.H. (2015) *J. Magn. Magn. Mater.*, **385**, 313–317.
79. Iwama, S. and Hayakawa, K. (1985) *Surf. Sci.*, **156** (Pt. 1), 85–89.
80. Kurt, R., Pitschke, W., Heinrich, A., Schumann, J., and Wetzig, K. (1997), the 16th International Conference on Thermoelectrics (ICT'97), Proceedings of ICT 97, pp. 303–306.
81. Lin, C., Du, N., and Li, X. (2003) *J. Chin. Soc. Corros. Prot.*, **23** (5), 261–264.
82. Egermeier, J. and Hill, R. (1987) Annual Technical Conference – Society of Vacuum Coaters, pp. 329–337.
83. Xin, Y., He, X.D., Li, M.W., and Sun, Y. (2008) *J. Comput. Theor. Nanosci.*, **5** (8), 1696–1699.
84. Misaki, T. and Ohsawa, J. (2004) *Electron. Commun. Jpn., Part II: Electron.*, **87** (1), 9–15.
85. Lee, C.-H. and Kim, S.-I. (2003) *Integr. Ferroelectr.*, **57**, 1265–1270.
86. Blackstock, J.J., Rostami, A.A., Nowak, A.M., McCreery, R.L., Freeman, M.R., and McDermott, M.T. (2004) *Anal. Chem.*, **76**, 2544–2552.
87. Yang, D.Y., Zhua, X.H., Wei, Z.R., Yang, W.Q., Li, L.Z., Yang, J., and Gao, X.Y. (2011) *J. Semicond.*, **32** (2), 023001.
88. Sivaramamoorthy, K., Bahadur, S.A., Kottaisamy, M., and Murali, K.R. (2010) *J. Alloys Compd.*, **503** (1), 170–176.
89. Fang, L., Chen, J., Xu, L., Su, W.N., Yu, Y., Xu, J., and Ma, Z.Y. (2013) *Chin. Phys. B*, **22** (9), 098802.
90. Ragachev, A.V., Yarmolenko, M.A., Rogachev, A.A., Gorbachev, D.L., and Zhou, B. (2014) *Appl. Surf. Sci.*, **303**, 23–29.
91. Huang, H.L., Cheng, S.Y., and Huang, B.H. (2009) *J. Optoelectron. Laser*, **20** (3), 355–358.
92. Wang, S., Xia, G., Shao, J., and Fan, Z. (2006) *Proc. SPIE Int. Soc. Opt. Eng.*, **6149**, 61491 L.
93. Murali, K.V., Antony, A., Ramachandran, M., and Jayaraj, M.K. (2003) *Proc. SPIE Int. Soc. Opt. Eng.*, **5280 II**, 600–607.
94. Shu, X., Xu, C., Tian, Z., and Shen, G. (2005) *Solid State Electron.*, **49** (12), 2016–2017.
95. Syed, M.G., Ahamed, B., Nagarethinam, V.S., Thayumanavan, A., Murali, K.R., Sanjeeviraja, C., and Jayachandran, M. (2010) *J. Mater. Sci.: Mater. Electron.*, **21** (12), 1229–1234.
96. Suthan Kissinger, N.J., Velmurugan, N., and Perumal, K. (2009) *J. Korean Phys. Soc.*, **55** (4), 1577–1581.
97. Farooq, M.U., Khan, M., Faraz, A., Maqsood, A., Ahmad, W., and Li, L. (2014) *Mater. Technol.: Adv. Perform. Mater.*, **29** (1), 29–35.
98. Ho, C., Kao, M.J., Peng, C.H., Kuo, C.G., and Huang, K.D. (2011) *J. Nanosci. Nanotechnol.*, **11** (8), 7491–7494.
99. Richter, K., Lorbeer, C., and Mudring, A.V. (2015) *Chem. Commun.*, **51**, 114–117.
100. Arshi, N., Lu, J.Q., Lee, C.G., Koo, B.H., and Ahmed, F. (2014) *JOM*, **66** (9), 1893–1899.
101. Katzer, D.S., Meyer, D.J., Storm, D.F., Nepal, N., and Wheeler, V.D. (2014) *J. Vac. Sci. Technol., B*, **32**, 02C117–02C122.
102. Yonekura, K., Fujiyoshi, T., Sueyoshi, T., Awaji, S., and Watanabe, K. (2013) *Phys. C: Supercond. Appl.*, **480**, 108–110.
103. Zhu, X.H., Yang, D.Y., Wei, Z.R., Sun, H., Wang, Z.G., and Zu, X.T. (2010) *J. Semicond.*, **31** (8), 083002.
104. Zhan, M., Wu, Z., and Shao, J. (2008) *Proc. SPIE Int. Soc. Opt. Eng.*, **6984**, 698422–698427.
105. Shang, S., Yi, K., Huang, J., Shao, J., and Fan, Z. (2005) *Proc. SPIE Int. Soc. Opt. Eng.*, **5774** (89), 385–388.
106. Zhang, X.S., Li, L., Zou, K.-S., and Tao, Y. (2004) *J. Optoelectron. Laser*, **15** (5), 549–553.
107. Vasiliev, A.G., Vasiliev, A.L., Zakharov, R.A., Orlikovsky, A.A., Horin, I.A., and Eindou, M. (2003) *Russ. Microlectron.*, **32** (5), 275–281.
108. Rivera, V.A.G., Ferri, F.A., Clabel, J.L.H., Silva, M.A.P., Nunes, L.A.O., SiuLi, M., and Marega, E. Jr., (2014) *J. Lumin.*, **148**, 86–191.

109. Chen, Z.M., Miao, D.G., Jiang, S.X., Guo, R.H., and Xin, B.J. (2015) *J. Mater. Sci.: Mater. Electron.*, **26**, 4113–4118.
110. Schiller, S. *et al.* (1990), the 4th International Conference on Vacuum Web Coating, Proceedings of the 4th International Vacuum Web Coating Conference, pp. 138–158.
111. Pourteimoor, S., Araghi, M.E.A., Riyazi, S., and Askari, A. (2014) *J. Mater. Sci.: Mater. Electron.*, **25**, 2325–2332.
112. Pourteimoor, S. and Azim-Araghi, M.E. (2014) *Mater. Sci. Semicond. Proc.*, **18**, 97–104.
113. Ang, P.C., Ibrahim, K., and Pakhuruddin, M.Z. (2015) *AIP Conf. Proc.*, **1657**, 040003–040008.
114. Ozin, G.A., Andrews, M.P., Francis, C.G., Huber, H.X., and Molnar, K. (1990) *Inorg. Chem.*, **29** (5), 1068–1073.
115. Nukeaw, J. and Tunhoo, B. (2010), the 3rd International Nanoelectronics Conference (INEC 2010), 2010 3rd International Nanoelectronics Conference Proceedings, pp. 660–661.
116. Tunhoo, B., Thiwawong, T., and Nukeaw, J. (2007) the 7th IEEE International Conference on Nanotechnology (IEEE-NANO 2007), Proceedings, vol. 4601350, pp. 994–997.
117. Bunshah, R.F., Jou, S., Prakash, S., Doerr, H.J., Isaacs, L., Wehrsig, A., Yeretzian, C., Cynn, H., and Diederich, F. (1992) *J. Phys. Chem.*, **96** (17), 6866–6869.
118. Brus, V.V., Maryanchuk, P.D., Ilashchuk, M.I., Rappich, J., Babichuk, I.S., and Kovalyuk, Z.D. (2015) *Sol. Energy*, **112**, 78–84.
119. Mori, T., Chen, C.J., Hung, T.F., Mohamed, S.G., Lin, Y.Q., Lin, H.Z., Sung, J.C., Hud, S.F., and Liu, R.S. (2015) *Electrochim. Acta*, **165**, 166–172.
120. Sun, H.T., Wang, X.P., Kou, Z.Q., Wang, L.J., Wang, J.Y., and Sun, Y.Q. (2015) *Chin. Phys. B*, **24** (4), 047701–047706.
121. Mahmood, A., Rashid, R., Aziz, U., Shahn, A., Ali, Z., Raza, Q., and Ashraf, T. (2015) *Prog. Nat. Sci.: Mater. Int.*, **25**, 22–28.
122. Ali, S.M., Hussain, S.T., Muhammada, J., Ashraf, M., Farooq, A., Imran, M., and AbuBakar, S. (2013) *Mater. Sci. Semicond. Process.*, **16**, 899–904.
123. Begam, M.R., Rao, N.M., Kaleemulla, S., Krishnamoorthi, C., Krishna, N.S., and Kuppan, M. (2014) *Appl. Phys. A*, **117**, 793–798.
124. Lodha, G.S., Pandita, S., Gupta, A., Nandedkar, R.V., and Yamashita, K. (1996) *Appl. Phys. A: Mater. Sci. Proc.*, **62** (1), 29–32.
125. Mohamed, R.B., Nasina, M.R., Shaik, K., Narayananellore, S.K., and Kuppan, M. (2014) *J. Supercond. Novel Magn.*, **27**, 2147–2152.
126. Tsunoda, T., Mukaida, M., and Imai, Y. (2001) *Thin Solid Films*, **381**, 296–302.
127. Bianchi, L. (1991) *JOM*, **43** (5), 45–47.
128. Maiti, N., Karmakar, P., Barve, U.D., and Bapat, A.V. (2008) *J. Phys.: Conf. Ser.*, **114**, 012049–012059.
129. Murakami, K. and Takai, M. (2015) *Microelectron. Eng.*, **132**, 74–82.
130. Huth, M., Porrati, F., Schwalb, C., Winhold, M., Sachser, R., Dukic, M., Adams, J., and Fantner, G. (2012) *Beilstein J. Nanotechnol.*, **3**, 597–619.
131. Wnuk, J.D., Rosenberg, S.G., Gorham, J.M., van Dorp, W.F., Hagen, C.W., and Fairbrother, D.H. (2011) *Surf. Sci.*, **605**, 257–266.
132. Schiller, S. *et al.* (1988) Electron beam coating, in *Surfacing Technologies Handbook* (ed. T.S. Sudarshan), Marcel Dekker, New York.
133. Schiller, S., Jaesch, G., and Neumann, M. (1983) *Thin Solid Films*, **110**, 149–164.
134. Schiller, S. and Neumann, M. (1987) the 1st International Vacuum Web Coating Conference, Proceedings of the 1st International Vacuum Web Coating Conference, pp. 113–128.
135. Thiede, A. *et al.* (2003), the 46th SVC Annual Technology Conference, Proceedings of the 46th Annual Technology Conference SVC, pp. 149–153.
136. Schiller, S. *et al.* (1987), the 5th Annual Conference EB Melting and Refining, Proceedings of the 5th Annual Conference EB Melting and Refining, pp. 111–130.
137. Hayashida, M., Malac, M., Bergen, M., and Li, P. (2014) *Ultramicroscopy*, **144**, 50–57.

138. Srenzel, O. (ed.) (2005) *The Physics of Thin Film Optical Spectra: An Introduction*, Springer, Berlin.
139. Birkholz, M. (2006) *Thin Film Analysis by X-Ray Scattering*, Wiley-VCH Verlag GmbH, Weinheim.
140. Bubert, H. and Jenett, H. (2002) *Surface and Thin Film Analysis: Principles, Instrumentation, Applications*, Wiley-VCH Verlag GmbH, Weinheim.
141. Milling, A.J. (1999) *Surface Characterization Methods: Principles, Techniques and Applications*, Marcel Dekker, Inc., New York.
142. Birdi, K.S. (2003) *Scanning Probe Microscopes: Applications in Science and Technology*, CRC Press, New York.
143. Fujiwara, H. (2007) *Spectroscopic Ellipsometry: Principles and Applications*, Maruzen Co. Ltd., Tokyo.
144. Tompkins, H.G. and Erene, E.A. (2005) *Handbook of Ellipsometry*, William Andrew Inc., New York.
145. Gong, J.B., Dai, R.C., Wang, Z.P., and Zhang, Z.M. (2014) *Sci. Rep.*, **5**, 9279–9283.

3
Nanostructures and Thin Films Deposited with Sputtering

Weiqing Yang

3.1
Introduction

As one of the important physical vapor deposition (PVD) techniques, the sputtering process involves the physical (not thermal) vaporization of atoms from a surface by momentum transfer from bombarding energetic atomic-sized particles. A schematic of the sputtering process is shown in Figure 3.1. The energetic particles are usually ions of a gaseous material accelerated in an electric field or radio frequency (rf) magnetic field – that is, direct current (DC) magnetic sputtering and rf magnetic sputtering [1]. In the early 1850s, Grove and Pulker first observed the sputtering phenomenon by using von Guericke-type oil-sealed piston vacuum pumps [1, 2]. Moreover, the terms *chemical sputtering* and *electrochemical sputtering* have been associated with the process whereby bombardment of the target surface with a reactive species produces a volatile species [1, 2]. This process is often termed as *reactive plasma etching* or *reactive ion etching* and is important in the patterning of thin films [1, 2]. Sputter deposition, which is often called *just sputtering* (a poor use of the term), is the deposition of particles whose origin is from the surface (target) being sputtered. Sputter deposition of films was first reported by Wright in 1877 and was feasible because only a relatively poor vacuum is needed for sputter deposition. Edison patented the sputter deposition process for depositing silver on wax photograph cylinders in 1904. Besides traditional thin film deposition, some recently developed nanomaterials can also be fabricated using the sputtering technique [1, 2].

Nowadays, nanostructures and nanomaterials as well as thin films dominate the research landscape in almost all areas of natural sciences and engineering. Moreover, plenty of applications have already been realized and commercialized in the past years with the use of materials with nanometer-sized dimensions. The mesoscopic properties of nanostructures are not only determined by quantum-sized

Advanced Nano Deposition Methods, First Edition. Edited by Yuan Lin and Xin Chen.

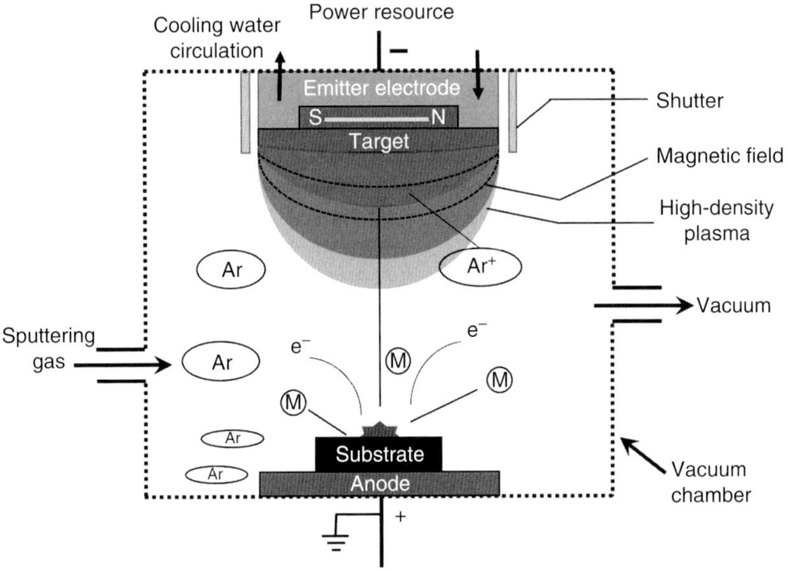

Figure 3.1 Sketch of sputtering work mechanism. (Mattox [1]. Reproduced with permission of Elsevier.)

effects but are actually dominated by surface effects due to the high surface-to-volume ratio. These are additional surface states, surface reconstruction, as well as Fermi-level pinning, band bending, and depletion in the nanomaterials. In any case, the mesoscopic properties are given by the interplay of three important parameters: the size, shape, and composition of the nanostructured material. In this book, we present the fabrication of functional nanowires and thin films by magnetic sputtering. Specifically, the growth and applications of ZnO nanowires, MoO_3 nanowires, and $BaTiO_3$ and SnO_2 thin films are investigated in this chapter.

3.2
Nanostructures with Sputtering

One-dimensional (1D) semiconductor nanowires and nanorods [3–9], such as MoO_3 [7, 9, 10], ZnO [11], SiC [12], and Cu_2S nanowires [13], owing to their surface effect, small-size effect, and quantum confinement effect, exhibit superior and unique optical, electronic, and thermal properties than those of their bulk counterparts. Sputtering is one of the important techniques to prepare the desired nanowires. Typically, we will introduce oxide nanowires such as Bi_2O_3, TiO_2, and MoO_3 nanowires and nitrides such as GaN and AlN core–shell nanorods by sputtering in this chapter.

3.2.1 Oxide Nanostructures

3.2.1.1 Needle-Shaped MoO_3 Nanowires

The unique field emission properties of 1D nanomaterials have been widely studied in the past few years owing to their prospective applications in large-area flat panel displays and other device development [9–13]. The key efforts for improving the field emission properties in nanomaterial device performance are to lower the turn-on and threshold voltages for field emission and to increase the field enhancement factor. These critical factors are related to the radius of curvature of the small tip for the field emission materials [9–13]. Usually, MoO_3 nanowires prepared by the thermal evaporation technique are misaligned flat tips [14–16] that are unfavorable to the corresponding field emission properties. Therefore, fabricating needle-shaped MoO_3 nanowires with preferred orientation on the surfaces of the substrates has become a critical issue in enhancing their field emission performance.

Recently, Yang *et al.* successfully fabricated the needle-shaped MoO_3 nanowires with preferred orientation on the Cu electrode by the rf magnetron sputtering technique. In Figure 3.2, the scanning electron microscopic (SEM) image shows that most of the MoO_3 nanowires are basically straight and up with needle-shaped tips. The lengths of the nanowires are close to 2 μm and their shapes are slab-sided and straightforward. In general, the tips of the nanobelts are needle shaped and the radii of their tips are in the range of 50–90 nm, which is remarkably in favor of the field emission application. It is interesting to find a broken tip at the lower left corner of Figure 3.2a, and its enlarged picture is shown at the upper left corner of the figure. The needle-shaped tip is usually cleaved in a flat surface parallel to the bottom of the slab-sided MoO_3 nanobelt. Figure 3.2b shows a schematic view of the crystal structures of MoO_3 nanobelts in which the MoO_6 octahedra have

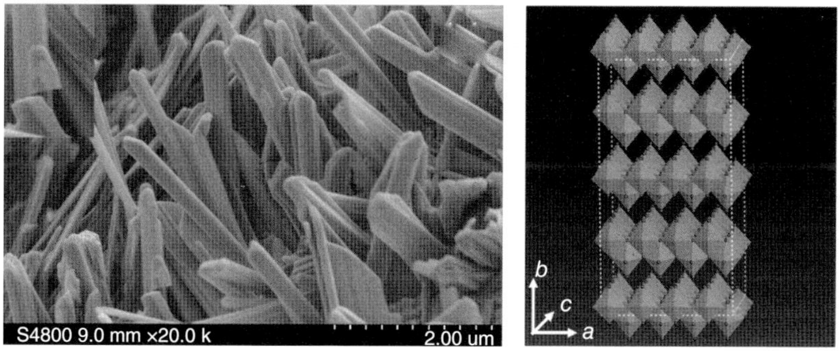

(a) (b)

Figure 3.2 (a) The SEM image of the MoO_3 nanobelts and (b) the crystal structure of MoO_3 nanowires (silver gray lozenge representing MoO_6 octahedra) [17]. (Yang *et al.* [20]. Reproduced with permission of Elsevier.)

the corners shared with each other to form a layer of MoO_3 in the *ac* plane. Thus, along the *b*-axis, the layers are stacked in a staggered arrangement with the van der Waals' force [17], which are much weaker than the interactions between the atoms within each layer. So the nanobelt should be more easy to break off along the transverse *ac* plane, further confirming that as-grown MoO_3 nanobelts should have the preferred *b*-axis orientation.

As we know, MoO_3 has an orthorhombic structure with the lattice constants of $a = 3.962$ Å, $b = 13.858$ Å, $c = 3.697$ Å, and *Pbnm* (62) space group (PDF# 05-0508) [18]. Figure 3.3a,b are, respectively, the typical XRD patterns of the post-annealing and pre-annealing MoO_3 nanobelt samples, which can be indexed with the orthorhombic structure mentioned earlier. It is worth noting that the structure of the sample becomes α-MoO_3 phase from polycrystalline $\alpha-\beta$ mixed MoO_3 phase after annealing [18]. Moreover, the α-MoO_3 nanobelts have a preferred orientation with the *b*-axis normal to the Cu electrode surface [18–20].

The sputtering process of MoO_3 nanowires is as follows: the MoO_3 nanobelts were deposited directly on the copper thin film electrode by the rf magnetron sputtering with the MoO_3 target (purity 99.99%). The working pressure was maintained at about 1.5×10^{-1} Pa with a flowing gas mixed with O_2 and Ar throughout the sputtering process. The flow rates of O_2 and Ar were, respectively, 5 and 25 sccm. The distance between the target and the substrate was 7.5 cm. MoO_3 nanobelts were uniformly deposited directly on the copper/quartz substrate at 200 °C with a sputtering power of 400 W for 10 min. Subsequently, the samples were annealed in air at 500 °C for 1 h to decrease the oxygen vacancies of the as-grown samples and to form the nanobelts with α-MoO_3 structure.

To understand the growth mechanism of the needle-shaped MoO_3 nanowires, the crystal structure of MoO_3 needs to be considered. Since the nanowires were

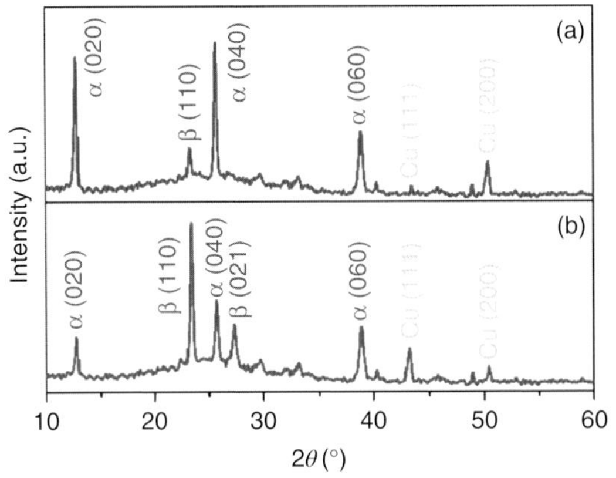

Figure 3.3 The XRD pattern of the post-annealing (a) and pre-annealing (b) MoO_3 nanowires/Cu/quartz [17]. (Yang *et al.* [20]. Reproduced with permission of Elsevier.)

grown at a low temperature with no metallic catalysts in the original growth process, a modified vapor–solid (VS) mechanism [16, 21, 22] should be dominant for the growth process of MoO_3 nanowires rather than the vapor–liquid–solid (VLS) growth mechanism [23, 24]. In a typical VS process, the molybdenum oxides (MoO_x) are sputtered from a solid source by bombarding with high energetic ions and directly deposited onto the very thin CuO film, which is the oxidation layer on the surface of the Cu electrode during the sputtering process. When the sputtered molybdenum oxide molecules with high mobility reach the vicinity of the substrate, they have enough energy to diffuse and nucleate on CuO surface at the deposition temperature [25, 26]. These foreign atoms and molecules usually follow the atomic hopping mechanism rather than the atomic exchange diffusion mechanism [27] to hop to the low-energy surface sites such as surface step terrace edge and ledge to form into small clusters or nucleation seeds.

According to the kinetics of whisker growth [21, 28, 29], CuO has a much lower surface energy than MoO_x. Thus, the MoO_x molecules/particles will prefer to nucleate in the flat MoO_3 islands along the (010) crystallographic plane on the CuO surfaces, as shown in Figure 3.4a. The newly arriving MoO_x molecules will accumulate on the newly formed MoO_3 island surfaces, resulting in fast growth along the *b*-axis direction on the MoO_3 surface area, as shown in Figure 3.4b. Continual accumulation of the incoming MoO_3 molecules on the MoO_3 island surfaces will enable fast stacking of MoO_3 layers to form nanobelts (Figure 3.4c). With the belts continually growing, the top surface areas are gradually reduced because of the Ehrlich–Schwoebel potential at the surface step terrace edges [30–32]. Therefore, a long nanobelt with a needle-shaped tip is formed along the *b*-axis direction, as shown in Figure 3.4d. But at this moment, the structure of the

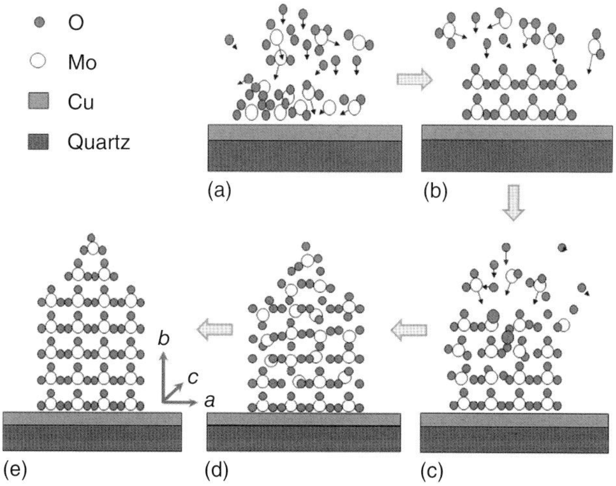

Figure 3.4 The proposed growth process (a–d) and post-annealing (e) of needle-shaped MoO_3 nanowires [17]. (Yang *et al.* [20]. Reproduced with permission of Elsevier.)

sample is a polycrystalline α–β mixed MoO_3 phase (Figure 3.3b). After annealing in air at 500 °C for 1 h, the final as-grown nanobelts (Figure 3.4e) have α-MoO_3 structure (Figure 3.3a) and have a preferred orientation with the *b*-axis normal to the Cu electrode surface. This mechanism is in good agreement with the previous observation as shown in Figure 3.2a, where the tips of MoO_3 nanowires have the needle shape.

In summary, needle-shaped MoO_3 nanowires with preferred (010) orientation normal to the surfaces of Cu-coated quartz substrates could be deposited by the rf magnetron sputtering technique. The investigation demonstrated that needle-shaped MoO_3 nanowires could be a promising candidate for electronic display devices.

3.2.1.2 Bi_2O_3 Nanowires

Bi_2O_3 nanowires, as a typical photocatalyst, exist in two crystal structures, that is, α-Bi_2O_3 and β-Bi_2O_3 [33]. Their band gaps are, respectively, 2.91 and 2.58 eV, which are dramatically lower than that of TiO_2 with the structures of anatase (3.32 eV). This suggests that excitation by visible light can generate electron–hole pairs more efficiently in Bi_2O_3 than in TiO_2 [34, 35]. One-dimensional Bi_2O_3 nanostructures should be efficient visible light-driven photocatalysts [36–40] owing to their high surface-to-volume ratio. Many methods, such as template synthesis, hydrothermal methods, the vapor transport process, and oxidative metal vapor transport deposition [41–44], were used for synthesizing 1D Bi_2O_3 nanostructures. However, it is difficult to use these synthetic methods to prepare 1D Bi_2O_3 nanostructures for large-scale applications because of the small yield and the high fabrication cost. In addition, owing to the complexity of Bi_2O_3 compounds, the phase-selective synthesis of Bi_2O_3 nanostructures is more difficult to be controlled.

Recently, Tien had used the rf magnetron sputtering technique to controllably prepare phase-selective α- and β-Bi_2O_3 nanocones on Si(001) substrate. Figure 3.5 shows that numerous 1D nanostructures were distributed over the entire Si (001) substrate, with diameters of roughly 70–130 nm and lengths of 1–3 μm. Figures 3.6 and 3.7 show that the high-density 1D nanostructures are highly oriented in the direction normal to the substrate, with a top diameter of less than 100 nm and a bottom diameter of more than 500 nm. The nanocones obtained have excellent chemical composition.

According to the possible mechanism of formation of Bi_2O_3 nanocones shown in Figure 3.8, the Bi seeding layer is crucial to the growth of bismuth oxide nanostructures [33]. The thin Bi seeding layer sputtered onto Si (001) substrate at room temperature was heated above its melting point (271.4 °C), and then Bi droplets formed due to surface tension. The mobility of the Bi atoms increased with temperature, causing large droplets to grow. The V-shaped notch above the grain boundaries resulted in a higher adsorption rate than the smooth surfaces. Driven by the high adsorption rate at the grain boundaries, the Bi and O atoms were adsorbed on specific sites with a high probability. As the reactive vapor species were supplied by the sputtering process, the growth occurred initially from the

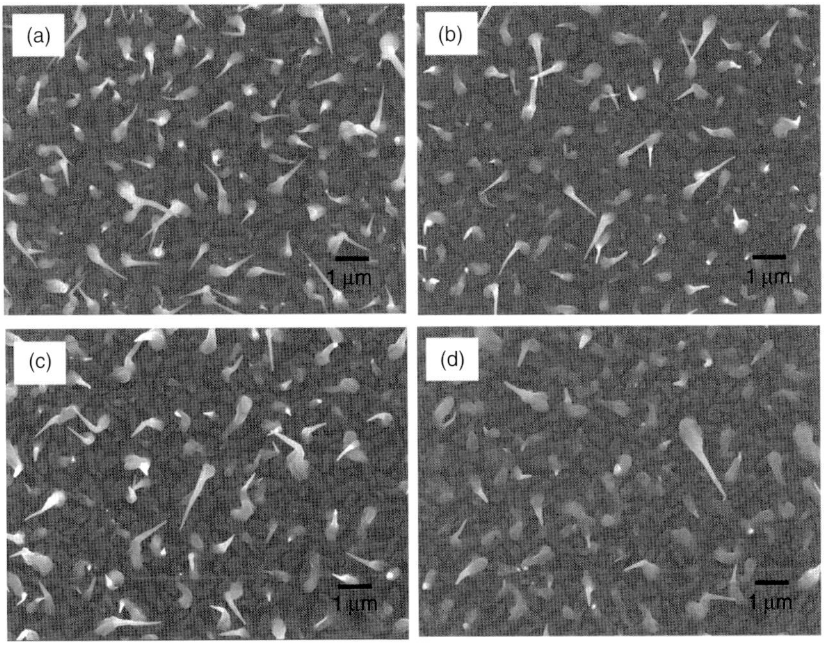

Figure 3.5 FESEM images of samples deposited on Si(001) with a Bi seeding layer at (a) 400 °C, (b) 450 °C, (c) 500 °C, and (d) 550 °C by rf magnetron sputtering. (Tien and Liou [33]. Reproduced with permission of Elsevier.)

grain boundaries of Bi nanoislands, and the Bi_2O_3 nanostructures continued to grow unidirectionally.

Compared with the growth mechanism of conventional VLS-grown oxide nanowires, the mechanism of the grain boundary-assisted growth of the Bi_2O_3 nanocones is that the grain boundaries serve as nucleation sites and facilitate the adsorption of Bi and O ions vapor onto the surface. This growth mechanism is driven by the reducing surface energy of the substrate, and the specific phase (α, β) is formed by the phase stability at different temperatures.

3.2.2 Nitride Nanostructures

3.2.2.1 Graphitic-C_3N_4 Nanocone Array

Since Liu and Cohen [45] predicted empirically that diamond-like β-C_3N_4 might have an exceptional hardness even higher than diamond, first-principles calculations have been applied to predict several allotropes of C_3N_4, such as α-C_3N_4, β-C_3N_4, cubic C_3N_4, and graphitic C_3N_4 (g-C_3N_4), and propose their wide band gap, high thermal conductivity, outstanding field emissivity better than elemental carbon structures, and nonlinear optical properties [46–49]. Since N_2 is difficult

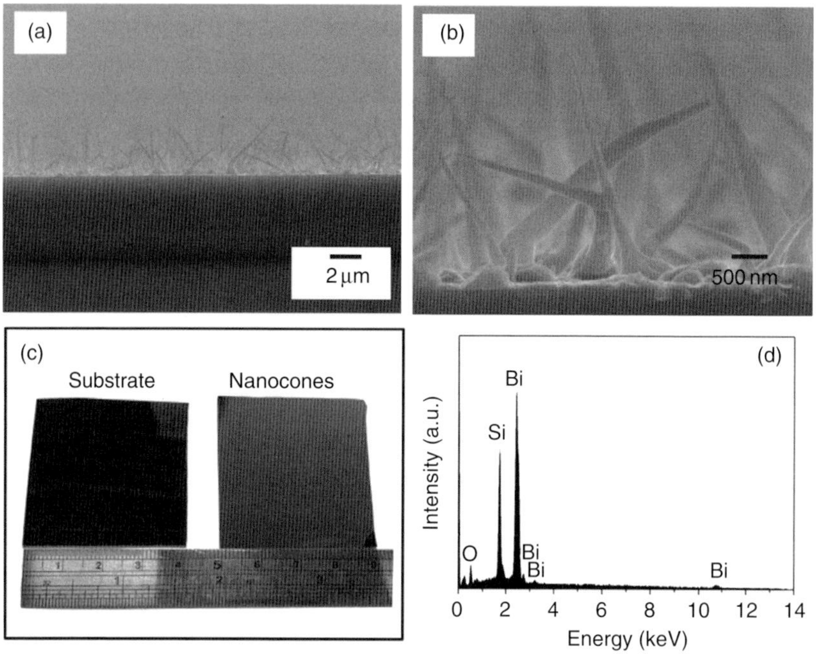

Figure 3.6 (a,b) Cross-sectional FESEM images of samples deposited with a Bi seeding layer at 500 °C. (c) Photograph of Bi_2O_3 nanocones grown on a 4 cm × 4 cm Si (001). (d) EDX spectrum of sample deposited at 500 °C. (Tien and Liou [33]. Reproduced with permission of Elsevier.)

to be dissociated and it is easy for carbon to form the graphite structure, the experimentally synthesized carbon nitride is limited to amorphous or nanosphere thin films with nitrogen deficiency and nanobelts with low nitrogen content [49–52]. Recently, Guan *et al.* [53] successfully prepared the vertically and neatly arranged crystalline g-C_3N_4 nanocone arrays on nickel-covered silicon (100) substrates supported on a graphite frame by using an abnormal glow discharge plasma sputtering reaction deposition method with a feed gas mixture of CH_4, N_2, and H_2. The corresponding SEM images are shown in Figure 3.9.

As shown in Figure 3.10, transmission electron microscopy (TEM), selected area electron diffraction (SAED), and energy dispersive X-ray spectroscopy (EDXS) data were used to further confirm the composition and structure of the nanocones grown. The nanocones displayed good mechanical stability and quite strong adhesion to the substrates for the samples.

Figure 3.11 presents the growth mechanisms of the three kinds of nanocones. Figure 3.11a–c lay out the growth processes of the black silicon, g-CNNCs, and diamond nanocones, respectively. In Figure 3.11, the growth processes of all nanocones could be roughly divided into two stages, that is, the production of building elements and the building up of nanocones. In the first stage, the

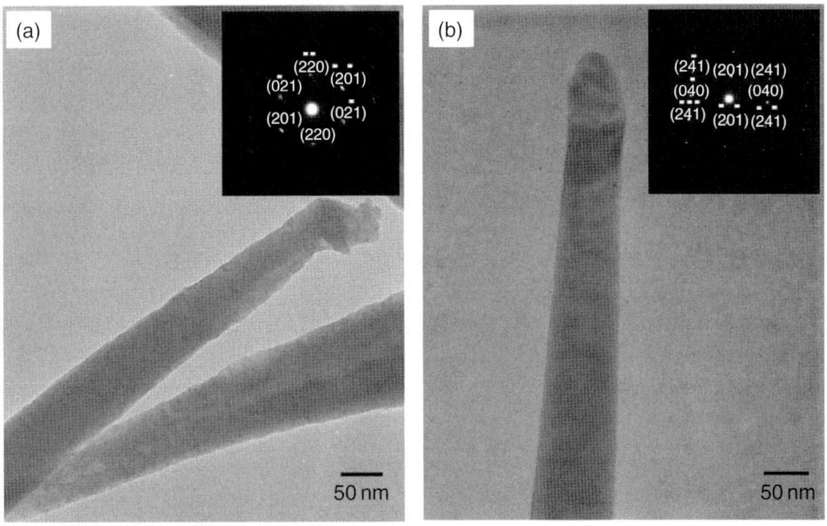

Figure 3.7 (a) TEM image of α-Bi_2O_3 nanocones. Inset shows SAED patterns from a single α-Bi_2O_3 nanocone grown at 400 °C. (b) TEM image of a β-Bi_2O_3 nanocone. Inset shows SAED patterns from a single β-Bi_2O_3 nanocone grown at 500 °C. (Tien and Liou [33]. Reproduced with permission of Elsevier.)

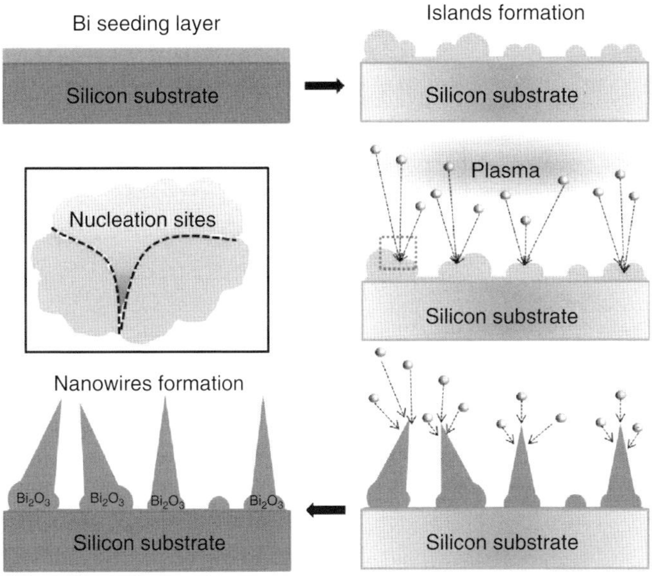

Figure 3.8 Model of the growth of Bi_2O_3 nanocones by magnetron sputtering. (Tien and Liou [33]. Reproduced with permission of Elsevier.)

Figure 3.9 FESEM images of the samples prepared on the scratched Si (100) wafers for 45 min, with a Ni catalyst layer of about 100 nm; a discharge current of 180 mA; a voltage of 350 V; and different $CH_4/(N_2 + H_2)$ ratios of (a) 1/10, (b) 1/20, (c) 1/40, (d) 1/100, (e) 1/150, and (f) 0, respectively. (Guan *et al.* [53]. Reproduced with permission of Elsevier.)

abnormal glow discharge produces a high-density plasma containing a large number of CH_n^- ($n \leq 3$), N^+, and H^+. In the second stage of Figure 3.11, the predeposited Ni layer will melt and fragment into separated spheres under the heating of the discharge plasma.

3.2.2.2 **InAlN Nanorods**

Group III nitride semiconductors such as InN, GaN, and AlN, as well as their ternary and quaternary alloys, have attracted great research interests lately because of their promising applications in high-performance devices, including ultraviolet/blue/green-light-emitting diodes and laser diodes that operate in the blue and ultraviolet spectrum, high-temperature diodes, transistors or solar cells, or high-frequency/high-power electronics used for high-electron-mobility transistors (HEMTs) and photodetectors [54–57].

Especially, increasing number of investigations on the growth of III-nitride nanorods have been developed rapidly and vigorously because of the obvious advantage of a large junction area that makes the resultant structures suitable for fabricating high-performance nanodevices, such as high-brightness light-emitting devices and high-sensitivity sensors [58–60].

Among the III-nitride nanomaterials, the vertically aligned $Al_{1-x}In_xN$ nanorod array with a controlling range of 0.7–6.2 eV is an excellent nanomaterial to fabricate broad-bandgap-range semiconductor nanoscale optoelectronics [61]. The conducting transition metal nitride $Ti_{1-y}Zr_yN$ acts as seed layers on sapphire substrates for resolving the bottom contact electrode for making devices [61]. In Figure 3.12, the well-distributed $Al_{1-x}In_xN$ nanorods with hexagonal shape are demonstrated in the cross-sectional and top-view SEM images. The rods have a uniform height of about 2.7 µm and diameter of about 70 nm at the upper

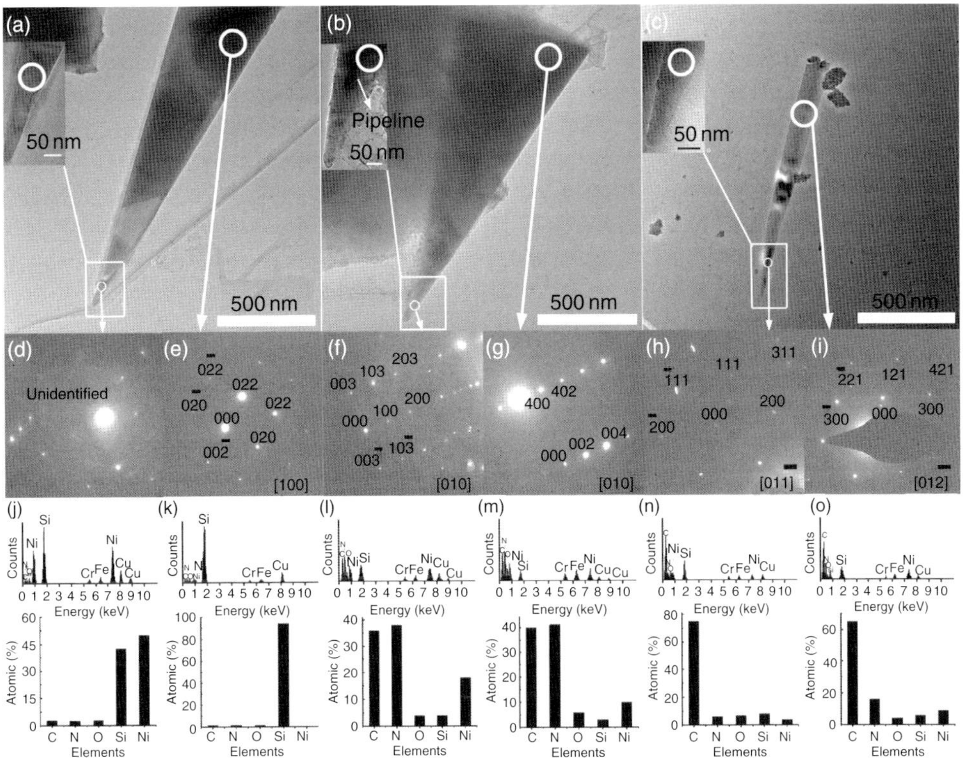

Figure 3.10 TEM images of the nanocones grown at the $CH_4/(N_2 + H_2)$ ratios of (a) 1/10, (b) 1/150, and (c) 0; corresponding SAED patterns taken from (d, f, h) bodies and (e, g, i) tops in (a–c); and (j–o) EDXS spectra and their analytical histograms at the nearby locations corresponding to subgraphs (d–i). (Guan *et al.* [53]. Reproduced with permission of Elsevier.)

part of the rods after 4 h of growth. The inset of Figure 3.12a clearly reveals that $Al_{1-x}In_xN$ nanorods were nucleated on the flat $Ti_{0.21}Zr_{0.79}N$ seed layer epitaxially grown on the sapphire substrate. The growth of nanorods depends on proper seed layer assistance.

Recently, Serban *et al.* successfully prepared Al-rich $In_xAl_{1-x}N$ core–shell nanorods on Si (111) substrates by using the DC magnetron sputtering technique [54]. The SEM images of core–shell nanorods are shown in Figure 3.13. The core–shell nanorods can be grown as a semiconductor core coated with a shell of another material, usually exhibiting a larger bandgap [54, 62–64]. These core–shell nanorods show a high-quality crystalline structure and uniform shells along the whole length of the nanorods and then could overcome the shadowing effect caused by directional growth [62, 64]. Moreover, spontaneously formed core–shell nanostructures can effectively avoid the generation of misfit

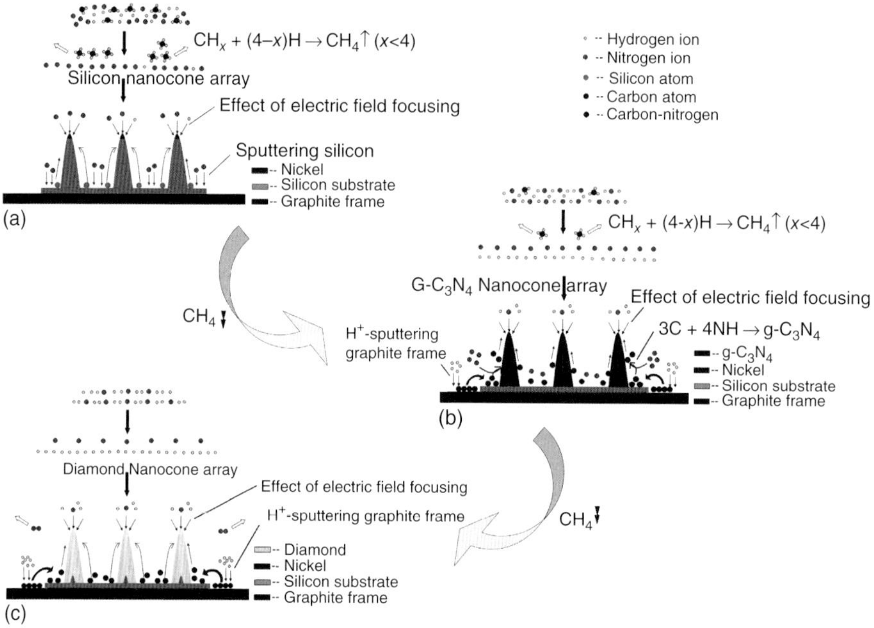

Figure 3.11 Schematic of the growth mechanisms of (a) silicon nanocones, (b) g-CNNCs, and (c) diamond nanocones. (Guan *et al.* [53]. Reproduced with permission of Elsevier.)

Figure 3.12 (a) Cross-sectional SEM image of the $Al_{1-x}In_xN$ nanorods grown with a $Ti_{0.79}Zr_{0.21}N$ seed layer assistance. The inset shows the interface of $Al_{1-x}In_xN$ nanorods/$Ti_{1-y}Zr_yN$ seed layer/sapphire. The scale shown in the inset is 60 nm. (b) Corresponding top-view image of the $Al_{1-x}In_xN$ nanorod arrays. (Hsiao *et al.* [61]. Reproduced with permission of Japan Society of Applied Physics.)

Figure 3.13 Top-view (a,c) and side-view (b,d) SEM images of the $In_xAl_{1-x}N$ grown at the various substrate temperatures (a,b) 700 °C and (c,d) 750 °C by DC magnetron sputtering technique [54]. (Serban *et al.* http://iopscience.iop.org/article/10.1088/0957-4484/26/21/215602/meta. Used under CC BY 3.0 http://creativecommons.org/licenses/by/3.0/.)

dislocations between the core and the shell because of surface relaxation or reconstruction of the rod core [65].

Ching-Lien Hsiao *et al.* successfully fabricated $In_xAl_{1-x}N$ nanospirals with tailored chirality by a novel curved-lattice epitaxial growth (CLEG) technique [66], the sketch of which is illustrated in Figure 3.14. The SEM images of as-grown samples are shown in Figure 3.2. The nanorods exhibit right-handed nanospirals comprised of five turns with a period of 200 nm, which are rather homogeneous in height (~1 μm), spiral diameter (~80 nm), and rod diameter (~60 nm) (Figure 3.15).

3.3 Thin Films Deposited with Sputtering

Compared to bulk materials, thin films will be used in applications with miniaturization of many devices. In addition, the unique physical and chemical properties of the films drastically deviate from the bulk properties because of the confined geometry and the properties of thin films. There are several methods to prepare

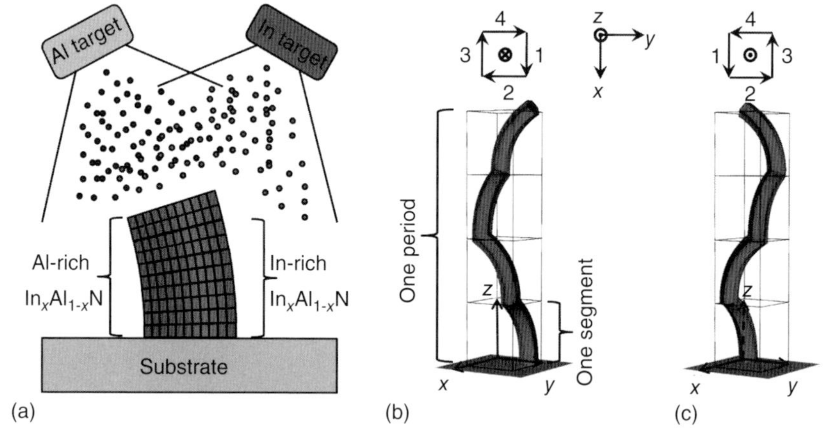

Figure 3.14 Schematic illustration of controlled curved-lattice epitaxial growth (CLEG) and the growth of one period of a nanospiral structure. (a) A curved rod segment of an $In_xAl_{1-x}N$ nanospiral (one-fourth of a spiral turn in this sample) controlled by CLEG. Dark gray to light gradient along the lateral direction of the curved rod segment indicates the compositional gradient from high to low Al content $In_xAl_{1-x}N$. (b,c) One period of a left- and right-handed nanospiral structure, respectively. One turn of a nanospiral is comprised of four curved rod segments, obtained by temporal control of the azimuthal orientation of the deposition fluxes. The arrow diagrams illustrate the sequence and direction of the curvature of each segment, as seen from the top, in the two cases of left- and right-handed nanospiral growth [66]. (Hsiao *et al.* http://pubs.acs.org/doi/abs/10.1021/nl503564k. Used under CC BY 4.0 https://creativecommons.org/licenses/by/4.0/.)

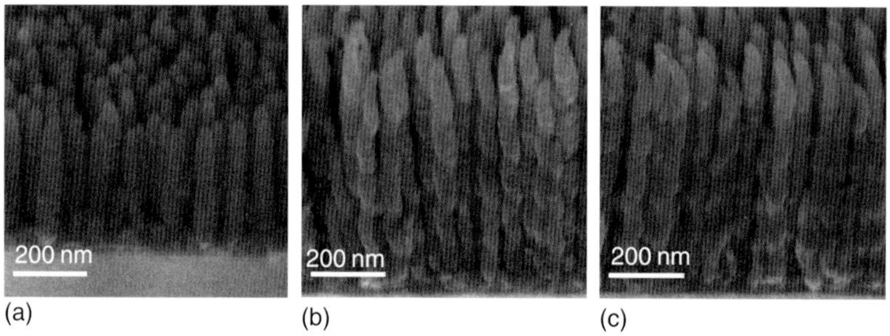

Figure 3.15 SEM images of $In_xAl_{1-x}N$ nanorods and nanospirals. (a) Straight nanorods. (b,c) Left- and right-handed nanospirals, respectively, comprising five turns at 200 nm pitch. The nanospiral's growth is homogeneous in both length (~1 μm) and diameter (60 nm) [66]. (Hsiao *et al.* http://pubs.acs.org/doi/abs/10.1021/nl503564k. Used under CC BY 4.0 https://creativecommons.org/licenses/by/4.0/.)

functional thin films. Among them, sputtering technique is one of the important techniques to prepare the desired thin films. Specifically, we will introduce functional alloy thin films, such as $LaNi_5$ and Ni–Mn–In, and composite metal oxide thin films by sputtering in this chapter.

3.3.1 Metal Alloy Thin Films

3.3.1.1 $LaNi_5$ Alloy Thin Films

Compared with $LaNi_5$ bulk and powder, $LaNi_5$ thin film is one of the most hydrogen-sensitive materials, owing to the $LaNi_5$ response mechanism that depends on electrical resistance [67, 68]. Han Tian *et al.* successfully prepared $LaNi_5$ thin films for hydrogen gas sensors by rf magnetron sputtering under various sputtering conditions [67]. As shown in Figure 3.16, the thicknesses of $LaNi_5$ thin film and Ni buffer layer are 820 and 400 nm, respectively. The cross-sectional diagram of the $LaNi_5$ thin film presents obvious columnar crystal formation, while in Figure 3.17, the grain sizes of $LaNi_5$ films are about 50 nm before annealing and increase to 100–200 nm after annealing, clearly indicating grain growth. Moreover, the gaps that promote the absorption of hydrogen on the

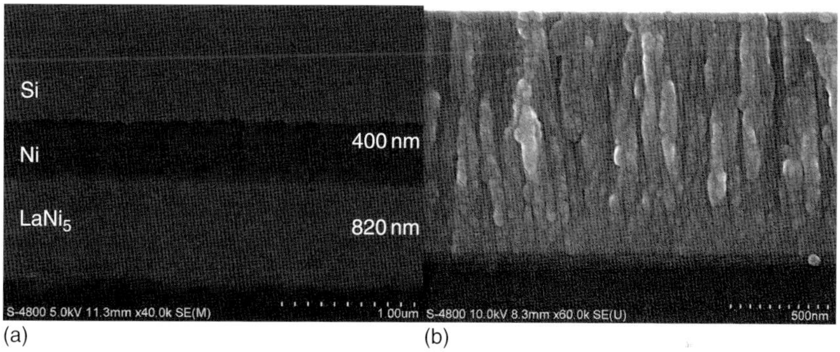

Figure 3.16 SEM images of sectional $LaNi_5$ thin films at (a) low magnification and (b) high magnification. (Tian *et al.* [67]. Reproduced with permission of Elsevier.)

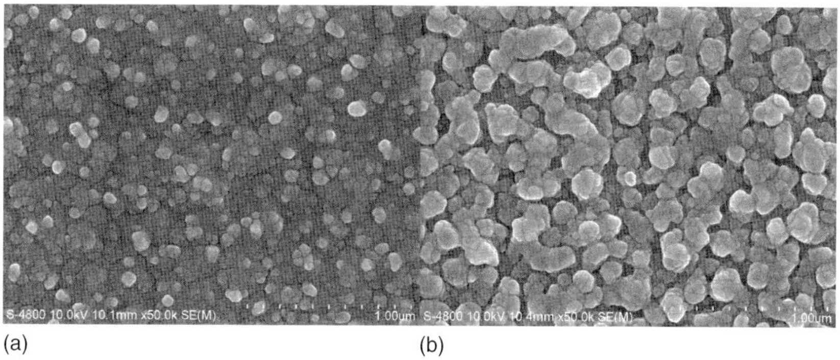

Figure 3.17 SEM images of $LaNi_5$ thin films: (a) non-annealed and (b) annealed at 700 °C. (Tian *et al.* [67]. Reproduced with permission of Elsevier.)

surface of $LaNi_5$ film become larger between grains. The as-grown alloy thin films by rf magnetron sputtering perform with excellent hydrogen-sensitive properties.

3.3.1.2 Ni–Mn–In Alloy Thin Films

Ni–Mn–In alloy thin films are a new kind of nanomaterial with multifunctional properties such as the large magnetocaloric effect [69, 70], magnetoresistance [71], magnetic field-induced strain [72], and magnetic shape memory effect [73]. Exchange bias (EB) was also reported in these alloys because of the coexistence of ferromagnetic (FM) and antiferromagnetic exchange interactions in martensitic phase [74]. The remarkable properties of these alloys are usually attributed to the occurrence of any solid–solid displacive, diffusionless first-order thermoelastic martensitic transformation (MT), which takes place from high-temperature austenitic phase (high symmetrical) to a low-temperature martensitic phase (low symmetrical). Recently, Davinder Kaur group reported the martensitic phase transformation of nanostructured Ni–Mn–In ferromagnetic shape memory alloy thin films by DC magnetron sputtering. As shown in Figure 3.18, the cross-sectional field emission scanning electron microscope (FESEM) and atomic force microscope (AFM) images show an increase in grain size with increase in film thickness. The average grain size values observed by FESEM for the samples was about 14–95 nm. The cross-sectional FESEM images present a clear interface between the substrate and the film, which indicates that there is no diffusion at the interface.

Figure 3.18 FESEM cross-sectional images and atomic force microscope images of samples [74]. (Sharma *et al.* [69, 70]. Reproduced with permission of Elsevier.)

3.3.2 Composite Metal Oxide Thin Films

3.3.2.1 $BiFeO_3/BaTiO_3$ Bilayer Thin Films

The interplay between magnetic and ferroelectric order parameters offers various applications in novel spintronics and magnetoelectric devices with fast speed and low power consumption [75, 76]. The interfacial effects play a crucial role in deciding the ferroelectric, magnetic, and magnetoelectric properties of bilayer thin films [77–80]. Bismuth ferrite, $BiFeO_3$ (BFO), is a novel multiferroic material with the simultaneous existence of room temperature antiferromagnetism (TN = 643 K) and ferroelectricity (TC = 1103 K). The use of 18 bilayer thin films of BFO with other perovskite ferroelectric materials such as $BaTiO_3$ (BTO) thin film has been found to be an effective way to reduce its leakage current and to improve ferroelectric polarization [78–82]. $BaTiO_3$ is one of the lead-free ferroelectric materials that has been very well studied in the solid solution form of $BiFeO_3-BaTiO_3$. Recently, Kotnala and coworkers [83] have prepared bilayer thin films of BFO/BTO on $Pt/Ti/SiO_2/Si(100)$ substrate with different thicknesses of BFO with an objective to understand the influence on

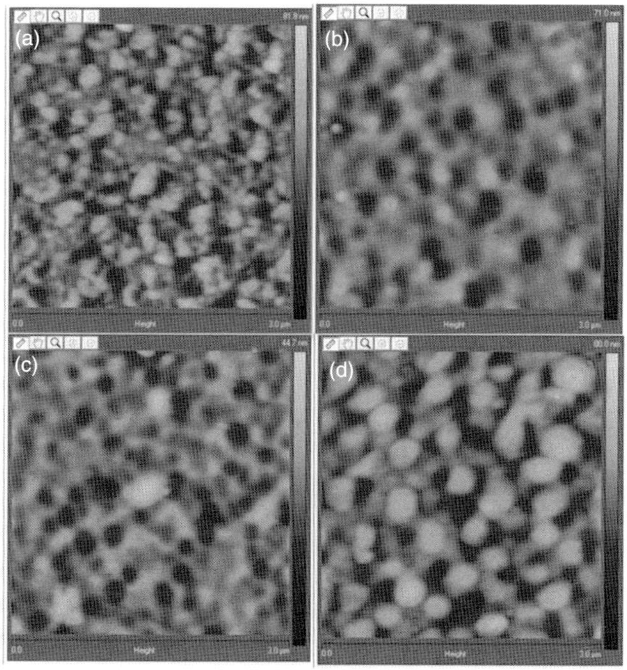

Figure 3.19 AFM images captured on 3 μm × 3 μm area of the different bilayer thin films. The thickness of BTO layer was kept fixed at 100 nm while BFO layer was deposited for different thicknesses of 50 nm (a), 100 nm (b), 150 nm (c), and 200 nm (d), respectively. (Gupta *et al.* [83]. Reproduced with permission of American Chemical Society.)

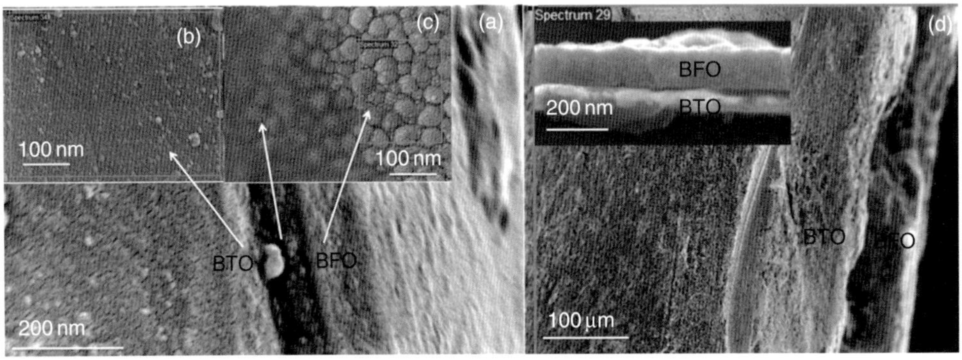

Figure 3.20 Cross-sectional FESEM image of bilayer thin film of BFO/BTO: (a) top view of film interface; inset (b) BTO layer; inset (c) interdiffusion and BFO layer; and (d) side-view bilayer interface. (Gupta *et al.* [83]. Reproduced with permission of American Chemical Society.)

the multiferroic properties and magnetoelectric coupling interaction mediated by the nature of interface between the two layers by using the rf magnetron sputtering technology. As shown in Figure 3.19, a dense, crack-free surface morphology without pinholes can be observed from atomic force microscope (AFM) images for all bilayer films. Homogeneous distribution of grains and grain boundaries irrespective of increasing thickness of BFO layer has been deduced from the AFM images, and their average grain sizes were calculated to be about 12–30 nm. The morphology and nature of bilayer interface were further analyzed by cross-sectional FESEM images as shown in Figure 3.20. Along with well-observed grains and grain boundaries of both BTO and BFO layers, an interdiffused layer of both phases was also observed at the bilayer interface.

3.4 Summary

In summary, sputtering technologies can be widely applied for the growth of various functional nanomaterials and thin films as well as the corresponding semiconductor processing for metal and compound depositions. Physical sputtering mechanism is well understood, and a wide variety of functional nanowires, nanobelts, and thin films are available. This PVD sputtering technology can be either in the area of applications, such as interconnection wiring on semiconductors, or in the realm of materials, such as the development of high dielectric or piezoelectric films.

Acknowledgments

The authors acknowledge the National Natural Science Foundation of China (No. 51202023) and the scientific and technological projects for distinguished young scholars of Sichuan Province (No. 2015JQ0013).

References

1. Mattox, D.M. (1998) *Handbook of Physical Vapor Deposition (PVD) Processing*, Noyes Publications, New Jersey.
2. Roth, J. (1983) Chemical sputtering, Chapter 3, in *Sputtering by Particle Bombardment II* (ed. R. Behrisch), Springer-Verlag.
3. Ali, K., Rouhollah, A., Ali, A.Z., and Alireza, Z.M. (2009) *J. Phys. Chem. C*, **113**, 19298–19304.
4. Dekker, C. (1999) *Phys. Today*, **52**, 22–28.
5. Pan, Z.W., Dai, Z.R., and Wang, Z.L. (2001) *Science*, **291**, 1947–1949.
6. Yao, J.N., Hashimoto, K., and Fujishima, A. (1992) *Nature (London)*, **355**, 624–626.
7. Yang, W.Q., Wei, Z.R., Gao, M., Chen, Y., Xu, J., Chen, C.L., and Lin, Y. (2013) *J. Alloys Compd.*, **576**, 332.
8. Wen, X.N., Yang, W.Q., Ding, Y., Niu, S.M., and Wang, Z.L. (2014) *Nano Res.*, **7**, 180.
9. Yang, W.Q., Chen, J., Zhu, G., Wen, X.N., Bai, P., Su, Y.J., Lin, Y., and Wang, Z.L. (2013) *Nano Res.*, **6**, 880.
10. Yoon, S.K., Jung, K.N., Jin, C.S., and Shin, K.H. (2012) *J. Alloys Compd.*, **536**, 179.
11. Zhang, Q., Zhang, Y., Yu, K., and Zhu, Z. (2008) *Vacuum*, **82**, 30.
12. Wu, Z.S., Deng, S.Z., Xu, N.S., Chen, J., Zhou, J., and Chen, J. (2002) *Appl. Phys. Lett.*, **80**, 3829.
13. Chen, J., Deng, S.Z., Xu, N.S., Wang, S., Wen, X., Yang, S., Yang, C., Wang, J., and Ge, W. (2002) *Appl. Phys. Lett.*, **80**, 3620.
14. Ali, K., Rouhollah, A., Ali, A.Z., and Alireza, Z.M. (2009) *J. Phys. Chem. C*, **113**, 19298.
15. Li, X.L., Liu, J.F., and Li, Y.D. (2002) *Appl. Phys. Lett.*, **81**, 4832.
16. Yan, B., Zheng, Z., Zhang, J.X., Gong, H., Shen, Z.X., Huang, W., and Yu, T. (2009) *J. Phys. Chem. C*, **113**, 20259–20263.
17. Brezesinski, T., Wang, J., Tolbert, S.H., and Dunn, B. (2010) *Nat. Mater.*, **9**, 146.
18. Mohamed, S.H. and Venkataraj, S. (2007) *Vacuum*, **81**, 636–643.
19. Swanson, F. (1954) Natl. Bur. Stand. (U. S.), Circ. **539**, III30.
20. Yang, W.Q., Wei, Z.R., Zhu, X.H., and Yang, D.Y. (2009) *Phys. Lett. A*, **373**, 3965–3969.
21. Dai, Z.R., Pan, Z.W., and Wang, Z.L. (2003) *Adv. Funct. Mater.*, **13**, 9–24.
22. Liu, J.G., Zhao, Y., and Zhang, Z.J. (2003) *J. Phys.: Condens. Matter*, **15**, L453–L461.
23. Duan, X.F. and Lieber, C.M. (2000) *Adv. Mater.*, **12**, 298–302.
24. Huang, M.H., Wu, Y.Y., Feick, H.N., Tran, N., Weber, E., and Yang, P.D. (2001) *Adv. Mater.*, **13**, 113–116.
25. Chen, C.L. and Tsong, T.T. (1993) *Phys. Rev. B*, **47**, 15852–15859.
26. Chen, C.L. and Tsong, T.T. (1990) *Phys. Rev. B*, **41**, 12403–12412.
27. Chen, C.L. and Tsong, T.T. (1990) *Phys. Rev. Lett.*, **64**, 3147–3150.
28. Sears, G.W. (1955) *Acta Metall.*, **3**, 361–366.
29. Blakely, J.M. and Jackson, K.A. (1962) *J. Chem. Phys.*, **37**, 428–430.
30. Voigtländer, B., Zinner, A., Weber, T., and Bonzel, H.P. (1995) *Phys. Rev. B*, **51**, 7583–7591.
31. Amar, J.G. and Family, F. (1996) *Phys. Rev. Lett.*, **77**, 4584–4587.
32. Hlawacek, G., Puschnig, P., Frank, P., Winkler, A., Draxl, C.A., and Teichert, C. (2008) *Science*, **321**, 108–111.
33. Tien, L.C. and Liou, Y.H. (2015) *Surf. Coat. Technol.*, **265**, 1.
34. Ho, C.-H., Tsai, M.-C., and Wong, M.-S. (2008) *Appl. Phys. Lett.*, **93**, 081904.

35. Ho, C.-H., Chan, C.-H., Huang, Y.-S., Tien, L.-C., and Chao, L.-C. (2013) *Opt. Express*, **21**, 11965.
36. Zhang, L.S., Wang, W.Z., Yang, J.O., Chen, Z.G., Zhang, W.Q., Zhou, L., and Liu, S.W. (2006) *Appl. Catal., A*, **308**, 105.
37. Zhou, L., Wang, W., Xu, H., Sun, S., and Shang, M. (2009) *Chem. Eur. J.*, **15**, 1776.
38. Brezesinski, K., Ostermann, R., Hartmann, P., Perlich, J., and Brezesinski, T. (2010) *Chem. Mater.*, **22**, 3079.
39. Sirota, B., Reyes-Cuellar, J., Kohli, P., Wang, L., McCarroll, M.E., and Aouadi, S.M. (2012) *Thin Solid Films*, **520**, 6118.
40. Hou, J., Yang, C., Wang, Z., Zhou, W., Jiao, S., and Zhu, H. (2013) *Appl. Catal., B*, **142**, 504.
41. Li, L., Yang, Y.W., Li, G.H., and Zhang, L.D. (2006) *Small*, **2**, 548.
42. Qiu, Y., Liu, D., Yang, J., and Yang, S. (2006) *Adv. Mater.*, **18**, 2604.
43. Wu, C., Shen, L., Huang, Q., and Zhang, Y.-C. (2011) *Mater. Lett.*, **65**, 1134.
44. Tien, L.-C. and Lai, Y.-C. (2014) *Appl. Surf. Sci.*, **290**, 131.
45. Liu, A.Y. and Cohen, M.L. (1989) *Science*, **245**, 841.
46. Teter, D.M. and Hemlev, R.J. (1996) *Science*, **271**, 53.
47. Zheng, W.T., Sun, C.Q., and Tay, B.K. (2003) *Solid State Commun.*, **128**, 381.
48. Hu, J., Cheng, W., Huang, S., Wu, D., and Xie, Z. (2003) *Appl. Phys. Lett.*, **89**, 261117.
49. Zhang, Y.F., Zhou, Z.H., and Li, H.L. (1996) *Appl. Phys. Lett.*, **68**, 634.
50. Yen, T.Y. and Chou, C.P. (1995) *Appl. Phys. Lett.*, **67**, 2801.
51. Stafstrom, S. (2000) *Appl. Phys. Lett.*, **77**, 3941.
52. Zhang, G.Y., Ma, X.C., Zhong, D.Y., and Wang, E.G. (2002) *J. Appl. Phys.*, **91**, 9324.
53. Guan, L.L., Xu, N., Liu, X.J., Zhao, Y., Li, H., Sun, J., Wu, J.D., and Ying, Z.F. (2014) *Carbon*, **79**, 578.
54. Serban, E.A., Persson, P.O.A., Poenaru, L., Junaid, M., Hultman, L., Birch, J., and Hsiao, C.L. (2015) *Nanotechnology*, **26**, 215602.
55. Mohammad, S.N. and Morkoc, H. (1996) *Prog. Quantum Electron.*, **20**, 516.
56. Lu, N. and Ferguson, I. (2013) *Semicond. Sci. Technol.*, **28**, 074023.
57. Jr Eddy, C.R., Nepal, N., Hite, J.K., and Mastro, M.A. (2013) *J. Vac. Sci. Technol., A*, **31**, 058501.
58. Barth, S., Hernandez-Ramirez, F., Holmes, J.D., and Romano, A. (2010) *Prog. Mater Sci.*, **55**, 563.
59. Wang, G.T., Li, Q., Huang, J., Wierer, J., Armstrong, A., Upadhya, P., and Prasan kumar, R. (2011), *ECS Transactions*, **35**, 3.
60. Kölper, C., Sabathill, M., Römer, F., Mandl, M., Strassburg, M., and Witzigmann, B. (2012) *Phys. Status Solidi A*, **209**, 2304.
61. Hsiao, C.L., Palisaitis, J., Junaid, M., Chen, R.S., Persson, P.O.A., Sandström, P., Holtz, P.O., Hultman, L., and Birch, J. (2011) *Appl. Phys. Express*, **4**, 115002.
62. Durand, C., Bougerol, C., Carlin, J.F., Rossibach, G., Godel, F., Eymery, J., Jouneau, P.H., Mukhtarova, A., Butte, R., and Grandjean, N. (2014) *ACS Photonics*, **1**, 38.
63. Webster, R.F., Cherns, D., Novikov, S.V., and Foxon, C.T. (2014) *Phys. Status Solidi C*, **11**, 417.
64. Kamimura, J., Kishino, K., and Kikuchi, A. (2012) *Phys. Status Solidi RRL*, **6**, 123.
65. Enzevaee, C., Gutkin, M.Y., and Shodja, H.M. (2014) *Philos. Mag.*, **94**, 492.
66. Hsiao, C.L., Magnusson, R., Palisaitis, J., Sandström, P., Persson, P.O.Å., Valyukh, S., Hultman, L., Järrendahl, K., and Birch, J. (2015) *Nano Lett.*, **15**, 294.
67. Tian, H., Chen, K.S., Liu, H.R., Xie, K., and Gu, H.S. (2015) *Appl. Surf. Sci.*, **331**, 35.
68. Zhu, K.G., Yao, W., Chen, Z.Q., and Zhang, L.D. (1997) *J. Chin. Electron Microsc. Soc.*, **16**, 445–446.
69. Sharma Akkera, H., Singh, I., and Kaur, D. (2015) *J. Alloys Compd.*, **642**, 53.
70. Pathak, A.K., Khan, M., Dubenko, I., Stadler, S., and Ali, N. (2007) *Appl. Phys. Lett.*, **90**, 262504.
71. Pathak, A.K., Gautam, B.R., Dubenko, I., Khan, M., Stadler, S., and Ali, N. (2008) *J. Appl. Phys.*, **103**, 07F315.
72. Sharma, V.K., Chattopadhyay, M.K., Chouhan, A., and Roy, S.B. (2009) *J. Phys. D: Appl. Phys.*, **42**, 185005.

73. Heczko, O. and Straka, L. (2003) *J. Appl. Phys.*, **94**, 7139.
74. Wang, B.M., Liu, Y., Xia, B., Ren, P., and Wang, L. (2012) *J. Appl. Phys.*, **111**, 043912.
75. Valencia, S., Crassous, A., Bocher, L., Garcia, V., Moya, X., Cherifi, R.O., Deranlot, C., Bouzehouane, K., Fusil, S., Zobelli, A., Gloter, A., Mathur, N.D., Gaupp, A., Abrudan, R., Radu, F., Barthélémy, A., and Bibes, M. (2011) *Nat. Mater.*, **10**, 753.
76. Liu, M., Obi, O., Lou, J., Chen, Y., Cai, Z., Stoute, S., Espanol, M., Lew, M., Situ, X., Ziemer, K.S., Harris, V.G., and Sun, N.X. (2009) *Adv. Funct. Mater.*, **19**, 1826.
77. Thiele, C., Dörr, K., Bilani, O., Rödel, J., and Schultz, L. (2007) *Phys. Rev. B*, **75**, 054408.
78. Eerenstein, W., Wiora, M., Prieto, J.L., Scott, J.F., and Mathur, N.D. (2007) *Nat. Mater.*, **6**, 348.
79. Lou, J., Liu, M., Reed, D., Ren, Y., and Sun, N.X. (2009) *Adv. Mater.*, **21**, 4711.
80. Hu, J.M., Nan, C.W., and Chen, L.Q. (2011) *Phys. Rev. B*, **83**, 134408.
81. Catalan, G. and Scott, J.F. (2009) *Adv. Mater.*, **21**, 2463.
82. Wu, J. and Wang, J. (2010) *J. Phys. Chem. C*, **114**, 19318.
83. Gupta, R., Chaudhary, S., and Kotnala, R.K. (2015) *ACS Appl. Mater. Interfaces*, **7**, 8472.

4
Nanostructures and Quantum Dots Development with Molecular Beam Epitaxy

Wen Huang

4.1
Introduction

Molecular beam epitaxy (MBE) is one of several methods to deposit single crystals. It was invented in the late 1960s at Bell Telephone Laboratories by Arthur and Cho [1]. MBE is widely used in the manufacture of semiconductor devices, including transistors for cellular phones and Wi-fi. MBE is a process for making compound semiconductor materials with great precision and purity. These materials are layered one on top of the other to form semiconductor devices such as transistors and lasers. These devices are used in such applications as fiber optics, cellular phones, satellites, radar systems, and display devices.

Semiconductor devices consist of various materials with differing electronic properties stacked up much like a many-layered cake. How electrons move through the various layers determines the behavior and function of the device. For example, a transistor acts like a switch by only allowing electrons to pass through when a voltage is applied. Other structures, used for light-emitting diodes and lasers, produce light as electrons pass through the layers. To make each layer using the MBE technique, ultrapure elements are delivered to a surface (called the *substrate wafer*) as a beam of gas. The atoms or molecules in the gas deposit on the substrate to form the growing solid layer. Typically, each element is delivered in a separately controlled beam, so the choice of elements and their relative concentrations may be adjusted for any given layer, thereby defining the composition and electrical characteristics of that layer. Each gas beam may be turned on and off rapidly with a shutter or a valve. The beam intensity (called the *flux*) is adjusted for precise control of layer composition.

MBE offers tremendous control over layer thickness. The growth rate is sufficiently slow so that layers that are only a few atoms thick can be produced reliably. Thicker layers are obtained with longer deposition times. Also, MBE growth produces complex structures of varying layers that are further processed to produce a range of electronic and optoelectronic devices, including high-speed transistors, light-emitting diodes, and solid-state lasers. MBE is a powerful

Advanced Nano Deposition Methods, First Edition. Edited by Yuan Lin and Xin Chen.

technique both for research into new materials and layer structures and for producing high-performance devices.

4.2 Technology of MBE

MBE is an epitaxial growth technology that is based on the interaction of beams of atoms and molecules of thermal energy on a heated crystalline surface under ultra-high vacuum (UHV) conditions. Different MBE variants have been developed; as for the gas-source molecular beam epitaxy (GS-MBE) of III–V semiconductors, group-V materials are hydrides such as arsine (AsH_3) or phosphine (PH_3); for metalorganic molecular beam epitaxy (MO-MBE), group-III materials are metalorganic compounds such as triethylgallium (TEGa) and trimethylindium (TMIn), while in solid-source molecular beam epitaxy (SS-MBE), group-III and group-V molecular beams are produced by evaporating or sublimating source materials contained in high-purity crucibles in the so-called effusion cells. The chambers are equipped with pump systems that give base pressures in the 10^{-11} Torr range; UHV environment is required in order to reduce the background pressure of contaminants that, otherwise, would be incorporated into the solid phase. Moreover, under UHV conditions, atoms and molecules cover collision-free paths toward the substrate; hence, the molecular beams can be interrupted by means of mechanical shutters with actuation times corresponding to those required to deposit less than an atomic layer of the material. As a result, MBE grown structures are characterized by (i) low level of unintentional dopants (in the low 1014 cm^{-3} for GaAs), (ii) atomically abrupt interfaces, a feature related also to the kinetics of growth that is discussed later, and (iii) controlled composition and doping profiles along the growth direction. These basic features paved the way for the preparation of advanced structures suitable for new electronic and optical devices, thus justifying the great effort made over the last 30 years to develop and improve the MBE growth technique [2–5].

Recently, based on the conventional MBE with complex structure, a relatively inexpensive simple laser-MBE technology was developed. Laser-MBE is a versatile technique for growing epitaxial structures with molecular beams under UHV condition, as shown in Figure 4.1. It has been proved to be a great success in growing chemistry-complicated compounds, such as high-Tc superconductors, colossal magnetoresistance (CMR), manganites, and ferroelectrics. A laser-MBE system can also be considered as a modified pulsed laser deposition (PLD) system employing UHV and reflected high-energy electron diffraction (RHEED) that can deposit films in a layer-by-layer mode. PLD has been proved to be an ideal instrument to deposit high-quality thin films whose stoichiometry is nearly the same as that of the target. To take advantage of both PLD and MBE, the laser-MBE system has been successfully developed by employing UHV conditions and RHEED. Compared with the conventional PLD system that is basically in an island growth mode, laser-MBE is an approach that uses the PLD

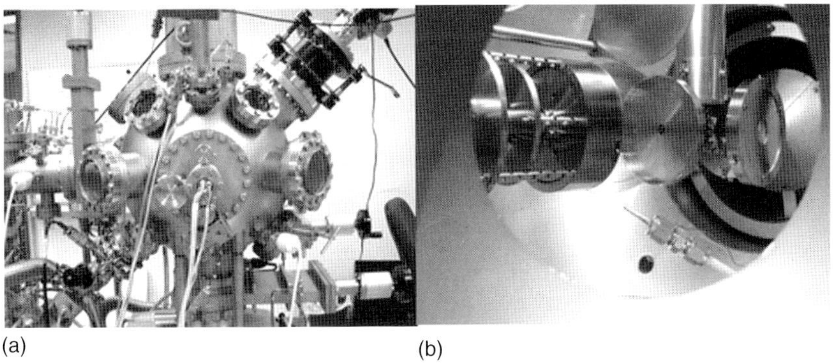

Figure 4.1 The setup of the laser-MBE (a) and the UH vacuum chamber of the LMBE (b). (Reproduced with kind permission of The Hong Kong Polytechnic University.)

to grow a layer-by-layer epitaxial thin film, which is similar to MBE. Figure 4.1 illustrates the setup of the system. Laser-MBE has several distinct advantageous features. For example, it can be used to deposit complex oxide epitaxial films that are almost impossible with conventional MBE. One particularly dramatic difference between crystal growth in conventional MBE and that in laser-MBE is the instantaneous deposition rate. In MBE, a typical growth rate might be only 1 ML/s. In laser-MBE, however, one can grow films at these rates, but it is also possible to grow films at instantaneous rates as high as several orders of magnitude faster. The average growth speed is limited only by the repetition rate of the laser. Hence, laser-MBE growth is an area of opportunity for a variety of fundamental kinetic studies that are difficult or impossible in MBE growth. Another advantage for industrial film fabrication is that this technique allows the fabrication of *c*-axis epitaxial films at relatively lower substrate temperatures than the standard MBE deposition. In addition, the laser-MBE system can be used for different film depositions without significant cross-contamination. This allows the system to serve different projects and to become a multiuser equipment.

4.2.1 The Physics of MBE

The main elements of a conventional MBE growth chamber are shown in Figure 4.2. Molecular beams, generated by effusion cells, impinge on a crystalline substrate mounted on a rotating holder; the substrate is held at a suitable temperature in order to activate the thermal migration of adsorbed species on the surface. The fluxes of molecular beams that depend on the temperature of the respective cells are measured by means of an ion gauge that can be placed in the substrate position. To ensure UHV conditions during growth, the chamber is provided with liquid nitrogen shrouds. The UHV environment allows the use of *in situ* analysis tools such as quadrupole mass analyzers (QMA) to monitor the vacuum environment and RHEED equipments to study the growing surface.

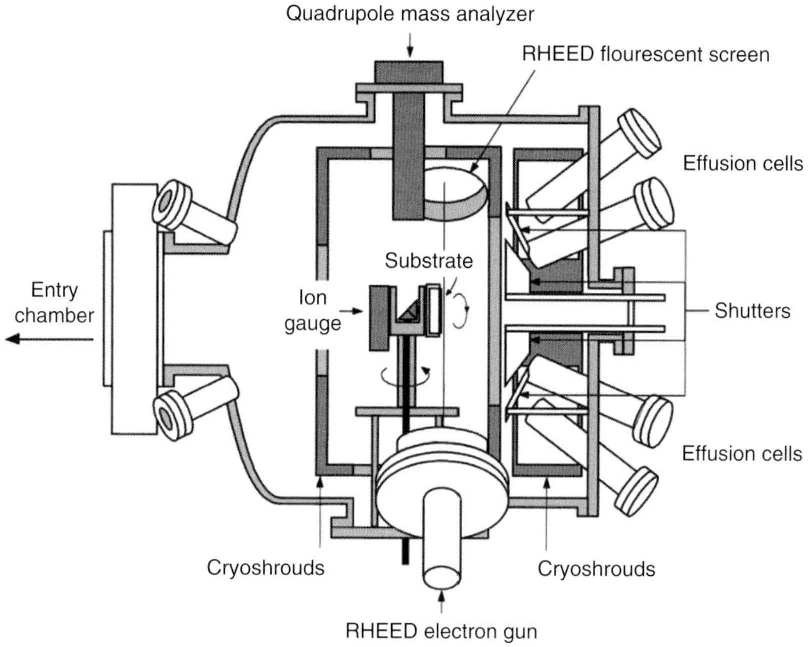

Figure 4.2 Schematic diagram of a solid-source MBE growth chamber. (Franchi *et al.* [6]. Reproduced with permission of Elsevier.)

Other characterization techniques, such as auger electron spectroscopy (AES), X-ray photoelectron spectroscopy (XPS), and secondary ion mass spectroscopy (SIMS), are available in the growth chamber or, more frequently, in separate chambers connected to the growth one by UHV modules.

Figure 4.3 illustrates the experimental setup of an laser molecular beam epitaxy (LMBE) system. A laser generated by a 200 keV laser gun serves as the evaporation source with a glancing angle onto the substrate surface; owing to the geometry of the system, the electron beam does not interfere with the molecular beams, thus allowing the real-time study of the growing surface. In the MBE/LMBE system, the RHEED proved to be a powerful characterization technique to gain insight into the growth process. The diffracted electrons of RHEED create a diffraction pattern on the fluorescent screen. The constructive interference of electrons diffracted by a crystalline epilayer gives rise to the well-known streaky diffraction patterns as shown in Figure 4.4. As experimentally proved by Cho [4, 5] and directly inferred from the analysis of the Ewald construction on a reciprocal two-dimensional (2D) lattice, the streaky pattern is typical of electron diffraction from an atomically flat surface and shows up a few tens of seconds after the beginning of growth. The presence of weaker, "fractional order" streaks between the "integer order" ones typical of bulk structures reveals that atoms on semiconductor surfaces arrange themselves in a configuration different from those of the corresponding bulk crystal. Surface reconstructions take place in order to reduce the surface energy by

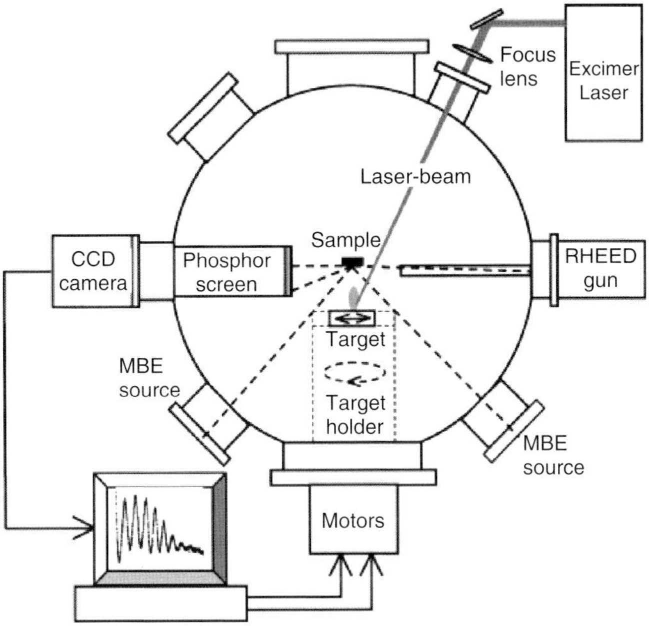

Figure 4.3 A diagram of laser-MBE system equipped with RHEED (including an electron gun and phosphor screen).

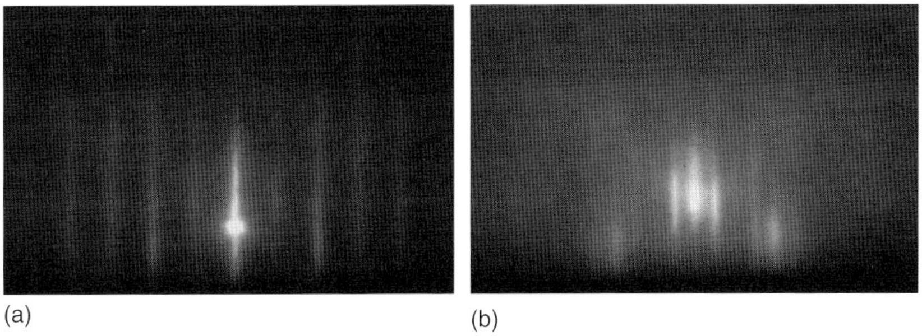

Figure 4.4 RHEED patterns during the MBE growth of an epitaxial GaAs layer deposited on a GaAs (100) substrate; the surface reconstruction is the As-stabilized (2 × 4) one. The azimuths are [1 −1 0] and [−1 −1 0] for the panels (a) and (b), respectively. (Franchi *et al.* [6]. Reproduced with permission of Elsevier.)

minimizing the number of dangling bonds as compared to that of unreconstructed surfaces. The reconstruction of surfaces has been studied by theoretical calculations and by a variety of experimental techniques, such as grazing incidence X-ray diffraction (XRD) and scanning.

Besides providing qualitative information about surface roughness and reconstruction, RHEED is used to study the time dependence of the intensity of diffracted streaks (Figure 4.5); after the beginning of growth or when growth is

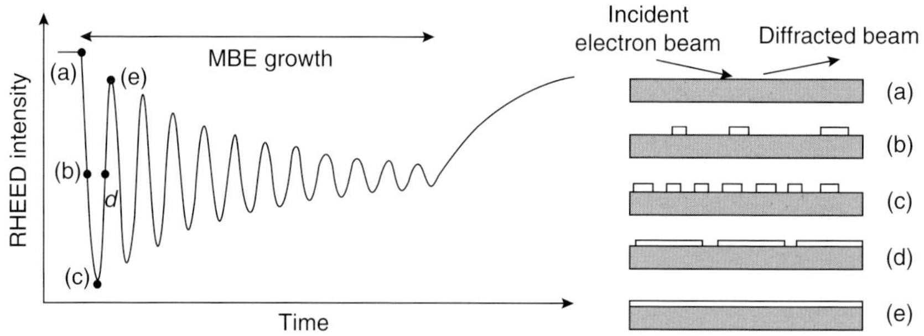

Figure 4.5 Different stages of layer-by-layer growth by nucleation of 2D islands and the corresponding intensity of the zero-order diffracted RHEED beam. See text. (Franchi *et al.* [6]. Reproduced with permission of Elsevier.)

resumed after an interruption, the intensity of the zero-order diffracted beam shows pronounced oscillations that damp down as growth proceeds. Then, as soon as the growth is interrupted, the intensity of the reflected beam increases almost to the initial value. The oscillatory behavior is related to the presence of 2D islands on the growing surface: by assuming that the height of islands is 1 ML and, then, much higher than the wavelength of the beam electrons, the island edges (steps) behave as scattering centers that reduce the intensity of the beam diffracted in the Bragg directions; therefore, the higher the surface density of steps, the lower the diffracted intensity. The notion that the islands are 1 ML high is confirmed by Monte Carlo simulations of the MBE growth process that will be described later. With reference to Figure 4.5, the oscillatory behavior may be explained as follows: after the beginning of growth (Figure 4.5a), the density of 2D islands and, then, of steps on the growing surface increases until it reaches a maximum value (Figure 4.5c); as a consequence, the diffracted intensity shows a minimum. As the growth goes on, 2D islands coalesce reducing the density of atomic steps (Figure 4.5d); correspondingly, the diffracted beam intensity increases until it reaches a maximum value when the coverage of the surface tends to be complete (Figure 4.5e). Thus, the period of oscillation corresponds to the time needed for the deposition of 1 ML of material; therefore, the RHEED intensity oscillations provide a real-time *in situ* tool to measure the deposition rate of the epitaxial growth.

4.2.2
MBE Growth Mechanisms

The growth of epitaxial (homogeneous or heterogeneous) thin films on a single crystal surface depends critically on the interaction strength between adatoms and the surface. Although it is possible to grow epilayers from a liquid solution, most epitaxial growth occurs via a vapor phase technique such as MBE. Basically, there are three kinds of growth modes during MBE growth according to the lattice mismatch between the film and the substrate: *3D growth*, *2D growth*, and

Stranski–Krastanov (SK) growth. In 3D growth mode, adatom–adatom interactions are stronger than those of the adatom with the surface, leading to the formation of 3D adatom clusters or islands. Therefore, 3D growth is also called *island growth*. Growth of these clusters, along with coarsening, will cause rough multi-layered films to grow on the substrate surface. Antithetically, during 2D growth, adatoms attach preferentially to surface sites resulting in atomically smooth, fully formed layers. This layer-by-layer growth is two dimensional, indicating that complete films form prior to the growth of subsequent layers. Stranski–Krastanov growth is an intermediary process characterized by both 2D layer and 3D island growth. Transition from the layer-by-layer to island-based growth occurs at a critical layer thickness, which is highly dependent on the chemical and physical properties, such as surface energies and lattice parameters, of the substrate and the film.

4.2.2.1 Two-Dimensional (2D) MBE Growth Mechanism

The growth mechanism that results in the oscillatory behavior of RHEED intensity (Figure 4.5) is usually referred to as *layer-by-layer* 2D growth mode. It takes place (i) since the incorporation of group-III adatoms is energetically more favorable at the edges of existing islands than on top of them and (ii) when the surface migration length l of group-III adatoms is shorter than the mean distance D between islands; under these conditions, part of the cations may reach the preferred sites for incorporation at the edges of existing islands, while others contribute to the formation of new islands even on top of not yet completed ones. A second 2D growth mode takes place when l is longer than D, that is, when the group-III adatoms may easily reach the edges of existing islands and, then, the islands laterally expand causing the apparent flow of their edges. This 2D growth mechanism is referred to as *step-flow* or *step-propagation mode*; under this condition, the surface density of steps is constant and, therefore, the RHEED intensity has stationary values, as opposed to the oscillating ones that stem from the periodic variation of step density when $\lambda < D$. Recently, advanced *in situ* imaging techniques such as scanning tunneling microscope (STM) and reflected electron microscopy (REM) gave the possibility of observing the different growth modes during the MBE growth of Si on Si (111) surfaces simply by changing the substrate temperature and, then, the migration length of the surface species with respect to the mean distance among islands.

The main result of MC simulations for compound semiconductor growth is the highlight of the leading role of cation surface migration length to obtain atomically smooth surfaces. Owing to the temperature dependence of the migration processes, at relatively low growth temperatures the migration length is too short and several atomic layers may be formed before the completion of the lower ones. On the other hand, at higher temperatures, the hopping rate is sufficiently high (in the order of 10^4 hops/s) and the migration length of cations is large enough for them to be prevalently incorporated at the edges of existing islands, thus assuring a smooth growth front. In choosing the MBE growth temperatures, it has to be considered that in order to have smooth interfaces, the growth temperature should be sufficiently high while, on the other hand, the probability of incorporation of vacancies and interstitials increases with increasing growth temperature; therefore, an optimum growth temperature window exists for the MBE process, centered at a temperature that is material dependent and that basically scales with the anion–cation binding energy.

The migration length of the adsorbed cations not only depends on the deposition temperature but also increases with decreasing surface population of adsorbed anions. Since the anion desorption rate is larger than that of the cations, after a growth interruption the surface anion population preferentially decreases; as a result, the cation migration length increases and the growth front smoothens. This is experimentally proved by the observation of the recovery of the RHEED reflected beam intensity usually observed after a growth interruption (Figure 4.5). Consistent with the results of Monte Carlo simulations, alternative growth techniques have been developed in order to improve the smoothness of the growth front. Migration enhanced epitaxy (MEE) and atomic layer molecular beam epitaxy (ALMBE) are MBE variants that allow the growth of smooth surfaces at temperatures lower than those of usual MBE growth; they are based on the alternated supply (in ML or sub-ML amounts) of group-V and group-III species to the substrate. During the group-III subcycle, the relatively low surface concentration of group-V species results in the enhancement of the cation migration rate, even at a low growth temperature. Since the reduced substrate temperature inhibits interdiffusion of species, these techniques may allow the growth of heterostructures with atomically flat interfaces and even sharper composition and doping profiles as compared to those of structures grown by conventional MBE at relatively higher temperatures.

4.2.2.2 Three-Dimensional (3D) MBE Growth Mechanism

The growth of lattice-mismatched heteroepitaxial structures follows mechanisms that are different from those described for 2D MBE growth. With reference to the growth on GaAs of the 7% lattice-mismatched InAs (Figure 4.6), during the initial stage of deposition, InAs grows following the layer-by-layer 2D mechanism with the in-plane lattice parameter matched to that of the underlying GaAs substrate (Figure 4.6d); the resulting strained layer is termed a *wetting layer* (WL). The corresponding RHEED pattern shows the arrangement of streaks typical of the InAs (100) reconstruction (Figure 4.6c). When the amount of deposited InAs exceeds the so-called critical coverage (q_c), the streaks are suddenly replaced by bright spots (Figure 4.6e), thus indicating that the 2D layer-by-layer growth mode is no longer followed and that growth proceeds by the formation of 3D islands (Figure 4.6f). q_c depends both on the growth conditions and on the lattice mismatch of the heterostructure: a typical value for InAs deposited on (001) GaAs is q_c Z 1.6 ML (w 0.5 nm). The well-known evolution described in Figure 4.6e,f is usually referred to as *2D-to-3D growth transition*; the nucleation of 3D islands homogeneously distributed on the growing surface can be evidenced by atomic force microscopy (AFM) (Figure 4.7). The islands are of nanometer size, with heights and diameters in the order of a few nanometers and a few tens of nanometers,

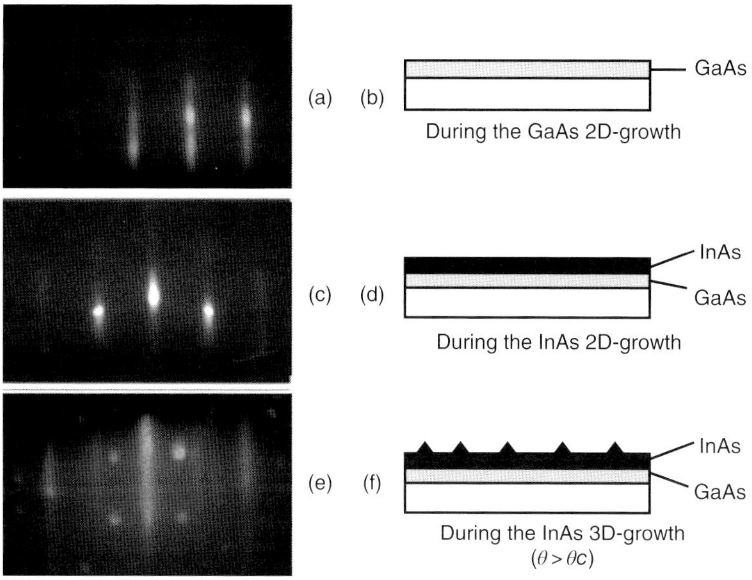

Figure 4.6 RHEED patterns (a,c,e) during the 2D deposition of GaAs (b), the 2D deposition of InAs (d), and the 3D deposition of InAs (f) on a (100) GaAs buffer layer, respectively. (Franchi *et al.* [6]. Reproduced with permission of Elsevier.)

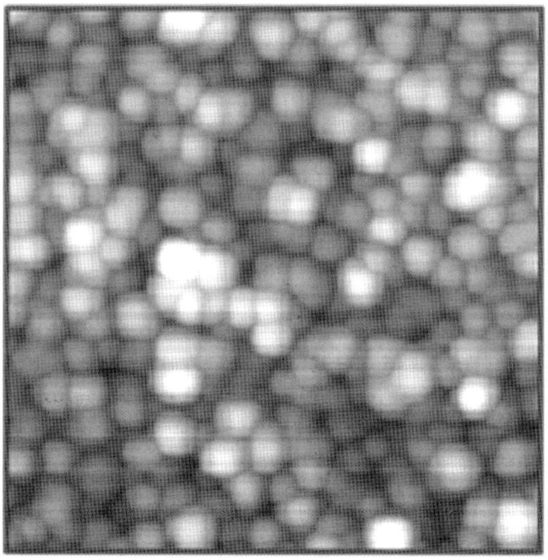

Figure 4.7 AFM image ($250 \times 250\,nm^2$) of InAs deposited on GaAs (100) with a coverage exceeding q_c. (Franchi *et al.* [6]. Reproduced with permission of Elsevier.)

respectively. Moreover, AFM analysis of the ensemble of islands reveals that their dimensions have reduced dispersions that, depending also on growth conditions, are in the order of 10%. The growth of self-assembled 3D islands on top of a 2D wetting layer pseudomorphically grown on the underlying epilayer is usually referred to as *Stranskie–Krastanow 3D growth* [6].

4.2.2.3 Stranskie–Krastanow 3D Growth Mechanism

Stranski–Krastanov growth (SK growth, also Stransky–Krastanov or Stranski–Krastanow) is one of the three primary modes by which thin films grow epitaxially at a crystal surface or interface. It is also known as *layer-plus-island growth,* which means a 2D-to-3D growth transition. The SK mode follows a two-step process: initially, complete films of adsorbates, up to several monolayers thick, grow in a layer-by-layer manner on a crystal substrate. Beyond a critical layer thickness, which depends on strain and the chemical potential of the deposited film, growth continues through the nucleation and coalescence of adsorbate "islands." This growth mechanism was first noted by Ivan Stranski and Lyubomir Krastanov in 1938. Since then, SK growth has been the subject of intense investigation, not only to better understand the complex thermodynamics and kinetics at the core of thin film formation but also as a route to fabricating novel nanostructures for application in the microelectronics industry. The driving force for the 2D-to-3D growth transition is the relaxation of the elastic energy that builds up as the thickness of mismatched epilayers increases.

As schematically shown in Figure 4.8, in a 3D island, the in-plane lattice parameter may continuously relax along the growth direction, hence allowing the relaxation of elastic energy without the formation of misfit dislocations at the heterointerfaces. This explanation has been experimentally verified by Grandjean

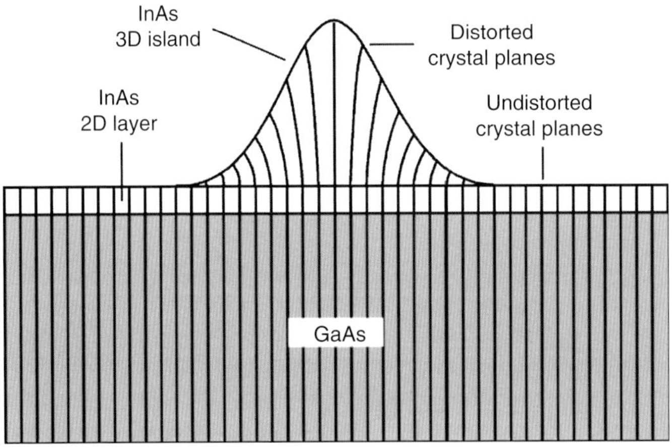

Figure 4.8 Schematic representation of the lattice distortion in a 3D island. (Franchi *et al.* [6]. Reproduced with permission of Elsevier.)

and Massies [3], who analyzed the distance between streaks or spots of RHEED patterns for the growth of $In_xGa_{1-x}As$ ($x = 0.35$ and $x = 0.53$) on GaAs substrates; the authors showed that the lattice parameter of the epilayer relaxes toward that of free-standing $In_xGa_{1-x}As$ after the 2D-to-3D growth transition. The occurrence of the 2D-to-3D transition has been observed in heteroepitaxial systems: III–V compounds such as arsenides, phosphides, antimonides, and nitrides; IV–IV compounds such as Ge/Si and SiGe/Si; II–VI compounds such as CdSe/ZnSe; and mixed-group compounds such as InAs/Si. It is worth noting that by growing lattice-mismatched systems, it is possible to obtain 3D self-assembled, coherent nanostructures without the use of sophisticated nanolithographic techniques.

High-resolution microscopic techniques have been used for the study of island sizes and shapes. By means of AFM, different shapes of transmission electron microscopy (TEM) and STM have been attributed to 3D islands, which vary from lenticular to conical or truncated conical, from pyramidal to truncated pyramidal with various facet orientations [6].

AFM images, as shown in Figure 4.7, are largely used in the analysis of size distributions and surface density of uncapped 3D islands. Although the evaluation of the height and density of 3D islands is quite straightforward, the measurement of the lateral island sizes and their shapes is affected by convolution effects between islands and probing tips that have sizes of the same order of magnitude. The convolution effects can be taken into account only if tip sizes and shapes are accurately known. Although AFM has been mainly used for the study of uncapped 3D island structures, it is worth remembering that cross-sectional AFM measurements have been recently performed on stacked layer structures, giving the possibility to obtain interesting information about the vertical ordering in quantum dot (QD) nanostructures.

4.3 Nanoheterostructures Fabricated by Molecular Beam Epitaxy

4.3.1 Semiconducting Oxide Heterostructures Grown by Laser Molecular Beam Epitaxy

Oxide-based ferromagnet/semiconductor heterostructures offer substantial advantages for spin electronics. Among them, semiconducting oxide heterostructures based on ZnO show high functional versatility. Efforts have been intensified in developing high-quality ZnO films and crystals on semiconductors for device applications. Epitaxial ZnO films have been reported to be fabricated on different substrates, such as GaN and sapphire, or by introducing Al_2O_3, NiO buffer layers. As it is known that optical and electrical properties of metal-oxide-based films show high dependence upon interface structure, oxygen stoichiometry, and crystal orientation, those substrates served for depositing ZnO films have similar structure or close lattice parameters compared to that of ZnO. On the other hand, GaAs has widely been used in the microwave and optoelectronic industries because of its highly saturated electron mobility, direct band gap, and various optical characteristics. It could be expected that the integration of ZnO with GaAs will open the way to the fabrication of new devices with novel functionalities.

A ZnO/p-GaAs (001) heterostructure was grown via an STO buffer by O_2 flowing LMBE system with an operation wave length of KrF ($\lambda = 248$ nm, energy density of 6 J/cm^2). In brief, initially approximately 50 nm STO was deposited on a chemically cleaned p-GaAs (001) substrate at 600 °C in a UHV chamber (1×10^{-6} Pa). Subsequently, the substrate was heated to 650 °C and a 200-nm-thick ZnO film was grown under O_2 partial pressure of 10 Pa. The film growth was monitored by *in situ* RHEED (Staib GmbH). Finally, the 100-nm-thick Au metal electrodes with 0.2 mm diameter were deposited for electrical measurements. The film structure was determined by XRD. The measurement of the current–voltage ($I-V$) curve was performed by using a Keithley 2400 electrometer. The dielectric properties including capacitance–frequency ($C-F$) and capacitance–voltage ($C-V$) characteristics were measured by using an LCR meter.

Figure 4.9 shows RHEED patterns of the epitaxial feature of ZnO films grown on GaAs substrates using STO as a buffer layer. It can be clearly seen that the diffraction spots of GaAs (001) in Figure 4.9a gradually faded away as the STO started to grow. Figure 4.9b shows the diffraction pattern with 25 unit cells of STO (about 10 nm), indicating that the film was coherent at the onset of the STO growth. The clear elongated bright spots suggest that a well-defined single-crystal STO layer forms on the GaAs substrate. For subsequent growth of the ZnO film on the

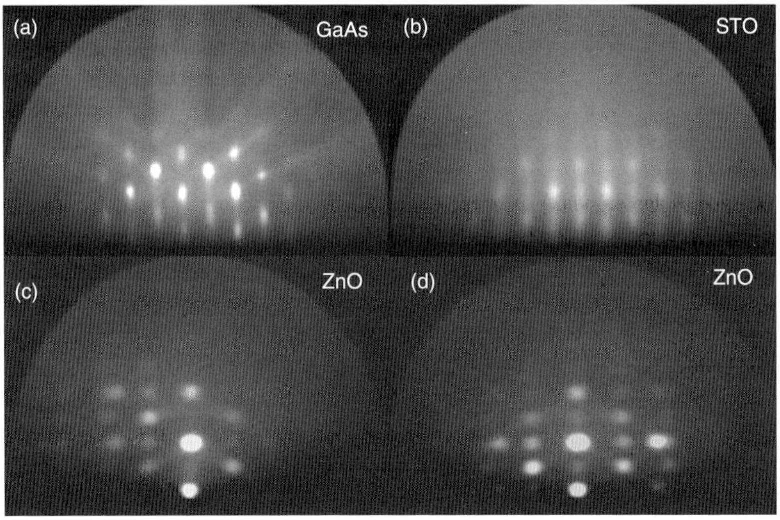

Figure 4.9 RHEED patterns for heteroepitaxial growth of ZnO thin films on GaAs via STO buffer layer with the laser repetition of 2 Hz and 2 J/cm^2 laser energy density at various thicknesses: GaAs substrate (a), 100 Å STO (b), 50 Å ZnO (c), and 2000 Å ZnO (d). (Huang *et al.* [7]. Reproduced with permission of American Institute of Physics.)

STO-buffered GaAs substrate, the diffraction pattern becomes weaker in contrast because of the interface transition from the perovskite to the wurtzite structure. As the deposition went to about 3 nm, several Bragg reflection spots gradually appeared as shown in Figure 4.9c, suggesting that the film grew with the island growth mode. With increasing ZnO film thickness, no unidentifiable RHEED patterns were detected and the quality of the diffraction pattern was enhanced as shown in Figure 4.9d, suggesting that the epitaxy and crystallinity of the ZnO film remained satisfactory via an STO interfacial layer.

Figure 4.10 shows a typical $\theta-2\theta$ XRD scan of the ZnO/STO/GaAs heterostructure. Only the (00l) peaks of the ZnO appear in the diffraction pattern, indicating that the ZnO exhibits basically *c*-axis oriented growth on the STO-buffered GaAs substrate. To further evaluate the in-plane alignment of the film, a Φ scan was performed (not shown here), which reveals that the ZnO film is epitaxially grown on the (001) surface of STO and has an in-plane film–substrate orientation relationship of [1−100]ZnO//[1−10]STO//[100]GaAs, which is consistent with the results from RHEED observation.

Figure 4.11a shows a schematic diagram of the Au/ZnO/STO/GaAs heterostructure. The heterostructure exhibits a variation in dielectric constant as a function of frequency at room temperature. The dielectric constant can be calculated to be 64 at 1 kHz. As shown in Figure 4.11a, the experimental dielectric constant value of the heterostructure was smaller compared to that of STO bulk and decreased significantly with increase in the measurement frequency. This reduced dielectric constant at room temperature may be attributed to the ZnO thin film with a small dielectric constant in this heterostructure. Figure 4.11b shows the capacitance–voltage curve at high frequency (1 MHz) between −6 and 6 V. There exists almost "zero" hysteresis in this $C-V$ curve with the maximum

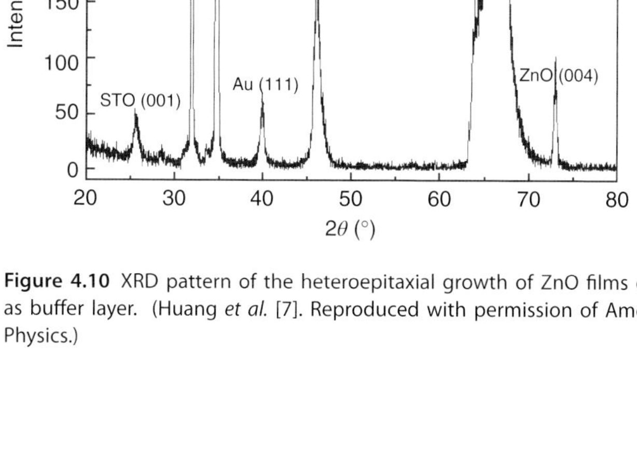

Figure 4.10 XRD pattern of the heteroepitaxial growth of ZnO films on GaAs (001) via STO as buffer layer. (Huang *et al.* [7]. Reproduced with permission of American Institute of Physics.)

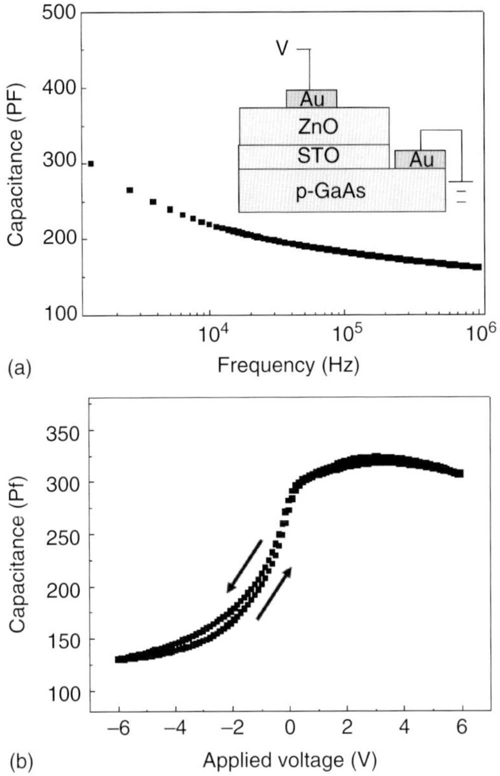

Figure 4.11 The capacitance of the Au/ZnO/STO/GaAs heterostructure as a function of frequency (a) and the inset shows the schematic of the Au/ZnO/STO/GaAs heterostructure and the capacitance–voltage response of the heterostructure under the bias voltage of 6 V at the frequency of 1 MHz (b). The arrows denote the direction of the bias sweeping. (Huang *et al.* [7]. Reproduced with permission of American Institute of Physics.)

shift less than 50 mV. The extremely low hysteresis, along with the sharp transition from depletion and accumulation, reveals the high quality of crystalline and interface properties of the heterostructure.

Figure 4.12 shows the typical I–V curve of the heterostructure at room temperature, where the resistance switching characteristic can be observed. The I–V curve reveals a strong nonlinear and asymmetric behavior. As shown in Figure 4.4, the bias voltage was swept as $-5 \rightarrow 0 \rightarrow +5 \rightarrow 0 \rightarrow -5$ V, and a forward-to-reverse bias ratio of 125 was obtained at the voltage of ± 4 V. It is apparent that the heterostructure exhibits a high resistance state when sweeping from 0 to +5 V and a low resistance state when sweeping from +5 to 0 V. We have also measured hysteresis I–V characteristics for different bias voltages (not show here). All I–V curves exhibit a similar hysteresis property. It is interesting to note that the junction shows a larger I–V loop with higher bias voltage.

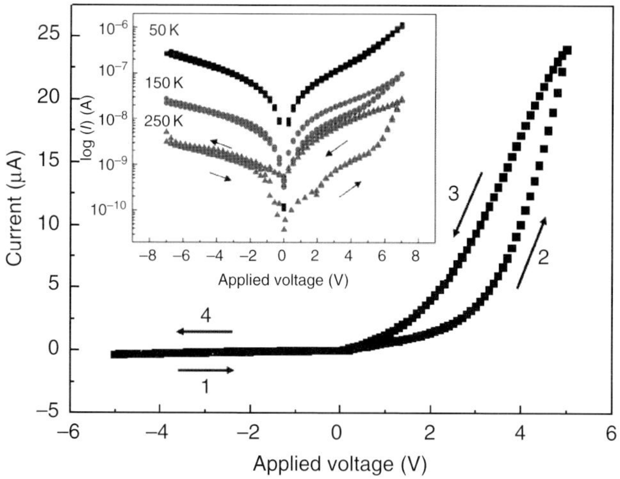

Figure 4.12 Current–voltage properties under the bias voltage of 5 V of the ZnO/STO/GaAs heterostructure at room temperature. The arrows indicate that the bias voltage was swept as −5 → 0 → +5 → 0 → −5 V, and the numbers from 1 to 4 indicate the direction of the sweep. The inset shows the current–voltage properties (in log scale) under the bias voltage of 7 V at various temperatures in the range of 50–250 K. (Huang *et al.* [7]. Reproduced with permission of American Institute of Physics.)

In earlier studies, reversible electroresistive (ER) switching effect has been found in some oxide systems, which is a claimed potentiality for nonvolatile memory applications. Recently, a lot of research has been focused on the switching properties of ZnO thin films. Although the mechanism is not well understood yet, it is recognized that the interface effect should play an important role in controlling the ER switching and other electric properties. For instance, in ZnO/$BaTiO_3$/ZnO heterostructure, the switching behaviors have been explained by coupling of the ferroelectric BTO and piezoelectric ZnO interface charges. In our case, the origin of the switching behavior seems to be quite complicated. According to earlier studies on ZnO/GaAs heterostructure, a diode-like rectifying $I-V$ characteristic without the switching behavior has been observed. Similar results have also been found in STO/GaAs and other STO/GaAs-related heterostructures. As a consequence, we consider that the switching behavior in the ZnO/STO/GaAs heterostructure could be attributed to the interface contact with a Schottky barrier between the ZnO/STO interface. It is known that interface properties in STO-based heterostructure could differ quite considerably. Here, we propose a scenario of charging effect at the interface through Fowler–Nordheim tunneling to defect states, that is, the acceptor trapping states present in the Schottky barrier because of impurities and structural defects. In the high resistance state, these traps are occupied by electrons. When a sufficiently high forward bias voltage is applied, electrons are detrapped, leaving resonant tunneling paths in the Schottky barrier; as a consequence, the resistance drops. A negative reverse bias voltage is able to produce retrapping, hence to switch back the device to the previous state. In other words, resonantly tunneling paths into the Schottky barrier would play the same role as filamentary percolative paths created by electroformation in insulating oxides [7].

In order to get more insights into the current transport mechanism in the Au/ZnO/STO/GaAs heterostructure, temperature-dependent $I-V$ measurements have been carried out in the temperature range of 50–300 K, as shown in the inset of Figure 4.12. While the temperature decreases, the hysteresis (in logarithmic scale) window gets smaller. At 50 K, the hysteresis disappears while the diode-like behavior is persistent. This result is quite different compared with the temperature-dependent switching behavior in the ZnO/BTO/ZnO system, which is caused by the disappearance of the charge coupling between the ferroelectric and the piezoelectric interface above the Curie temperature of BTO due to the phase transition from its tetragonal ferroelectric to cubic paraelectric phase in the BTO material. By contrast, in our case, the disappearance of the hysteresis can be considered as disappearance of the Schottky barrier at a low temperature. This is somewhat similar to the temperature-dependent $I-V$ curves in other systems of metal-oxide-semiconductor (MOS) structures, such as Al/HfO_2/SiO_2/p-Si, TaN/ZrO_2/SiO_2/p-Si. These temperature-dependent $I-V$ behaviors always show similar performances by inducing an interfacial layer. Therefore, these reversible electroresistive switching effects in our heterostructure could also be considered as the interface contact with a Schottky barrier between ZnO and the STO buffer layer [7].

4.3.2 Strain-Induced Magnetic Anisotropy in Highly Epitaxial Heterostructure by LMBE

Magnetic anisotropy is the directional dependence of a material's magnetic properties. The magnetic moment of magnetically anisotropic materials will tend to align with an "easy axis," which is an energetically favorable direction of spontaneous magnetization. The two opposite directions along an easy axis are usually equivalent, and the actual direction of magnetization can be along either of them.

In contrast, a magnetically isotropic material has no preferential direction for its magnetic moment unless there is an applied magnetic field. Magnetic anisotropy is a prerequisite for hysteresis in ferromagnets. Without it, a ferromagnet is superparamagnetic. As recently verified by experiments, magnetic anisotropy can be effectively modulated by epitaxial strain induced by lattice mismatch.

Among ferromagnetic materials, cobalt ferrite, $CoFe_2O_4$ (CFO), has attracted much interest due to its broad magnetic properties and promising applications in microelectronic and microwave devices as well as in sensitive sensors. Cobalt ferrite ($CoFe_2O_4$) thin films were epitaxially grown on (001) $SrTiO_3$ (STO) and (001) MgO by laser MBE. Microstructural studies indicate that the $CoFe_2O_4$ grown on (001) STO with compressive strain are *c*-oriented by island growth mode with rough surface morphology, whereas the films on (001) MgO with tensile strain become *c*-oriented with layer-by-layer mode. Magnetic property studies reveal that the compressive strained $CoFe_2O_4$ films on (001) STO can significantly enhance out-of-plane magnetization (190 emu/cm^3) with a large coercivity (3.8 kOe). In contrast, the tensile-strained $CoFe_2O_4$ films on (001) MgO reveal weak magnetic anisotropy. These results suggest that strong magnetic anisotropy is highly dependent on the lattice misfit-induced strain.

RHEED patterns showing the epitaxial features of CFO films grown on STO and MgO substrates are shown in Figure 4.13. It is evident that the films on different substrates have completely different growth natures. In other words, the epitaxial behavior of a CFO film is strongly dependent on the lattice mismatch between the films and the substrates and their crystal structures. For the CFO film on MgO, the in-plane stress is tensile since the film has a smaller lattice parameter than the substrate. During the entire growth process, the RHEED patterns remain streaky (Figure 4.13a–c), suggesting that the growth is by 2D layer-by-layer mode. With

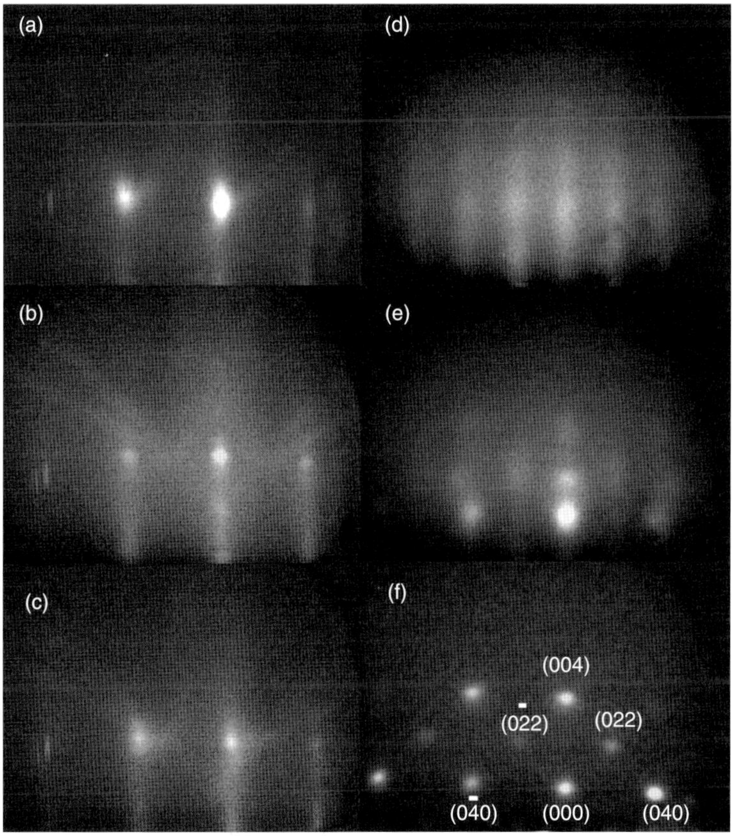

Figure 4.13 RHEED patterns for heteroepitaxial growth of $CoFe_2O_4$ thin films on MgO(100) and STO(100) with the laser repetition of 3Hz and 2J/cm^2 laser energy density at various time: 0 (a), 10s (b), 60s (c), 300s (d), 600s (e), and laser off (f). (Huang *et al.* [8]. Reproduced with permission of American Institute of Physics.)

film thickness increasing, the diffraction streaks change only slightly, because of a subtle lattice misfit-induced strain. However, the CFO film on STO undergoes a strong in-plane compressive stress with a large lattice misfit of 7.4%. As seen in Figure 4.13d–f, the growth on STO starts from an initial streaky pattern and gradually changes to a spotty pattern, suggesting that the CFO growth on STO is an island growth mode. Furthermore, a remarkable satellite pattern was found when the film thickness increases, which may result from the large lattice distortion.

Figure 4.14 shows the XRD $\theta-2\theta$ scans of CFO films on both MgO and STO substrates. Only the (00l) peaks of the CFO films appear in the diffraction patterns indicating that the CFO films are basically c-axis oriented. The CFO peaks on (001) MgO and STO are shifted by −0.12° and 0.36° from the bulk material, respectively. This result indicates that the lattice parameters of the CFO films are highly dependent on the substrates. The c-axis lattice parameter of the CFO film on MgO was therefore determined to be 0.833 and 0.8418 nm on STO substrates. These results further suggest that the CFO film on MgO is tensile strain, while the CFO film on STO is compressive strain. Rocking measurements of the CFO films on both STO and MgO substrates reveal that the full-width half maxima were found to be 0.91° and 0.43°, respectively, suggesting that CFO films have good single crystallinity.

In order to understand the strain effect on the CFO films grown on both compressive and tensile strains, AFM and MFM were employed to study the surface morphology and magnetic domain structures. Figure 4.15a is an AFM image of the CFO films on (001) MgO showing that the as-grown films have very smooth surface morphology with the root mean square roughness (R_{rms})

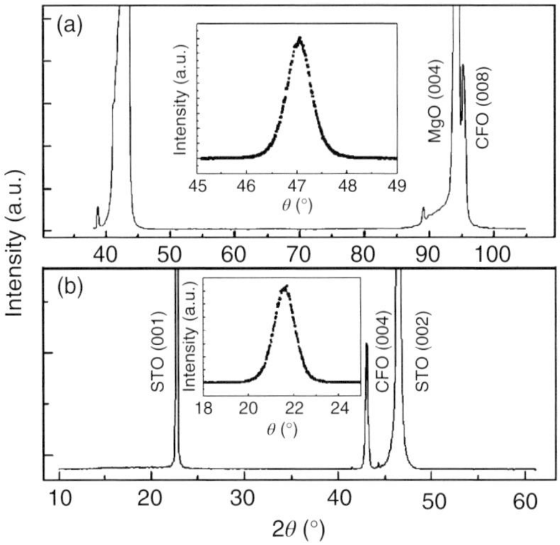

Figure 4.14 XRD pattern and rocking curve of heteroepitaxial growth $CoFe_2O_4$ films on MgO (100) in which separate peaks from (004) MgO substrate peak ($2\theta \approx 93.6°$) to $CoFe_2O_4$ film peak ($2\theta \approx 94.112°$) are clearly distinguishable (a) and the XRD pattern and rocking curve of heteroepitaxial growth $CoFe_2O_4$ films on STO (100) (b). (Huang *et al.* [8]. Reproduced with permission of American Institute of Physics.)

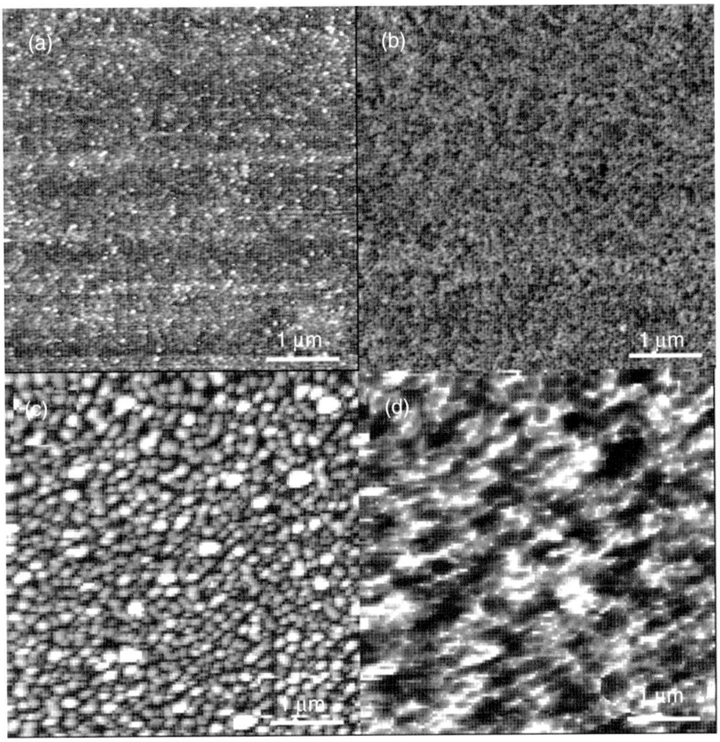

Figure 4.15 AFM and MFM images of the $CoFe_2O_4$ thin films on MgO (100) (a,b) and STO (100) (c,d), respectively. (Huang *et al.* [8]. Reproduced with permission of American Institute of Physics.)

value of only 0.94 nm and the maximum peak to valley roughness (R_{p2v}) value of 2.0 nm. The small root mean square roughness value of the as-grown CFO films can be considered to be the result of the layer-by-layer growth mode with subtle tensile strain. Figure 4.15b is the MFM image of the same film showing the out-of-plane domain image recorded in zero fields at room temperature. The weak contrast image implies a small magnetization perpendicular to the film. However, there are some remarkably different surface morphologies and magnetic domain structures for CFO films on STO. As seen in Figure 4.15c, the surface morphology of CFO on STO appears to consist of grains with various shapes, which is mainly attributed to the island growth mode and the in-plane compressive strain since both *a*- and *b*-axes are stressed by the large lattice mismatch between the film and the substrate. The values of R_{rms} and R_{p2v} of this film are 3.0 and 28.0 nm, respectively. The observed domain image (Figure 4.3d) with high contrast indicates that the domains consist of a cluster-like structure with various shapes as well, where the magnetization is confined up and down. Since the domain rotation can occur near the coercivity of the CFO, a larger lattice distortion in the perpendicular direction may result in a higher coercivity, which is consistent with the VSM measurement results.

In order to fully understand the strain effect on magnetic properties of the highly epitaxial CFO films, VSM studies were performed to achieve magnetic hysteresis loops in both perpendicular and parallel directions to the film planes. As seen in Figure 4.16, both films exhibit good hysteresis behaviors. For the CFO films on MgO, the remanent magnetization (Mr) for both perpendicular and parallel directions is found to have the same value of 75 emu/cm^3, but the coercive fields (Hcs) for the perpendicular and parallel directions are of 2.5 and 2.1 kOe, respectively. The similar magnetization values with little difference in coercive fields in perpendicular and parallel directions indicate that the CFO films on MgO have isotropic magnetic property. In contrast, the hysteresis loops of CFO on STO exhibit strong

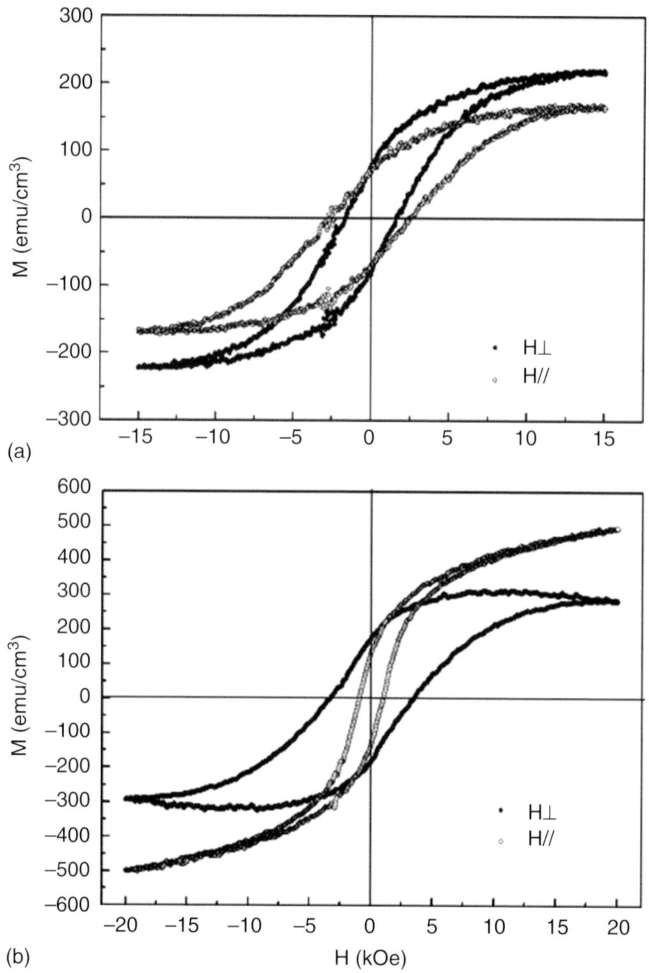

Figure 4.16 Parallel and perpendicular hysteresis loops of the $CoFe_2O_4$ thin films on MgO (100) (a) and STO (100) (b), respectively. (Huang *et al.* [8]. Reproduced with permission of American Institute of Physics.)

magnetic anisotropy. The Mr and Hc of the CFO film on STO in perpendicular and parallel directions were found to be 190 and 150 emu/cm^3 and 3.8 and 1 kOe, respectively. The large remanent magnetization difference between the perpendicular and parallel directions suggests that the CFO films with compressive strain are strongly anisotropic. It should be noted that the magnetic anisotropy in CFO material can only be seen from the residual strained structures, which were fabricated by mechanical milling. This is somewhat similar to the compressive strain-induced anisotropic behavior. The evolution and mechanisms of the generation of the large magnetic anisotropy are not well understood yet. However, strain-induced strong ferromagnetic and ferroelectric behaviors have been found in many ferroelectric thin films and related epitaxial materials such as (La, Ca)MnO_3, (Pb, Sr)TiO_3, and many others. Therefore, the high perpendicular coercivity of CFO film on STO can be considered to be the in-plane compressive strain induced by the lattice misfit between the film and the substrate. Furthermore, changes in the saturation magnetization values, $Ms_{\perp}$ and $Ms_{\parallel}$, in the films indicate that the lattice misfit-induced strains can strongly alter the magnetic properties of the as-grown CFO films [8].

4.4 Quantum Dots Development with Molecular Beam Epitaxy

QDs are structures where carriers are confined in all three spatial directions and where zero-dimensional systems of carriers are realized. In order to achieve this regime, self-assembled islands are embedded into layers of semiconductors with larger energy gap and appropriate band discontinuities.

When the sizes of the 3D islands are comparable to the de Broglie wavelengths of carriers lB and to their mean free paths, the carriers are quantum confined in levels inside the QDs. This is usually the case for semiconductors, since lB is of the order of a few tens of nanometers. The band profile of the InAs/GaAs QD structure along an arbitrary direction is schematically shown in Figure 4.17, where the confined ground levels for electrons and holes are represented. The difference between the confined ground levels for electrons (holes) and the bottom (top) of the conduction band (valence band) of the QD material is often referred to as the *confinement energy* for electrons *Ee* (holes E_h).

MBE has been widely used to grow QD structures. The spontaneous formation of MBE QD ensembles based on the mismatched heterostructure has attracted tremendous interest in both fundamental research and practical application. An important parameter affecting the optical performance is the homogeneity in both size and shape for the QD ensemble, which is always improved if the QDs are laterally ordered or periodically aligned along some specific crystallographic directions on the growing surface. In addition, long-range ordering of QDs is of critical importance in future micro- and optoelectronics devices that exploit the properties of a single or a well-defined number of nanostructures [9, 10]. Combining the pre-patterning with self-assembly, long-range lateral alignment of In(Ga)As

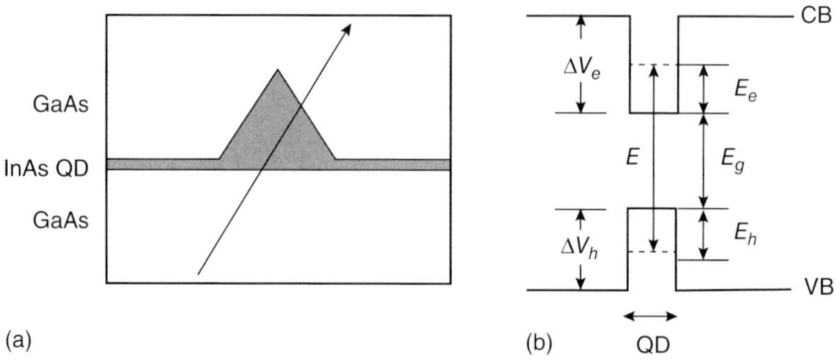

Figure 4.17 Schematic diagram of conduction band (CB) and valence band (VB) profiles for an InAs/GaAs QD structure (b) along an arbitrary direction shown by the arrow (a). The confined ground levels for electrons and holes are indicated. The fundamental optical transition has an energy E. The energy gap of the QD material is indicated by E_g, while E_e and E_h are the confinement energies for electrons and holes, respectively. DV_e and DV_h are the band discontinuities for CB and VB, respectively. (Franchi *et al.* [6]. Reproduced with permission of Elsevier.)

QDs on GaAs(001) was obtained [11, 12], whereas some contamination or crystal defects may be induced in the extrinsic techniques. Therefore, spontaneous lateral alignment of QDs on the growth plane or among the neighboring layers in a multilayer structure has been studied theoretically and experimentally [13, 14]. Figure 4.18 shows the AFM image of InAs dot arrays on InGaAs/GaAs [15].

Extensive work has demonstrated the short-range ordering of QDs along the elastically soft 100 directions in both the In(Ga)As/GaAs (001) and Ge(Si)/Si

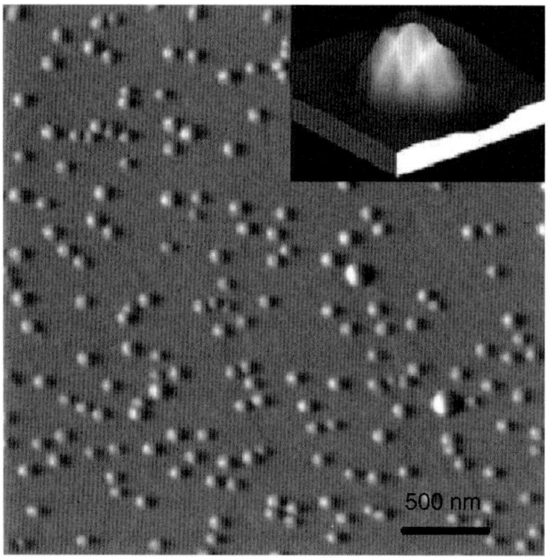

Figure 4.18 AFM image of InAs dot arrays on InGaAs/GaAs. (Inset) A single InAs quantum dot. (Reproduced with kind permission of Northwestern University.)

(001) systems, which was attributed to the elastic energy interaction between the neighboring islands via the strained wetting layer [16–18]. Alignment along the 110 directions on the (001) substrate was also observed in the InGaAs/GaAs system, which was explained in terms of the anisotropic diffusion of the adatoms [19]. In addition to the lateral alignment within the growth plane, vertical alignment of QDs in the multilayer structure along the growth direction was also observed in the InAs/GaAs(001) or Ge/Si systems [20, 21]. Such vertical alignment of QDs is due to the strain field provided by the underlying QDs in the multilayer structure [22]. In material systems with comparatively large elastic anisotropy, the strain field induced by the burying dots results in oblique alignment of QDs along the growth direction, and, furthermore, 3D QD crystals with fcc-like stacking were obtained in the $PbSe/Pb_{1-x}Eu_xTe$ multilayer structure [23, 24]. In the vertical or oblique alignment of QD multilayers, both size uniformity and spatial ordering of QDs could be invariably improved [6, 25–27]. Some other self-organizing phenomena, such as mismatch dislocation networks and atomic steps or terrace on the vicinal substrates, were also used to control the nucleation sites of QDs to get the ordered QD arrays. Experimental results have definitely proved that the MBE technique is a feasible technique to fabricate QD nanostructures.

4.5 Summary

In summary, we presented the basic features of MBE/LMBE growth technology and aspects of the main properties of nanostructures based on oxide and III–V semiconductors. 2D and 3D growth mechanisms were discussed, highlighting the underlying physical processes and their dependence on growth and design parameters, with a particular emphasis given to the 3D growth mechanism, which takes place in the deposition of highly lattice-mismatched structures and allows the preparation of self-assembled nanoislands. In order to discuss the effect of growth parameters on basic properties of these nanostructures, we presented the most relevant results obtained in the study of the nucleation and growth of 3D islands. Then, we showed that by embedding the nanoislands in suitable structures, the charge carriers undergo a complete quantum confinement in the so-called QDs. We focused our attention on aspects of the morphological and electronic properties of nanostructures and on the relevant research dedicated to the dependence of nanostructures on design and growth parameters. The understanding of this dependence is required to tune their electric properties. We also gave a brief presentation of other QD devices that, while rather behind the QD laser on the path toward industrial applications, are of great interest for the future of the nanoelectronic revolution in different fields such as computing and processing of information.

Acknowledgments

The authors acknowledge the Open Fund of IPOC (BUPT) Grant No. IPOC2013B006 and the grants from the Fundamental Research Funds for the Central Universities of China (Grant No. ZYGX2013J040, ZYGX2012J032).

References

1. Bimberg, D., Grundmann, M., and Ledentsov, N.N. (1999) *Quantum Dot Heterostructures*, John Wiley & Sons, Ltd., Chichester.
2. Grundmann, M. (ed.) (2002) *Nano-Optoelectronics Concepts, Physics and Devices*, Springer, Berlin.
3. Grandjean, N. and Massies, J. (1993) *J. Cryst. Growth*, **134**, 51.
4. Cho, A.Y. and Arthur, J.R. (1975) *Prog. Solid State Chem.*, **10**, 157.
5. Cho, A.Y. (1985) in *The Technology and Physics of Molecular Beam Epitaxy*, vol. **1** (ed. E.H.C. Parker), Plenum Press, New York.
6. Franchi, S. *et al.* (2003) *Prog. Cryst. Growth Charact. Mater.*, **47**, 166–195.
7. Huang, W., Hao, J.H., and Dai, J.Y. (2010) *Appl. Phys. Lett.*, **97**, 162905.
8. Huang, W., Zhu, J., Zeng, H.Z., Wei, X.H., Zhang, Y., and Li, Y.R. (2006) *Appl. Phys. Lett.*, **89**, 262506.
9. Schmidt, O.G. and Eberl, K. (2001) *IEEE Trans. Electron Devices*, **48**, 1175.
10. Schmidt, O.G., Denker, U., Dashiell, M., Jin-Phillipp, N.Y., Eberl, K., Schreiner, R., Grabeldinger, H., Schweizer, H., Christiansen, S., and Ernst, F. (2002) *Mater. Sci. Eng., B*, **89**, 101.
11. Kiravittaya, S., Rastelli, A., and Schmidt, O.G. (2005) *Appl. Phys. Lett.*, **87**, 243112.
12. Heidemeyer, H., Denker, U., Muller, C., and Schmidt, O.G. (2003) *Phys. Rev. Lett.*, **91**, 196103.
13. Stangl, J., Holy, V., and Bauer, G. (2004) *Rev. Mod. Phys.*, **76**, 725.
14. Shchukin, V.A., Ledentsov, N.N., Kopev, P.S., and Bimberg, D. (1995) *Phys. Rev. Lett.*, **75**, 2968.
15. Xie, Q.H., Madhukar, A., Chen, P., and Kobayashi, N.P. (1995) *Phys. Rev. Lett.*, **75**, 2542.
16. Ledentsov, N.N. *et al.* (1996) *Solid-State Electron.*, **40**, 785.
17. Shchukin, V.A. and Bimberg, D. (1998) *Appl. Phys. A*, **67**, 687.
18. Capellini, G., De Seta, M., Evangelisti, F., Zinovyev, V.A., Vastola, G., Montalenti, F., and Miglio, L. (2006) *Phys. Rev. Lett.*, **96**, 106102.
19. Heinrichsdorff, F., Krost, A., Grundmann, M., Bimberg, D., Kosogov, A., and Werner, P. (1996) *Appl. Phys. Lett.*, **68**, 3284.
20. Ledentsov, N.N. *et al* (1996) *Phys. Rev. B*, **54**, 8743.
21. Le Thanh, V., Yam, V., Boucaud, P., Fortuna, F., Ulysse, C., Bouchier, D., Vervoort, L., and Lourtioz, J.M. (1999) *Phys. Rev.*, **B60**, 5851.
22. Marchetti, R., Montalenti, F., Miglio, L., Capellini, G., De Seta, M., and Evangelisti, F. (2005) *Appl. Phys. Lett.*, **87**, 261919.
23. Quek, S.S. and Liu, G.R. (2003) *Nanotechnology*, **14**, 752.
24. Springholz, G., Holy, V., Pinczolits, M., and Baucer, G. (1998) *Science*, **282**, 734.
25. Tersoff, J., Terchert, C., and Lagally, M.G. (1996) *Phys. Rev. Lett.*, **76**, 1675.
26. Pinczolits, M., Springholz, G., and Bauer, G. (1999) *Phys. Rev. B*, **60**, 11524.
27. Shiryaev, S.Y., Jensen, F., Hansen, J.L., Petersen, J.W., and Larsen, A.N. (1997) *Phys. Rev. Lett.*, **78**, 503.

5
Carbon Nanomaterials and 2D Layered Materials Development with Chemical Vapor Deposition

Taisong Pan

5.1
Introduction

Low-dimensional materials, including one-dimensional (1D) nanotubes and two-dimensional (2D) nanomaterials, have been intensively studied in the last 20 years. Since the paper "Helical microtubules of graphitic carbon" by Sumio Iijima [1] was published in 1991, the development of low-dimensional materials has strongly promoted nanotechnology and its wide applications. Due to their several unique and excellent physical and chemical properties, low-dimensional materials have been used in many fields, such as high-performance sensors, transistors, and microelectromechanical systems [2–5]. With the continuously expanding applications, the large-scale synthesis of high-quality, low-dimensional materials becomes a very important topic. During the development of the low-dimensional materials, several different techniques are used to prepare them. The carbon nanotubes (CNTs) can be synthesized by arc discharge, laser ablation, and plasma torch. Mechanical and chemical exfoliations have been used to obtain the 2D crystals, including graphene and metal dichalcogenides. However, these methods have their drawbacks such as low yield, small crystal size, and difficulties in setting up the equipment. The ideal technique for the massive production of low-dimensional materials should be easy to set up, low cost, and quality controllable. Chemical vapor deposition (CVD) is a promising candidate that satisfies these demands and has been very popular for the growth of the low-dimensional materials.

CVD can be described as a solid-state material synthesis technique that uses the chemical reaction of vapor-phase precursors. By flowing the precursor gases into a chamber containing a heated substrate, the precursors react chemically with themselves or catalysts around the hot substrate and then the targeted solid-state materials are deposited on the surface of the substrate. In practice, CVD can be categorized into many formats. Based on the ways the chemical reactions can be initiated, CVD can be classified as thermal CVD (heat), plasma-enhanced CVD (plasma), photoinitiated CVD (ultraviolet light), and so on.

Advanced Nano Deposition Methods, First Edition. Edited by Yuan Lin and Xin Chen.

CVD is not a brand-new technique and has been studied for more than 100 years. As early as 1880, a patent describing a CVD process to improve carbon fiber filaments was issued [6]. Soon after that, in 1890, Mond process, a well-known CVD method for obtaining pure nickel from $Ni(CO)_4$, was used to refine the nickel ore [7]. Since the invention of transistor in the late 1940s, the semiconductor revolution started. CVD has been intensively researched and developed by the electronic industry in the next decades [8]. It has become a critical and indispensable technique during the fabrication of semiconductor devices. It can be used to refine and purify the silicon to supply the fundamental material of the microelectronic devices. Some key materials required during the fabrication of the devices, such as silicon dioxide and silicon nitride, are also relied on the corresponding CVD techniques.

In this chapter, an overview of the CVD techniques for synthesizing CNTs, graphene, and metal dichalcogenides will be presented. We will summarize some relevant growth mechanisms. The methods to control the structure, morphology, and doping of the materials will also be introduced.

5.2 Carbon Nanotube Synthesis by Chemical Vapor Deposition

5.2.1 Overview of CVD Process of Carbon Nanotube Growth

CNT can be treated as the folded structure of a carbon atomic layer. Depending on the wall number of CNT, it can be categorized into single-walled carbon nanotube (SWCNT) and multiwalled carbon nanotube (MWCNT), as shown in Figure 5.1a [9]. Two years after Iijima's discovery of the nanotubes grown by the arc discharge method, the CVD technique for the synthesis of CNT was reported for the first time [11]. Generally, the CVD process of CNT growth can be described as follows. First, the carbon source in gas phase flows into the CVD chamber (usually a quartz tube). Then, certain kind of energy sources, such as heat or plasma, is used to initiate the reaction. Since CVD often uses hydrocarbons such as methane and acetylene as carbon sources, the hydrocarbon will break and the pure carbon atoms can be obtained. Finally, the carbon atoms diffuse on the heated substrate coated with a catalyst layer, for example, silicon dioxide with copper particles on its surface, and CNTs will finally form on the substrate. The schematic of the thermal CVD CNT growth process is shown in Figure 5.1b.

Although controversies still exist on the growth mechanism of CNT, there is a well-accepted one [10, 12]. During the CVD process, the carbon atoms produced by the hydrocarbon are dissolved into the heated metal catalyst on the surface of substrate. Then, when the carbon solubility reaches its limit in the catalyst, the dissolved carbon content will crystallize and the CNTs gradually form. The dynamic balance of carbon content dissolved in the catalyst particle keeps the

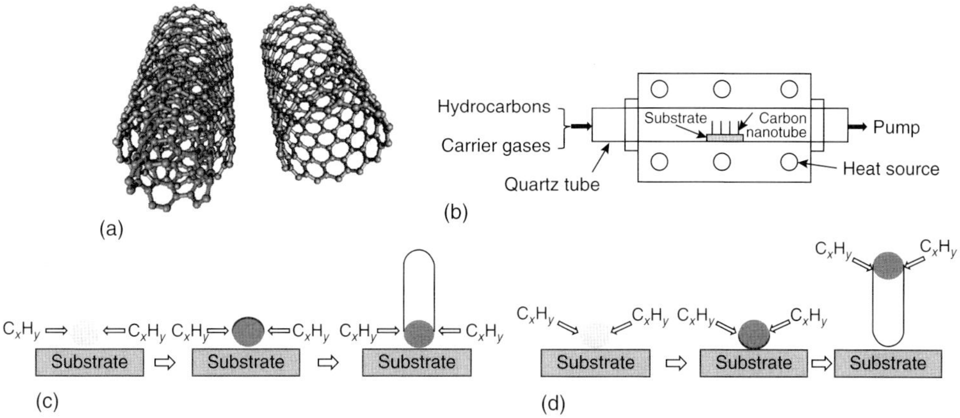

Figure 5.1 (a) Illustrations of the structure of MWCNT (left) and SWCNT (right). Schematic of (b) a thermal CVD equipment for CNT growth, (c) the base-growth mechanism, and (d) the tip growth mechanism. (Panels (a, b): Zhao and Stoddart [9]. Reproduced with permission of American Chemical Society. Panels (c, d): Chen *et al.* [10]. Reproduced with permission of Wiley.)

growth continuing for a long time during the effective supply of the carbon source precursor.

The interaction between the substrate and the catalyst is very important in determining the growth mode of CNT. After the dissolved carbon atoms diffuse in the catalyst, CNT precipitation occurs. When the wettability of the catalyst on the substrate is good, the CNT is unable to push the catalyst particle off the substrate. Thus, the carbon precipitates emerge out from the top of the catalyst particle and form the CNT. This is called *base-growth* mode (Figure 5.1c) [13]. If the contact angle of catalyst on the substrate is small enough, in other words, the wettability is poor, another growth mode called *tip growth* mode will happen (Figure 5.1d) [14]. The carbon precipitates will aggregate at the bottom of the catalyst particle, and the particle will be at the top of the as-grown CNT. Before the catalytic activity of the particle ceases due to the complete carbon coverage on particle, the CNT will keep growing with the catalyst "cap."

The substrate and the catalyst are the two important elements during the CVD growth of CNT. The main consideration on choosing the substrate is stability, especially chemical and thermal stability, during the high temperature of the CVD process. Commonly used substrates are some oxide substrates such as Al_2O_3 and SiO_2. Metals and silicon are not preferred because they are easy to transform into some alloys or silicides in the high-temperature environment and there may be diffusion of catalyst material in these bulk substrates. The properties of the catalyst have significant influences on the diameter, chirality, and growth rate of CNT. Thus, the preparation of the catalyst plays an important role in the growth of CNT. The catalyst, typically one kind of the transition metals (Fe, Co, Ni), is not a continuous thin film of several hundred nanometers, but clusters of the metal particles due to the growth mechanism of CNT introduced above [15]. Several methods have been used to prepare the catalyst. Physical vapor deposition techniques, such as evaporation and sputtering, are widely used to deposit the catalyst metal particles on substrates. Chemical-solution-based techniques, which use metal salts or organometallic precursors, are also applied to the preparation of the catalyst particles. For example, the catalyst particles can be obtained by supplying organometallic precursors, such as ferrocene, to the chamber during the growth process [16, 17]. In this so-called floating catalyst method, organometallic precursors are injected into the quartz tube and decomposed by heat. The metal particles generated by the thermal decomposition will diffuse to the surface of the substrate and become the catalyst clusters.

5.2.2 Control of Carbon Nanotube Structure

The controllable growth of CNT is a very important issue in the application of CNT and has been intensively studied. The control of the as-grown CNT structure can be achieved by adjusting the growth conditions (temperature, growth time, gas flow, etc.) and the catalyst structure (distribution, particle size, etc.).

As some applications of CNTs in complex electronic devices require the uniformity of the diameter of CNTs, it is a key issue in the CVD technique to control the diameter of CNTs. The size of the catalyst particles has been experimentally observed to have a strong influence on the diameter of CNTs. Dai *et al.* reported that there is a positive correlation between the CNT diameter and the particle size [18]. As shown in Figure 5.2a–f, Lieber *et al.* [19] synthesized CNTs using iron nanoclusters with average diameters of 3, 9, and 13 nm and obtained CNTs with diameters of $2.6 + 0.8$, $7.3 + 2.2$, and $11.7 + 3.2$ nm, respectively. Thus, the uniformity of the catalyst particle size is a necessary work to ensure the narrow distribution of the diameters of the as-grown CNTs. The temperature in CVD process is also very important in determining the diameter of CNTs. The influence of temperature on the diameter of CNTs is associated with the aggregation of the catalyst. When temperature is high enough, the nanoparticles will diffuse on the substrate surface. Due to Ostwald ripening, the particles with larger size will attract other particles and form particle clusters. The clusters that are too small in the CVD process are unable to grow CNTs because they lose their catalytic activity quickly due to the covered amorphous carbon. Furthermore, the absence of sufficient carbon supply will make the clusters of very large size fail to grow CNTs [21]. Only the catalyst clusters with proper size will be able to grow CNTs during the CVD process. Different temperatures lead to different size distributions of the particle clusters and different diameter ranges of the as-grown CNTs. The control of CNT diameter can also be achieved by choosing the suitable gas composition. For example, CNTs with larger diameters can be obtained by adding NH_3 during the CVD process [22].

As shown in Figure 5.2g,h, Hata *et al.* [20] have indicated that by using very small catalyst particles that have a diameter of 1–2 nm, SWCNT, which has more attractive electronic properties than the MWCNT grown with larger catalyst particles,

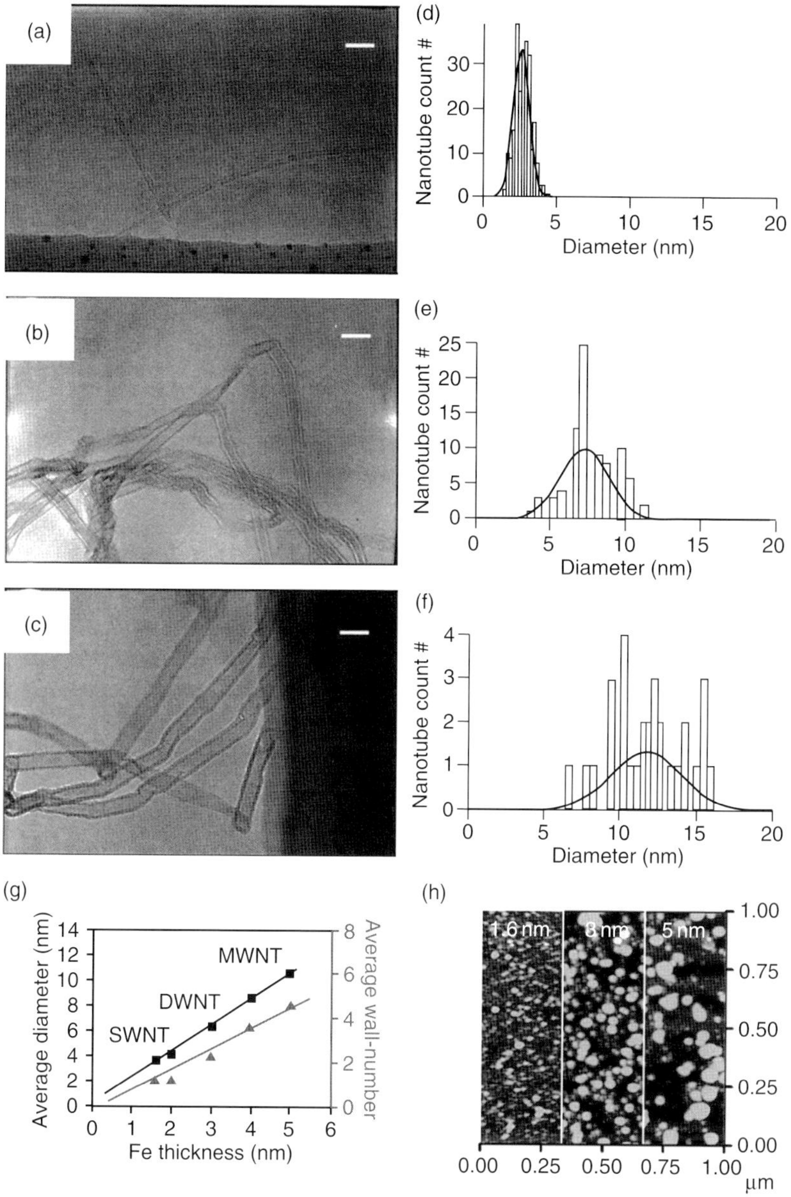

Figure 5.2 (a–c) Transmission electron microscopy (TEM) images of carbon nanotubes grown using iron nanoclusters with average diameters of 3, 9, and 13 nm (from (a) to (c)). Scale bars: 20 nm. (d–f) Histograms of nanotube diameters corresponding to the samples at the left. The solid lines correspond to Gaussian fits. (g) The relation between as-grown CNT average diameter, average wall number, and iron catalyst thin film thickness. (h) The surface atomic force microscopic (AFM) images of iron catalyst thin film with thickness of 1.6, 3, and 5 nm. (Panels (a–f): Cheung *et al.* [19]. Reproduced with permission of American Chemical Society. Panels (g, h): Zhao *et al.* [20]. Reproduced with permission of American Chemical Society.)

can be expected to be synthesized. The SWCNTs grown by CVD contain both metallic and semiconducting SWCNTs when no intentional control method is applied. However, when applying the SWCNTs to the electronic devices, metallic and semiconducting SWCNTs need to be used separately to ensure the successful operation of the device. Thus, how to obtain the SWCNTs with certain transport properties becomes a very interesting topic with respect to CVD technique. It has been reported that introducing methanol to the carbon gas source will result in the preferred growth of semiconducting CNTs [23]. The percentage of semiconducting SWCNTs can be more than 95% with the ethanol/methanol composite gas, whereas only 66% of semiconducting SWCNTs can be obtained with the ethanol-only gas. Although the decomposition temperature of methanol is too high (more than 900 °C) to supply carbon in the growth process, the OH radical from methanol can selectively etch the metallic SWCNT during the growth with carbon supplied by ethanol. Similarly, water vapor has been observed to be a good tool to etch the metallic SWCNT [24]. Other physical methods are also proposed to achieve the selective growth of SWCNT. For example, the electric field is used to prepare metallic SWCNT [25] and ultraviolet beam is used to prevent the growth of semiconducting SWCNT [26], as shown in Figure 5.3a,b.

5.2.3 The Alignment of Carbon Nanotube Array

The aligned CNT array is an important and versatile nanostructure of CNTs. In this nanostructure, the CNTs are aligned vertically or horizontally to the substrate during the CVD process and become a "forest" of CNTs. Several strategies, including electric field, substrate modification, and gas flow control, have been applied to synthesize the CNT array.

The morphology and lattice structure of the substrate is an important factor in the synthesis of both MWCNT and SWCNT arrays. In the early stage of aligned CNT array research, it is observed that vertically aligned MWCNT array can be grown with a growth template such as porous silicon [27, 28] or anodic aluminum

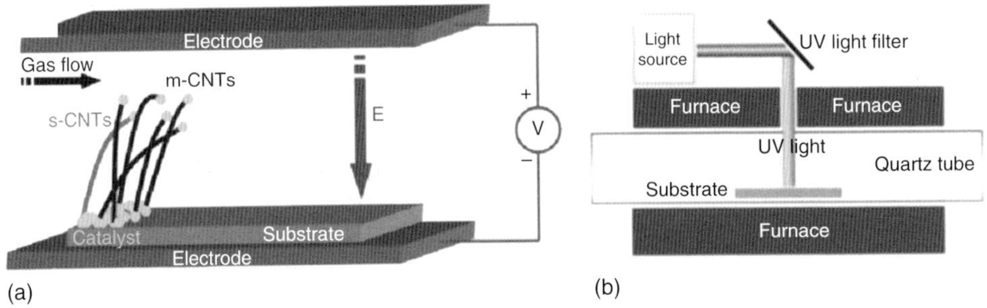

Figure 5.3 Schematic of (a) selective growth of metallic SWCNT with electric field and (b) selective growth of semiconducting SWCNT with ultraviolet beam. (Panel (a): Peng *et al.* [25]. Reproduced with permission of Elsevier. Panel (b): Hong *et al.* [26]. Reproduced with permission of American Chemical Society.)

oxide (AAO) substrates [29]. The porous morphology of the substrate will constrain the direction of the CNT growth and promote the growth of the aligned array, as shown in Figure 5.4a,b. Similar result can also be archived when the catalyst is deposited on silicon dioxide with carefully selected parameters. For the growth of SWCNT array, it has been showed that the single crystal substrate with anisotropic lattices or miscut steps is an appropriate candidate for guiding the alignment of CNTs [32, 33]. For example, the parallel aligned SWCNT array can be grown with *a*- and *r*-plane sapphire substrates, but not on *c*- and *m*-planes, as indicated by Figure 5.4c,d [30].

The electric field-directed alignment is to apply an electric field along the desired growth direction [34–36]. When an electric field is applied between two ends of CNT, the charges in CNT will relocate and the CNT will be polarized. Thus, the electric field force promotes the alignment of CNTs and directs the growth of CNT array.

At a critical temperature point, the catalyst and CNT will be with a floating state due to the weak interaction between the CNT/catalyst and the substrate at a high temperature. In this situation, the influence of gas flow direction on the alignment of CNTs will be significant. Moreover, since the catalyst is free from the substrate, the lifetime and the activity of the catalyst are enhanced. The catalyst with better performance offers the opportunity to grow the ultralong CNT array [37]. Few-walled aligned CNTs of 55 cm long have been prepared with the gas flow-directed method, which are shown in Figure 5.4e,f [31].

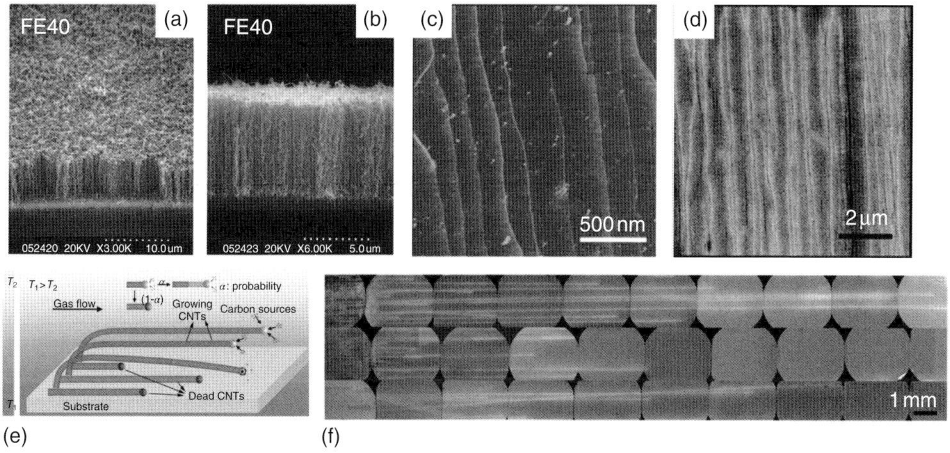

Figure 5.4 (a,b) Scanning electron microscope (SEM) images of well-aligned CNT array grown on porous silicon, (c) AFM image, (d) SEM image of aligned CNTs grown on the *a*-plane sapphire substrate, (e) schematic of tip growth mode of the aligned ultralong CNTs, and (f) mosaic SEM images of aligned ultralong CNTs. (Panels (a, b): Sohn *et al.* [27]. Reproduced with permission of American Institute of Physics. Panels (c, d): Han *et al.* [30]. Reproduced with permission of American Chemical Society. Panels (e, f): Zhang *et al.* [31]. Reproduced with permission of American Chemical Society.

5.3 Graphene Synthesis by Chemical Vapor Deposition

5.3.1 Overview of CVD Process of Graphene Synthesis

Since A. K. Geim and K. S. Novoselov exfoliated the graphene from the bulk graphite and reported on *Science* in 2004 [38], the graphene has already become one of the most popular nanomaterials. Graphene, which can be treated as one layer of bulk graphite, is a 2D nanomaterial with layered honeycomb structure. Due to its outstanding electronic, photonic, and optoelectronic performance [4, 39, 40], the preparation methods of graphene have been widely investigated. Several methods, including mechanical exfoliation, epitaxial on SiC, and chemical exfoliation, are proved to be effective in achieving the growth of graphene. However, since the CVD technique was reported to successfully synthesize graphene in 2008 [41], it has become the most commonly used method, especially when large-area graphene is required for the applications.

The mechanism of graphene growth with CVD is similar but not completely the same with the CVD growth of CNTs. The entire process can be separated into several steps. First, the hydrocarbon gas diluted by hydrogen is fed into the CVD chamber, and the hydrocarbon molecules are deposited on the catalyst substrate. Then, the carbon atoms obtained by the catalytic breaking of hydrocarbon molecule will diffuse on and into the catalyst substrate. When the temperature of CVD process decreases, the carbon content in the catalyst substrate exceeds the solubility limit and precipitates on the surface of catalyst substrate. The nucleation and reconstruction of precipitated carbon atoms lead to the formation of carbon atomic layers – graphene.

Since graphene is a continuous atomic layer, the catalysts used in the synthesis are metal foils or crystals without additional supporting substrate. Although many metals, including Pt [42], Ru [43], Ir [44], Co [45], and Pd [46], are experimentally proved to be capable for the growth of graphene, Ni and Cu are the most commonly used catalyst metal materials in today's CVD synthesis of graphene due to their low cost. Here, we will focus on these two major metal substrates.

Both single crystal and polycrystalline Ni have been used in growing graphene [47]. As Ni has a relatively high carbon solubility [48], the carbon atoms will diffuse into the substrate. When the substrate is cooled down, the solubility decreases and the carbon atoms will diffuse out to the surface of the Ni substrate. As shown in Figure 5.5a,b, depending on the crystal structure of Ni, the growth behavior will be a little different. In the case of single crystal Ni(111) substrate, since the lattice mismatch of Ni(111) to the densely packed hexagonal lattice of graphene is small, the precipitated carbon is expected to form continuous monolayer graphene on the smooth single crystal surface. But when the carbon atoms precipitate on the surface of polycrystalline Ni, the carbon will form multilayers of graphene in the grain boundary regions, which leads to poorer graphene quality than that of the graphene grown on single crystal Ni(111). Although high-quality graphene can be obtained by using Ni(111) single crystal substrates, the high price of single crystal Ni substrates limits the applications of Ni-based CVD graphene growth.

Cu offers the opportunity to synthesize large and high-quality graphene while keeping the cost low. Since Cu has much lower carbon solubility than Ni [49], most of the carbon atoms will diffuse on the surface of Cu substrate instead of dissolving in the substrate, as indicated in Figure 5.5c. After the carbon atoms diffused on the surface form the first layer of graphene, the metal is isolated from the hydrocarbon gas and the supply of carbon atoms is unable to continue due to the interrupted surface catalytic decomposition of hydrocarbon. This self-limiting behavior, which has been experimentally observed by Ruoff *et al.* by isotopic labeling [50], prevents the formation of graphene multilayers and is helpful for the synthesis of large-area monolayer graphene. The self-limiting surface catalysis mechanism does not require single crystal substrate to ensure surface smoothness and avoid the growth of multilayers in the grain boundary area. Thus, the polycrystalline Cu is a low-cost choice for the growth of the large-area monolayer graphene.

5.3.2 Control of Graphene Quality

The pretreatment of the substrate is a very important step in the CVD process of graphene growth. Generally, the substrate will be annealed in a reducing atmosphere, such as hydrogen, at a high temperature (about 1000 °C) for 30 min before the growth of graphene [51–53]. There are several purposes of the pretreatment. First, for both Ni and Cu, an oxide layer exists on the surface when they are contacted with air [54]. Obviously, the oxide layer on the substrate surface can significantly reduce the catalytic activity and make it difficult to grow uniform large-scale graphene on the substrate. The second is to increase the grain size of the substrate, reduce the impurities, and modify the substrate morphology for better graphene quality. As shown in Figure 5.6, McCarty *et al.* [55] observed the initial nucleation stage of graphene on Cu(111) by low-energy electron microscopy (LEEM). Most islands were found to nucleate at atomic steps or step bunches. It has been observed that with less grain boundaries, defects, and impurities of the substrate [55–57], the nucleation rate of graphene

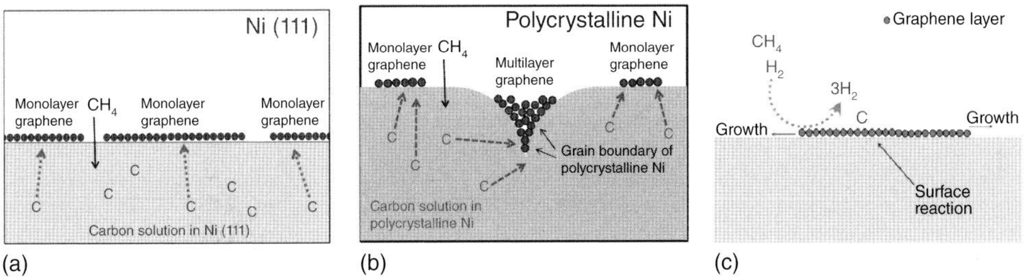

Figure 5.5 Schematic of graphene growth mechanisms on (a) single crystal Ni(111), (b) polycrystalline Ni, and (c) Cu. (Panels (a, b): Zhang *et al.* [47]. (Reproduced with permission of American Chemical Society. Panel (c): Zhang *et al.* [41]. Reproduced with permission of American Chemical Society.)

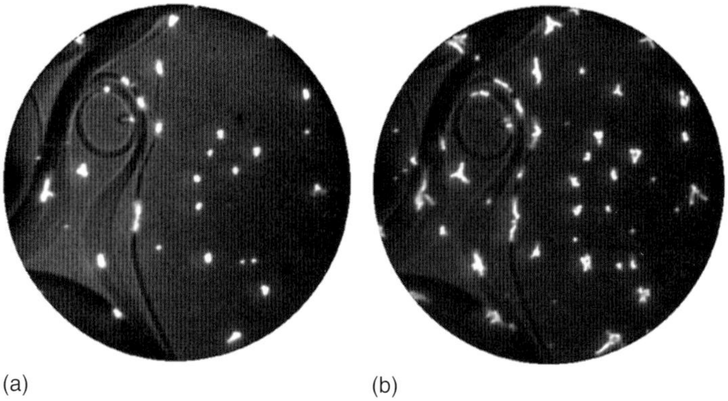

Figure 5.6 Low-energy electron microscopy (LEEM) images of graphene growing on Cu(111) at 815 °C. (a) After initial nucleation and (b) after secondary nucleation following an increase in the carbon flux. Graphene is bright, while dark stripes are Cu step bunches. The field of view is 20 μm. (Nie *et al.* [55]. Reproduced with permission of American Institute of Physics.)

considerably decreases and the quality of the as-grown graphene improves. To obtain the monolayer graphene with less grain boundaries, the nucleation density of graphene should be controlled in a certain range.

The temperature is also a key factor in the CVD process. Depending on the type of metal substrates used in the growth, the control strategy for temperature is a little different. For the Ni substrate, as it has relatively high carbon solubility, noticeable amount of carbon diffused in the bulk will precipitate during the cooling process with the reduced carbon solubility [58]. As indicated by Fick's laws of diffusion, a high cooling rate is essential to avoid the growth of the few-layered graphene. But for the Cu substrate, due to the much lower carbon solubility and the surface reaction mechanism, slow cooling rate in high-vacuum CVD will lead to the growth of bilayered graphene, whereas monolayer growth will be achieved in atmospheric pressure CVD (APCVD). An important temperature point for the Cu substrate is the melting point of Cu: 1084 °C. It has been experimentally observed that when the temperature is higher than 1084 °C, the liquid Cu phase can significantly suppress the nucleation process and large-scale single-domain graphene can be obtained [59].

The pressure of gases, including partial pressure of each gas content and the total pressure, was observed to have significant influence on the as-grown morphology of graphene. As shown in Figure 5.7, Zhang *et al.* [60] systematically studied the effects of total and partial pressures on the graphene morphology in the methane/hydrogen low-pressure CVD (LPCVD). When the ratio of methane and hydrogen is 1:12.5, the morphology of graphene grains changed from irregular small flakes to four-lobed grain and to six-lobed flowers with gradually increased pressure from 80 to 200 mTorr. However, when the pressure continued to increase to 300 mTorr, the six-lobed flower flakes turned to irregular flakes again. Their results also indicate that increasing the total pressure and high ratio of methane has similar effect on the morphology of as-grown graphene.

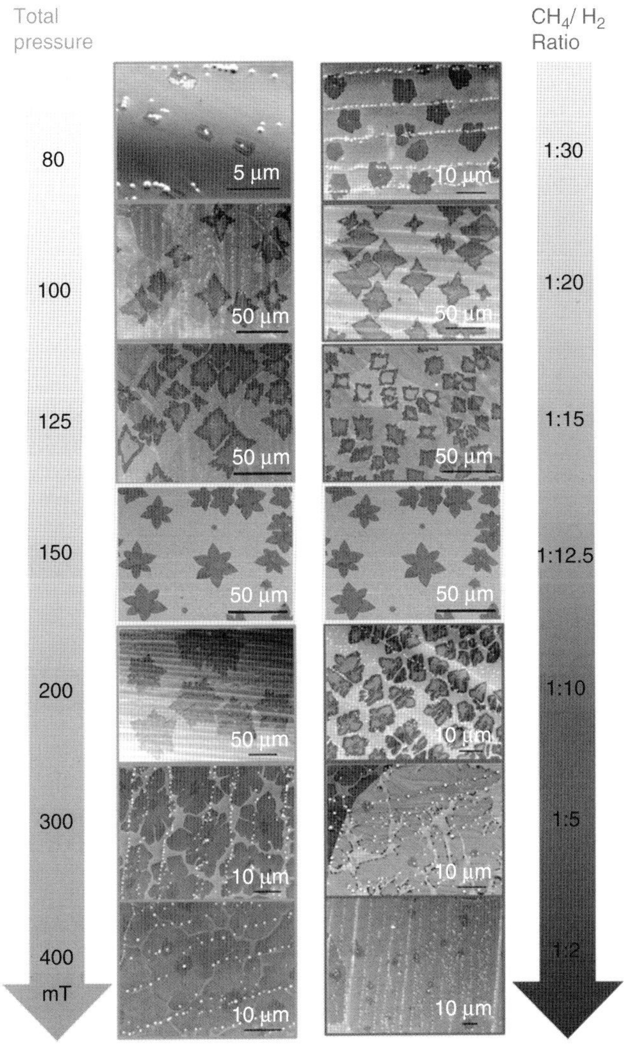

Figure 5.7 The graphene flakes grown by methane/hydrogen low-pressure CVD with different total pressure and methane ratio. (Zhang *et al.* [60]. Reproduced with permission of American Chemical Society.)

5.4 Metal Dichalcogenide Synthesis by Chemical Vapor Deposition

5.4.1 Overview of CVD Process of Metal Dichalcogenides

In recent years, another type of layered materials, transition metal dichalcogenides (TMDs), has received much attention due to their graphene-like 2D structures and diverse electronic and optoelectronic properties. The common formula of TMDs is MX_2, which usually consists of a transition metal M of groups 4–10 and a chalcogen X, as indicated in Figure 5.8a [61]. As shown in Figure 5.8b, the monolayer TMDs have two polymorphs: trigonal prismatic 1H (D_{3h}) and octahedral 1T (D_{3d}) phases. The different combination of the metal with the chalcogen makes the TMDs exhibit various properties. Semiconductors such as MoS_2 and WS_2, metals such as WTe_2 and $TiSe_2$, and insulators such as HfS_2 can all be found in the TMDs. Since the TMDs are also layered crystals, similar synthesis methods with graphene have been used, including mechanical exfoliation, chemical exfoliation, and CVD. With the successful utilization of the CVD technique in graphene synthesis, the CVD technique has been extended to the preparation of TMD layered crystal. As TMDs have numerous compositions and major work on CVD synthesis of TMDs is currently focused on MoS_2, the following discussion will be mainly about the preparation of MoS_2.

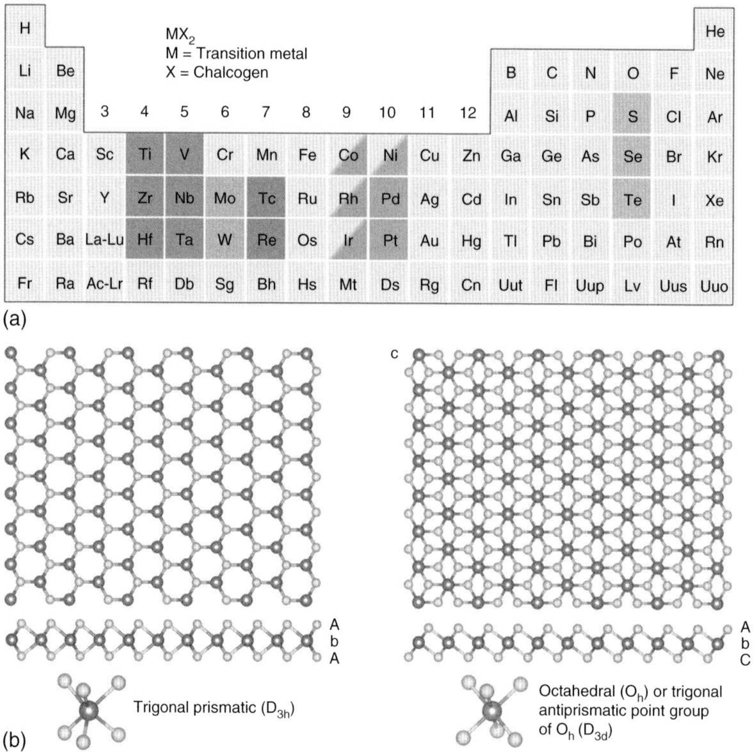

Figure 5.8 (a) The highlighted elements, including transition metals and three chalcogen elements, are able to form layered TMDs. Partial highlights for Co, Rh, Ir, and Ni indicate that not all the TMDs formed by these elements are layered structure. (b) *c*-axis and [10, 12–19, 21] sectional view of trigonal prismatic and octahedral phase monolayer TMD. Dark gray and light gray represent metal and chalcogen, respectively. AbA and AbC represent two different stacking sequences. (Chhowalla *et al.* [61]. Reproduced with permission of Nature Publishing Group.)

Unlike the graphene, TMD is a chemical compound involving at least two different elements. Thus, different routes have been applied for the synthesis of the MoS_2 using CVD. Generally, it can be categorized into two different routes [62]. The first is a one-step method, which uses the gas-phase reaction of Mo and S. As indicated by Figure 5.9a, Lin *et al.* [63] placed MoO_3 and sulfur powders in the CVD chamber with a substrate on the top of MoO_3. During heating, MoO_3 reacts with the vapor phase of sulfur and then forms MoS_2 on the substrate. The mechanism for this method is not completely clear, but two possible mechanisms have been proposed. One mechanism is that the suboxide MoO_{3-x} will diffuse on the substrate during heating and react with sulfur on the substrate. Another mechanism holds the view that MoO_{3-x} and sulfur directly react in the vapor phase and the formed MoS_2 clusters will nucleate and grow into flakes on the substrate. The second method is a two-step method wherein Mo-based precursor is deposited on the substrate and then annealed at a certain temperature to sulfurize or decompose the precursor into MoS_2. Several different types of precursors can be used in the two-step method. Li *et al.* [64] used ammonium thiomolybdate [$(NH_4)_2MoS_4$] as the precursor and deposited it on the substrate by dip coating. Then, as shown in Figure 5.9c, the precursor on the substrate was annealed at about 500 °C in reductive Ar/H_2 mixed gas to let the precursor convert into MoS_2. In the final stage, the as-grown MoS_2 was annealed again at about 1000 °C to improve the crystallinity. Lou *et al.* [65] also obtained few-layered MoS_2 by sulfurizing the Mo thin film deposited on SiO_2 substrate with e-beam evaporation, as indicated by Figure 5.9b. The diffusion of sulfur in Mo thin film led to the nucleation and growth

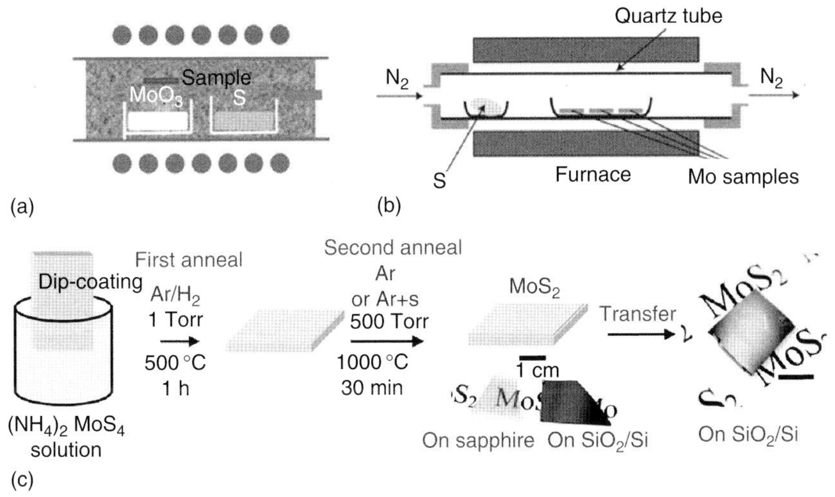

Figure 5.9 (a) Schematic of one-step MoS_2 growth method with gas reaction of MoO_3 and S. (b) Schematic of two-step MoS_2 growth method by sulfurizing the Mo thin film. (c) Schematic of two-step MoS_2 growth method by $(NH_4)_2MoS_4$ decomposition. (Panel (a): Lee *et al.* [63]. Reproduced with permission of American Chemical Society. Panel (b): Chhowalla *et al.* [61]. Reproduced with permission of Nature Publishing Group. Panel (c): Liu *et al.* [64]. Reproduced with permission of American Chemical Society.)

of MoS_2. However, it is still difficult to synthesize uniform large-scale MoS_2 with the two-step method due to some limitations such as insufficient diffusion and sulfurization on the substrate and nonuniformity of the deposited precursor. Both routes can be used to synthesize other TMDs, such as WS_2 [66, 67], $MoSe_2$ [68, 69], and WSe_2 [70], with corresponding precursors or oxides, but the one-step route is more promising in the synthesis of large uniform TMD crystals due to the limitations of the two-step route.

5.4.2
Growth Control of Metal Dichalcogenides in Chemical Vapor Deposition

The effect of substrates on the growth of MoS_2 was observed in many experiments of MoS_2 synthesis with CVD, especially in the one-step method described above. The SiO_2/Si substrates are widely used in the one-step method due to their compatibility with Si-based electronics. By pretreating the SiO_2/Si substrate with reduced grapheme oxide (rGO), perylene-3,4,9,10-tetracarboxylic dianhydride (PTCGA), or perylene-3,4,9,10-tetracarboxylic acid tetrapotassium (PTAS), a graphene-like surface will be created on the SiO_2/Si substrate [71]. With such a surface, nucleation and layered growth will be promoted. Instead of chemical modification of SiO_2/Si substrate surface, Lou *et al.* [72] successfully used the patterned SiO_2/Si substrate with uniform rectangular SiO_2 micropillar array to promote nucleation and layered growth. They found that scratches, dust particles, rough areas, and vicinity edges are preferred sites for the nucleation of triangular MoS_2 domains. Moreover, taking advantage of the strong interaction between MoS_2 and some single crystal substrates, the morphology of the as-grown MoS_2 crystal can be modulated. As shown in Figure 5.10a–c, the fractal growth of monolayer MoS_2 has been successfully achieved on $SrTiO_3$ (001) (STO) substrates using the one-step method [73]. The interaction between STO and MoS_2 affected the diffusion of the MoS_2 clusters and led to the diffusion-limited aggregation process.

In the two-step method, the parameters of annealing play key roles in determining the thickness and morphology of the as-grown MoS_2. Jiao *et al.* [74] synthesized the MoS_2 flakes with MoO_2 microplates using the two-step method. By controlling the sulfurization time to 0.5, 1, 3, and 6 h, MoS_2 flakes with thicknesses of one layer, one to two layers, bilayer, and three to four layers can be obtained, respectively (Figure 5.10d,e). When $(NH_4)_2MoS_4$, another commonly used precursor, is used in the two-step growth method, additional annealing stage in Ar/S gases after the decomposition of $(NH_4)_2MoS_4$ in reductive Ar/H_2 gases has been demonstrated to be able to improve the crystal quality due to the suppression of oxygen doping and an increase in the domain size [64]. As regards the growth with MoO_3 precursor, it has been shown that the pressure of CVD chamber, which reflects the concentration of sulfur, heavily influenced the morphology of MoS_2 [72]. As shown in Figure 5.10f–k, when the pressure was about 0.5–1 kPa, the formation of oxisulfides and hexagonal isolated MoS_2 islands was observed due to the lack of enough sulfur supply. The pressure ranging from 1 to 10 kPa led to the growth of triangles and increased domain areas with the sufficient supply of

Figure 5.10 (a–c) The dendritic MoS_2 flakes grown on STO under growth temperature, 730, 780, and 880 °C. (d,e) The MoS_2 flakes synthesized by two-step method with sulfurization time, 0.5 and 6 h. (f–k) The MoS_2 flakes grown by two-step route at 0.5–1, 1–4, 4–10, 10–40, 40–80, and 80–120 kPa, respectively. (Panels (a–c): Zhang *et al.* [73]. Reproduced with permission of American Chemical Society. Panels (d, e): Wang *et al.* [74]. Reproduced with permission of American Chemical Society. Panels (f–k): Najmaei *et al.* [72]. Reproduced with permission of Nature Publishing Group.)

both sulfur and MoO_3 precursor. Lower evaporation of MoO_3 in 10–40 kPa made the flakes become large but isolated. If the pressure kept increasing to more than 120 kPa, the size of triangle flakes would shrink and the star-shaped flakes were observed.

5.5 Summary

In this chapter, the principles and some growth control methods of CVD techniques used in synthesizing CNT, graphene, and TMDs are briefly introduced. CVD of low-dimensional nanomaterials is easy to set up but involves expertise from many fields, including thermodynamics, crystallography, and surface physics. With the fast-growing applications of these low-dimensional nanomaterials, CVD is more and more widely used not only by science research society but also by industries. Novel CVD techniques for the low-dimensional nanomaterials will continue to emerge and become the building block of future electronics.

References

1. Iijima, S. (1991) *Nature*, **354**, 56–58.
2. Terrones, M. (2003) *Annu. Rev. Mater. Res.*, **33**, 419–501.
3. Schwierz, F. (2010) *Nat. Nanotechnol.*, **5**, 487–496.
4. Geim, A.K. and Novoselov, K.S. (2007) *Nat. Mater.*, **6**, 183–191.
5. Wang, Q.H., Kalantar-Zadeh, K., Kis, A., Coleman, J.N., and Strano, M.S. (2012) *Nat. Nanotechnol.*, **7**, 699–712.
6. Jones, A.C. and Hitchman, M.L. (2009) *Chemical Vapour Deposition: Precursors, Processes and Applications*, Royal Society of Chemistry.
7. Kodas, T.T. and Hampden-Smith, M.J. (2008) *The Chemistry of Metal CVD*, John Wiley & Sons, Inc.
8. Park, J.H. and Sudarshan, T.S. (2000) *Chemical Vapor Deposition*, Surface Engineering Series, vol. **2**, ASM International.
9. Zhao, Y.-L. and Stoddart, J.F. (2009) *Acc. Chem. Res.*, **42**, 1161–1171.
10. Chen, Y., Zhang, Y., Hu, Y., Kang, L., Zhang, S., Xie, H., Liu, D., Zhao, Q., Li, Q., and Zhang, J. (2014) *Adv. Mater.*, **26**, 5898–5922.
11. Endo, M., Takeuchi, K., Igarashi, S., Kobori, K., Shiraishi, M., and Kroto, H.W. (1993) *J. Phys. Chem. Solids*, **54**, 1841–1848.
12. Kumar, M. and Ando, Y. (2010) *J. Nanosci. Nanotechnol.*, **10**, 3739–3758.
13. Baker, R.T.K. and Waite, R.J. (1975) *J. Catal.*, **37**, 101–105.
14. Baker, R.T.K., Barber, M.A., Harris, P.S., Feates, F.S., and Waite, R.J. (1972) *J. Catal.*, **26**, 51–62.
15. Dupuis, A.-C. (2005) *Prog. Mater Sci.*, **50**, 929–961.
16. Cheng, H.M., Li, F., Sun, X., Brown, S.D.M., Pimenta, M.A., Marucci, A., Dresselhaus, G., and Dresselhaus, M.S. (1998) *Chem. Phys. Lett.*, **289**, 602–610.
17. Liu, H., Li, S., Zhai, J., Li, H., Zheng, Q., Jiang, L., and Zhu, D. (2004) *Angew. Chem. Int. Ed.*, **43**, 1146–1149.
18. Dai, H., Rinzler, A.G., Nikolaev, P., Thess, A., Colbert, D.T., and Smalley, R.E. (1996) *Chem. Phys. Lett.*, **260**, 471–475.
19. Cheung, C.L., Kurtz, A., Park, H., and Lieber, C.M. (2002) *J. Phys. Chem. B*, **106**, 2429–2433.
20. Zhao, B., Futaba, D.N., Yasuda, S., Akoshima, M., Yamada, T., and Hata, K. (2009) *ACS Nano*, **3**, 108–114.
21. Lu, C. and Liu, J. (2006) *J. Phys. Chem. B*, **110**, 20254–20257.
22. Zhu, Z., Jiang, H., Susi, T., Nasibulin, A.G., and Kauppinen, E.I. (2011) *J. Am. Chem. Soc.*, **133**, 1224–1227.
23. Ding, L., Tselev, A., Wang, J., Yuan, D., Chu, H., McNicholas, T.P., Li, Y., and Liu, J. (2009) *Nano Lett.*, **9**, 800–805.
24. Zhou, W., Zhan, S., Ding, L., and Liu, J. (2012) *J. Am. Chem. Soc.*, **134**, 14019–14026.
25. Peng, B., Jiang, S., Zhang, Y., and Zhang, J. (2011) *Carbon*, **49**, 2555–2560.
26. Hong, G., Zhang, B., Peng, B., Zhang, J., Choi, W.M., Choi, J.-Y., Kim, J.M., and Liu, Z. (2009) *J. Am. Chem. Soc.*, **131**, 14642–14643.
27. Sohn, J.I., Lee, S., Song, Y.-H., Choi, S.-Y., Cho, K.-I., and Nam, K.-S. (2001) *Appl. Phys. Lett.*, **78**, 901–903.
28. Xu, D., Guo, G., Gui, L., Tang, Y., Shi, Z., Jin, Z., Gu, Z., Liu, W., Li, X., and Zhang, G. (1999) *Appl. Phys. Lett.*, **75**, 481–483.
29. Hu, W., Gong, D., Chen, Z., Yuan, L., Saito, K., Grimes, C.A., and Kichambare, P. (2001) *Appl. Phys. Lett.*, **79**, 3083–3085.
30. Han, S., Liu, X., and Zhou, C. (2005) *J. Am. Chem. Soc.*, **127**, 5294–5295.
31. Zhang, R., Zhang, Y., Zhang, Q., Xie, H., Qian, W., and Wei, F. (2013) *ACS Nano*, **7**, 6156–6161.
32. Ding, L., Yuan, D., and Liu, J. (2008) *J. Am. Chem. Soc.*, **130**, 5428–5429.
33. Ismach, A., Segev, L., Wachtel, E., and Joselevich, E. (2004) *Angew. Chem. Int. Ed.*, **43**, 6140–6143.
34. Zhang, Y., Chang, A., Cao, J., Wang, Q., Kim, W., Li, Y., Morris, N., Yenilmez, E., Kong, J., and Dai, H. (2001) *Appl. Phys. Lett.*, **79**, 3155–3157.
35. Dai, H. (2002) *Acc. Chem. Res.*, **35**, 1035–1044.
36. Ural, A., Li, Y., and Dai, H. (2002) *Appl. Phys. Lett.*, **81**, 3464–3466.

37. Huang, S., Cai, X., and Liu, J. (2003) *J. Am. Chem. Soc.*, **125**, 5636–5637.
38. Novoselov, K.S., Geim, A.K., Morozov, S.V., Jiang, D., Zhang, Y., Dubonos, S.V., Grigorieva, I.V., and Firsov, A.A. (2004) *Science*, **306**, 666–669.
39. Bonaccorso, F., Sun, Z., Hasan, T., and Ferrari, A.C. (2010) *Nat. Photonics*, **4**, 611–622.
40. Novoselov, K.S., Geim, A.K., Morozov, S.V., Jiang, D., Katsnelson, M.I., Grigorieva, I.V., Dubonos, S.V., and Firsov, A.A. (2005) *Nature*, **438**, 197–200.
41. Zhang, Y., Zhang, L., and Zhou, C. (2013) *Acc. Chem. Res.*, **46**, 2329–2339.
42. Sutter, P., Sadowski, J.T., and Sutter, E. (2009) *Phys. Rev. B*, **80**, 245411.
43. Sutter, P.W., Flege, J.-I., and Sutter, E.A. (2008) *Nat. Mater.*, 7, 406–411.
44. Coraux, J., N'Diaye, A.T., Busse, C., and Michely, T. (2008) *Nano Lett.*, **8**, 565–570.
45. Varykhalov, A. and Rader, O. (2009) *Phys. Rev. B*, **80**, 035437.
46. Kwon, S.-Y., Ciobanu, C.V., Petrova, V., Shenoy, V.B., Bareño, J., Gambin, V., Petrov, I., and Kodambaka, S. (2009) *Nano Lett.*, **9**, 3985–3990.
47. Zhang, Y., Gomez, L., Ishikawa, F.N., Madaria, A., Ryu, K., Wang, C., Badmaev, A., and Zhou, C. (2010) *J. Phys. Chem. Lett.*, **1**, 3101–3107.
48. Massalski, T.B., Okamoto, H., Subramanian, P.R., Kacprzak, L. (Eds) (2002) *ASM Handbook: Alloy Phase*, ASM International, Materials Park, OH.
49. López, G.A. and Mittemeijer, E.J. (2004) *Scr. Mater.*, **51**, 1–5.
50. Li, X., Cai, W., Colombo, L., and Ruoff, R.S. (2009) *Nano Lett.*, **9**, 4268–4272.
51. Lee, Y., Bae, S., Jang, H., Jang, S., Zhu, S.-E., Sim, S.H., Song, Y.I., Hong, B.H., and Ahn, J.-H. (2010) *Nano Lett.*, **10**, 490–493.
52. Bae, S., Kim, H., Lee, Y., Xu, X., Park, J.-S., Zheng, Y., Balakrishnan, J., Lei, T., Ri Kim, H., Song, Y.I., Kim, Y.-J., Kim, K.S., Ozyilmaz, B., Ahn, J.-H., Hong, B.H., and Iijima, S. (2010) *Nat. Nanotechnol.*, **5**, 574–578.
53. Li, X., Cai, W., An, J., Kim, S., Nah, J., Yang, D., Piner, R., Velamakanni, A., Jung, I., Tutuc, E., Banerjee, S.K., Colombo, L., and Ruoff, R.S. (2009) *Science*, **324**, 1312–1314.
54. Butt, M.Z. (1983) *J. Mater. Sci. Lett.*, **2**, 1–2.
55. Nie, S., Wofford, J.M., Bartelt, N.C., Dubon, O.D., and McCarty, K.F. (2011) *Phys. Rev. B*, **84**, 155425.
56. Hu, B., Ago, H., Orofeo, C.M., Ogawa, Y., and Tsuji, M. (2012) *New J. Chem.*, **36**, 73–77.
57. Wofford, J.M., Nie, S., McCarty, K.F., Bartelt, N.C., and Dubon, O.D. (2010) *Nano Lett.*, **10**, 4890–4896.
58. Reina, A., Jia, X., Ho, J., Nezich, D., Son, H., Bulovic, V., Dresselhaus, M.S., and Kong, J. (2009) *Nano Lett.*, **9**, 30–35.
59. Geng, D., Wu, B., Guo, Y., Huang, L., Xue, Y., Chen, J., Yu, G., Jiang, L., Hu, W., and Liu, Y. (2012) *Proc. Natl. Acad. Sci. U.S.A.*, **109**, 7992–7996.
60. Zhang, Y., Zhang, L., Kim, P., Ge, M., Li, Z., and Zhou, C. (2012) *Nano Lett.*, **12**, 2810–2816.
61. Chhowalla, M., Shin, H.S., Eda, G., Li, L.-J., Loh, K.P., and Zhang, H. (2013) *Nat. Chem.*, **5**, 263–275.
62. Ji, Q., Zhang, Y., Zhang, Y., and Liu, Z. (2015) *Chem. Soc. Rev.*, **44**, 2587–2602.
63. Lee, Y.-H., Yu, L., Wang, H., Fang, W., Ling, X., Shi, Y., Lin, C.-T., Huang, J.-K., Chang, M.-T., Chang, C.-S., Dresselhaus, M., Palacios, T., Li, L.-J., and Kong, J. (2013) *Nano Lett.*, **13**, 1852–1857.
64. Liu, K.-K., Zhang, W., Lee, Y.-H., Lin, Y.-C., Chang, M.-T., Su, C.-Y., Chang, C.-S., Li, H., Shi, Y., Zhang, H., Lai, C.-S., and Li, L.-J. (2012) *Nano Lett.*, **12**, 1538–1544.
65. Zhan, Y., Liu, Z., Najmaei, S., Ajayan, P.M., and Lou, J. (2012) *Small*, **8**, 966–971.
66. Gutiérrez, H.R., Perea-López, N., Elías, A.L., Berkdemir, A., Wang, B., Lv, R., López-Urías, F., Crespi, V.H., Terrones, H., and Terrones, M. (2013) *Nano Lett.*, **13**, 3447–3454.
67. Zhang, Y., Zhang, Y., Ji, Q., Ju, J., Yuan, H., Shi, J., Gao, T., Ma, D., Liu, M., Chen, Y., Song, X., Hwang, H.Y., Cui, Y., and Liu, Z. (2013) *ACS Nano*, **7**, 8963–8971.
68. Kong, D., Wang, H., Cha, J.J., Pasta, M., Koski, K.J., Yao, J., and Cui, Y. (2013) *Nano Lett.*, **13**, 1341–1347.

69. Lu, X., Utama, M.I.B., Lin, J., Gong, X., Zhang, J., Zhao, Y., Pantelides, S.T., Wang, J., Dong, Z., Liu, Z., Zhou, W., and Xiong, Q. (2014) *Nano Lett.*, **14**, 2419–2425.
70. Huang, J.-K., Pu, J., Hsu, C.-L., Chiu, M.-H., Juang, Z.-Y., Chang, Y.-H., Chang, W.-H., Iwasa, Y., Takenobu, T., and Li, L.-J. (2014) *ACS Nano*, **8**, 923–930.
71. Huang, X., Zeng, Z., and Zhang, H. (2013) *Chem. Soc. Rev.*, **42**, 1934–1946.
72. Najmaei, S., Liu, Z., Zhou, W., Zou, X., Shi, G., Lei, S., Yakobson, B.I., Idrobo, J.-C., Ajayan, P.M., and Lou, J. (2013) *Nat. Mater.*, **12**, 754–759.
73. Zhang, Y., Ji, Q., Han, G.-F., Ju, J., Shi, J., Ma, D., Sun, J., Zhang, Y., Li, M., Lang, X.-Y., Zhang, Y., and Liu, Z. (2014) *ACS Nano*, **8**, 8617–8624.
74. Wang, X., Feng, H., Wu, Y., and Jiao, L. (2013) *J. Am. Chem. Soc.*, **135**, 5304–5307.

6 Nanostructures Development with Atomic Layer Deposition

Hulin Zhang

6.1 Introduction

In recent years, plenty of novel techniques have been introduced to synthesize low-dimensional nanostructures. Atomic layer deposition (ALD) has emerged as a thin film deposition technique and has been widely applied. In a typical operation of ALD, a film is grown on a substrate material by exposing its surface to the alternating gaseous molecules. Compared with chemical vapor deposition (CVD), the precursors are never present simultaneously in the ALD reactor, but they are inserted as a series of sequential, nonoverlapping pulses. In each pulse, the precursor molecules react with the surface in a self-limiting routine. If all the reactive sites on the surface are consumed, the reaction will finish at once. Hence, the maximum amount of material deposited on the surface after a single exposure to all of the precursors (a so-called ALD cycle) is dependent on the precursor–surface interactive property [1]. Thin film materials with high uniformity and excellent properties can be obtained on arbitrary composites and large-scale substrates.

Through the sequential, self-limiting surface reactions, ALD is capable to meet the requirements of atomic layer control and conformal deposition. A schematic showing the continuous, self-limiting surface interactions during ALD is depicted in Figure 6.1 [2]. Mainly, ALD operations are based on the binary reaction sequences. Two surface reactions occur and deposit a binary compound film. The reactions can only deposit a finite number of surface species due to the finite surface sites. If each of the two surface reactions is self-limiting, the two surface reactions may proceed in a sequential manner to grow a thin film with atomic scale.

Although ALD is an active and a hot research field in materials science and engineering, it is motivated by the semiconductor engineering. Atomic-level control of nanoscale thin film deposition is essential for semiconductor miniaturization. Especially, semiconductor miniaturization has induced very high aspect nanostructures, which required to be coated conformally. Among all the nanofilm deposition methods, ALD might be the best to achieve the optimal conformality on

Advanced Nano Deposition Methods, First Edition. Edited by Yuan Lin and Xin Chen.

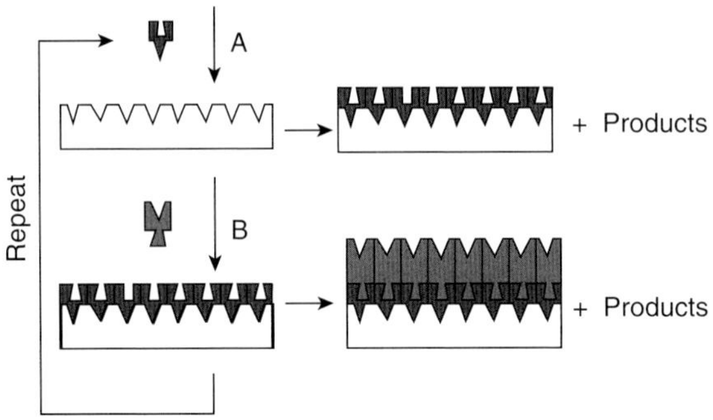

Figure 6.1 Schematic representation of ALD using self-limiting surface chemistry and an AB binary reaction sequence. (George *et al.* [2]. Reproduced with permission of American Chemical Society.)

high aspect nanostructures. The necessity for successive and pinhole-free films in semiconductor nanodevices has also promoted the development of ALD. ALD has been included for high dielectric constant gate oxides in the MOSFET structure and for copper diffusion barriers in back-end interconnects by the International Technology Roadmap for Semiconductors (ITRS) [3]. Moreover, ALD has encountered challenging requirements in other areas including the high-quality dielectric deposition for fabricating trench capacitors in dynamic random-access memory (DRAM) [4].

The term of ALD was first put forward under the name "molecular layering" (ML) in the early 1960s by Dr S.I. Koltsov from the Leningrad Technological Institute (LTI). These ALD experiments were performed under the scientific supervision of a member of the USSR Academy of Sciences, Prof. V.B. Aleskovskii, who first proposed the original concept of the ALD process in his doctoral thesis published in 1952 [5–7]. The foremost pioneer of ALD was Tuomo Suntola and coworkers, who demonstrated some of the first ALD processes as early as August/September 1974. They made the scientific idea a true thin film deposition technology and took that into an industrial use and worldwide awareness. The ALD reactors were also exploited by Suntola and coworkers, leading to the implementation of the ALD technology into an industrial level. At that time, this technique was named as atomic layer epitaxy (ALE) [8, 9]. The improving ALD technology was significantly useful to the manufacturing of thin film electroluminescent (TFEL) flat-panel displays, which served as the original motivation for developing the ALD technology owing to the requirement of the high-quality dielectric and luminescent films on large-area substrates. ALD has attracted increasing attention gradually in the mid-1990s and 2000s, especially focused on silicon-based microelectronics. It is considered as an excellent film deposition method with great potential for preparing very thin, conformal films under the control of the thickness and composition of the nanofilms at the atomic level [3, 10–13]. A major driving force for the recent interest is the prospective in the microelectronic nanodevice fabrication. The first of a series of yearly ALD meetings was held in 2001 and has continued through the present date. In 2004, the European SEMI award was given to Dr Tuomo Suntola for inventing the ALD technology and introducing it extensively [9].

With respect to ALE or ALD, several earlier reviews have been published [3, 12, 13]. Most of them have considered the application of ALD to microelectronics and nanotechnology [13–15]. The aim of this chapter is to give a brief and deep summary on the basis of typical ALD in nanostructure synthesis. The nanofilm as one important nanomaterial has been widely utilized in various nanodevices, microelectronics, and other fields. For continued improvement, the future for nanostructures will embrace as yet undefined paradigms. Certainly, ALD will almost be a new routine because of its ability to control deposition on the atomic scale and to deposit uniformly on very high aspect ratio structures.

6.2 Reaction Mechanisms

As an important and effective nanofilm deposition approach, ALD is quite similar to the common deposition techniques, such as CVD, molecular beam epitaxy (MBE), and so on. In the operation of the binary reactions, two reactants are present and react to form a single product or several products. The binary reaction is denoted simplistically as one reactant + another reactant → product [1, 2], as shown in Figure 6.1. When a solid surface is exposed to gaseous molecules, the gas species will adsorb on the surface because of the various molecular interactions that can attract the gas molecules to the surface. In ALD, enough time must be allowed in each reaction procedure so that a full adsorption density can be obtained [3]. The adsorption process can be expressed as a function of the rate of molecules arriving at the surface and the fraction of the molecules that undergo adsorption.

A pivotal difference between ALD and other deposition techniques lies in the method in which the reactants are exposed to the substrate. In ALD, the reactants are individually exposed to the surface, allowing for a consecutive layering process to cycle. Although several parallel reactions occur during the depositing, the only requirement for a reaction mechanism to be viable for ALD is that the reactants must be supplied separately for a binary reaction. Among the significant reaction mechanisms or techniques, thermal ALD, catalytic ALD, and metal ALD via elimination chemistry are the most useful in nanofilm or nanostructure preparation. Each reaction mechanism is feasible and advantageous for various sets of chemical precursors.

6.2.1 Thermal ALD

Thermal ALD can be used to deposit a wide variety of binary metal oxides, such as Al_2O_3. The deposition of Al_2O_3 is fairly common and will herein be discussed. In order to modify existing surface properties and conquer the activation energy inherent to chemical reactions, applying heat is an effective way to facilitating surface reactions which contribute to the formation of thin films stepwise. The most common temperature to deposit Al_2O_3 using ALD is 177 °C [10]. It is usually performed using trimethylaluminum (TMA) and H_2O. The first report of Al_2O_3 deposited by ALD using TMA and H_2O could date back to the late 1980s and early 1990s [16, 17]. More recent work in the semiconductor industry is using TMA and ozone for Al_2O_3 deposition by ALD [18, 19]. This review will focus on Al_2O_3 ALD using TMA and H_2O. TMA becomes exposed and chemisorbed on the substrate surface, and any remaining TMA not chemisorbed is purged out of the chamber by a high vacuum pump. Subsequently, CH_4 molecules as a by-product of the reaction and the $-CH_3$ groups from the TMA are reacted with and replaced by oxygen from introduced H_2O vapor. The last step in the production of a single Al_2O_3 monolayer is the desorption of H_2O by purging. In the previous research, it have been found the evidence of the loss of $AlOH^*$ species and the gain of $AlCH_3$ species during the reaction with TMA [10]. Contrarily, the loss of $AlCH_3$ species and the gain of $AlOH^*$ is verified in the reaction with H_2O. The aforementioned reaction schemes can be described in the chemical equations as following:

Primary reactions at surface:

$$AlOH^* + Al(CH_3)_3 \rightarrow AlOAl(CH_3)_2^* + CH_4 \tag{6.1}$$

$$AlCH_3^* + H_2O \rightarrow AlOH^* + CH_4 \tag{6.2}$$

Overall ALD reaction for Al_2O_3:

$$2Al(CH_3)_3 + 3H_2O \rightarrow Al_2O_3 + 3CH_4 \quad \Delta H = -376\ \text{kcal} \tag{6.3}$$

This is one of the highest reaction enthalpies encountered for any ALD reaction.

In situ FTIR research has been performed to study the surface chemistry of Al_2O_3 ALD [20–22]. It is clear that the loss of $AlOH^*$ species and concurrent gain of $AlCH_3^*$ species are seen during the TMA reaction in the FTIR difference spectra. Similarly, the loss of $AlCH_3^*$ species and the concurrent gain of $AlOH^*$ species is observed during the H_2O reaction. The gas-phase reaction products during Al_2O_3 ALD have also been investigated with the quadrupole mass spectrometry (QMS) [23, 24]. When $Al(CH_3)_3$ and D_2O served as the reactants, CH_3D was obtained as the main product, as expected from the surface chemistry for Al_2O_3 ALD. By repeating the surface reactions, Al_2O_3 growth is extremely linear with the reaction time [25]. Spectroscopic ellipsometry and quartz crystal microbalance (QCM) measurements have been used to characterize the growth process of Al_2O_3 ALD. The resulting Al_2O_3 ALD films are smooth and extremely conformal to the underlying substrate, which have indicated the excellent conformality of Al_2O_3 ALD films, as displayed in Figure 6.2 [26, 27].

Al_2O_3 films deposited using ALD have revealed the consecutive and pinhole-free properties by their preferable electrical performance. The curves of current–voltage for Al_2O_3 ALD films with different thicknesses on *n*-Si (100) reveal very similar electrical behavior compared to the thermal SiO_2 films [28].

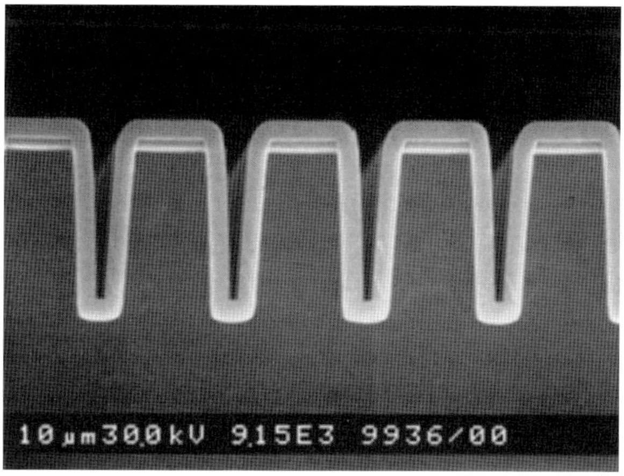

Figure 6.2 Cross-sectional SEM image of an Al_2O_3 ALD film with a thickness of 300 nm on a Si wafer with a trench structure. (Ritala *et al.* [26, 27]. Reproduced with permission of Wiley.)

The Al_2O_3 ALD films have a desirable dielectric constant and display very low electron leakage. Increases in the current density versus applied potential occur as a result of Fowler–Nordheim tunneling. This characteristic is in accordance with the absence of any defects or pinholes in the Al_2O_3 ALD film. Al_2O_3 ALD films can be utilized as gate oxides and for passivating semiconductor surfaces due to the excellent electrical properties.

Plenty of ALD processes in addition to Al_2O_3 ALD have been carried out at low temperatures. It is certified ALD can be used to deposit a wide variety of binary metal oxides such as Al_2O_3, TiO_2, ZnO, ZrO_2, HfO_2, and Ta_2O_5. These low-temperature ALD systems have been reported in a recent review and are important for ALD on organic and biological substrates [29].

6.2.2 Catalytic ALD

The catalysis is significantly important in delivering reliable SiO_2 films using the ALD technique. Although this reaction has a reasonable negative reaction enthalpy, surface reactions leading to the formation of SiO_2 are generally very slow and only occur at exceptionally high temperatures without catalysts.

The elementary reactions at surface:

$$SiOH^* + SiCl_4 \rightarrow SiOSiCl_3^* + HCl \tag{6.4}$$

$$SiCl^* + H_2O \rightarrow SiOH^* + HCl \tag{6.5}$$

Overall ALD reaction:

$$SiCl_4 + 2H_2O \rightarrow SiO_2 + 4HCl \tag{6.6}$$

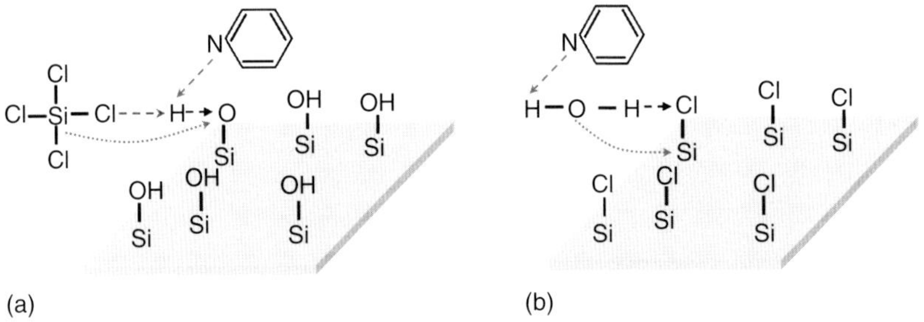

Figure 6.3 Proposed principles for Lewis base catalysis of SiO_2 ALD during (a) the $SiCl_4$ reaction and (b) the H_2O reaction.

It is fortunate that SiO_2 ALD can be catalyzed using other molecules. The catalysts employed for the SiO_2 ALD are commonly Lewis bases, such as pyridine or NH_3 (ammonia) [30, 31], which must be present during both the $SiCl_4$ and H_2O surface reactions. In addition, some studies have demonstrated that Lewis bases can also catalyze SiO_2 ALD using other silicon precursors such as $Si(OCH_2CH_3)$ [32].

The operation principle for the catalysis is considered to be the hydrogen bonding between the Lewis base and either the $SiOH^*$ surface species during the $SiCl_4$ reaction or the H_2O reactant during the H_2O reaction [29–31]. A schematic illustrating this mechanism is displayed in Figure 6.3. The hydrogen bonding between the Lewis base and the $SiOH^*$ surface species weakens the SiO–H chemical bond and makes the oxygen a stronger nucleophile. This hydrogen bonding interaction enhances the proportion of the nucleophilic attack by the oxygen on the Si atom in the $SiCl_4$ reactant [30]. Likewise, the hydrogen bonding between the Lewis base and the H_2O molecule makes the oxygen in H_2O an intense nucleophile on the Si atom in the $SiCl^*$ surface species.

The Lewis base catalyst is very effective to ALD. Without pyridine as the Lewis base catalyst, SiO_2 ALD using $SiCl_4$ and H_2O as the reagents requires reaction temperatures above 325 °C and reactant exposures at 10^3 Torr [33]. With pyridine added as the Lewis base catalyst, SiO_2 ALD can occur at temperatures nearly close to room temperature with reactant exposures of 10^{-2} Torr [31]. A variety of techniques such as QCM, ellipsometry, and profilometry have measured very linear SiO_2 ALD at 32 °C in a viscous flow ALD reactor with a growth per cycle of about 1.35 Å [31]. The resulting SiO_2 films have reasonable dielectric properties, although they are inferior to thermal SiO_2 oxide films.

In other ALD systems, Lewis bases also indicate similar catalysis, where the MOH^* surface hydroxyl is acidic media and can transfer a proton into liquid water. The Lewis base will hydrogen bond strongly to these acidic hydroxyl groups and make the oxygen more nucleophilic. Good candidates are metal oxides that are known to have their isoelectric points in water at pH below 7. These metal oxides include SnO_2, TiO_2, and ZrO_2 [34, 35].

6.2.3 Metal ALD

Metal ALD can also be carried out using thermal chemistry independently [14]. Metal ALD based on thermal chemistry has been demonstrated for various metal nanofilms. Typically, three main types of metal ALD using thermal chemistry have been successful. These ALD surface chemistries are based on fluorosilane elimination, combustion chemistry, and hydrogen reduction, respectively.

Metal ALD via elimination reaction routine mainly occurs when metals functionalized with halogens are reacted with silicon precursors. Common metals deposited using fluorosilane elimination reactions are W and Mo because the respective elimination reactions for these metals are highly exothermic [36, 37]. For W ALD, Si–H and W–F bonds exist on the material surface prior to the final purging process, and a linear deposition rate of W has been confirmed. The typical growth rate per cycle for tungsten ALD is about 5 Å with the typical reaction temperature ranging from 177 to 325 °C. Two surface reactions, as well as an overall ALD reaction for tungsten ALD, are illustrated here [37].

Elemental reactions at surface:

$$\mathrm{WSiF_2H^* + WF_6 \rightarrow WWF_5^* + SiF_3H} \tag{6.7}$$

$$\mathrm{WF_5^* + Si_2H_6 \rightarrow WSiF_2H^* + SiF_3H + 2H_2} \tag{6.8}$$

Overall ALD reaction:

$$\mathrm{WF_6 + Si_2H_6 \rightarrow W + SiF_3H + 2H_2} \quad \Delta H = -181\,\mathrm{kcal} \tag{6.9}$$

A multitude of other metals can be deposited by ALD via the reactions if their reaction sequences are based on fluorosilane elimination. This surface chemistry is supported by *in situ* FTIR studies of the absorbance from Si–H and W–F stretching vibrations on surface species during the WF_6 and Si_2H_6 exposures [37]. The absorbance from Si–H stretching vibrations decreases, while the absorbance from W–F stretching vibrations increases concurrently during WF_6 exposures. The absorbance from W–F stretching vibrations decreases, while the absorbance from Si–H stretching vibrations increases concurrently during Si_2H_6 exposures. The surface chemistry for W ALD leads to the linear deposition of W ALD films versus AB cycles, as shown in Figure 6.4 [36].

Some catalytic metals can be deposited using combustion chemistry. In this process, the organic ligands of the organometallic metal precursors react with oxygen to produce CO_2 and H_2O as combustion products. Ru and Pt were the first metal ALD systems that were deposited using combustion chemistry [38]. The Ru precursor was $Ru(C_5H_5)_2$, and the Pt precursor was $(CH_3C_5H_4)Pt(CH_3)_3$. The oxygen precursor was O_2. Ru ALD was accomplished at temperature ranging from 275 to 400 °C, and the growth per cycle was 0.5 Å around at temperature of about 375 °C. Pt ALD was previously reported at 300 °C, and the growth per cycle was also 0.5 Å [38].

The mechanism of Ru ALD and Pt ALD was explored using QMS studies [2]. The QMS investigations observed H_2O and CO_2 during both the organometallic

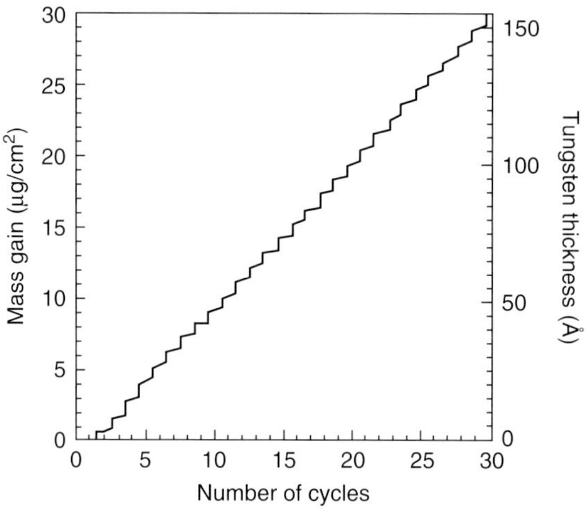

Figure 6.4 Mass gain monitored by quartz crystal microbalance for 30 cycles during W ALD with WF_6 and Si_2H_6 as the reactants. (Fabreguette *et al.* [36]. Reproduced with permission of Elsevier.)

precursor and O_2 exposures. Plenty of attention has been devoted to developing metal ALD based on various organic metallic precursors and H_2 reduction. The first efforts focused on Cu ALD using CuCl and H_2 as the reagents [39]. It was proved that Cu nanofilms could be grown on tantalum substrates. CuCl is problematic as a copper source. Other studies on Cu ALD use Cu-2,2,6,6-tetramethyl-3,5-heptanedionate and H_2 as the reactants. In addition, area-selective ALD was achieved on platinum-seeded substrates [40].

Cu ALD has also been achieved using another copper precursor, Cu (II)-hexafluoroacetylacetonate, and various organic reduction agents such as methanol, ethanol, and so on [41, 42]. Very good conformal Cu ALD nanofilms were prepared on trenched substrates. Moreover, the ALD of other metals has been demonstrated using organic reducing agents. Pd ALD has been demonstrated using $Pd(hfac)_2$ and formalin [43]. H_2 was also reported to be effective as the reducing agent for Pd ALD after the nucleation of the Pd ALD films using formalin as the reducing agent [43]. Alternative approaches to metal ALD have focused on depositing a metal oxide and then reducing this metal oxide with H_2 or other reducing agents. In this method, the fairly unreactive metal surface is circumvented following H_2 reduction. During metal oxide ALD, there is an oxide or MOH species at the surface to react with the organic metallic reactants. This approach has been verified for the growth of NiO nanofilms by using Ni acetylacetonate, O_3, and the subsequent reduction to Ni films using H_2 [44]. A similar technique was developed using bis(cyclopentadienyl) nickel and H_2O as the reagents for NiO growth and then hydrogen radicals for reduction to Ni metal [45]. Furthermore, there are persistent efforts to expand the range of metal precursors that can facilitate metal ALD using H_2 reduction chemistry. Many new families of metal precursors have been studied for metal ALD including the metal acetamidinates [46].

6.3
Nanostructures Based on ALD

ALD, as a unique nanofilm deposition and nanostructure synthesis method based on saturative surface reactions, has been widely used in nanoscale material preparation. Using ALD technology, atomic-level control of nanofilm composition and thickness, perfect step coverage, and large-area uniformity can be achieved [27, 47]. Owing to the self-limiting adsorption principle of ALD and the benefits it brings along, ALD is rather insensitive to the surface topology and suits very well for depositing conformal nanofilms. The potential of ALD for nanotechnology has been recognized already some time ago [15]. In catalyst applications, only one monolayer of catalyst molecule is necessary, and recently a real fluidized bed ALD reactor has been invented for deposition of thicker layers on nanoparticles [48]. The reactor has been used for coating of BN, ZrO_2, and SiO_2 nanoparticles with alumina film [26, 49, 50]. Coating of nanotubes, especially carbon nanotubes (CNTs), by ALD has attracted a lot of attention [51]. When multiwalled CNTs applied between the coaxial tube walls were filled, CNTs have served as templates for oxide nanotubes [52]. The deposition on CNT is not easy because of the inertness of the surface, but functionalization with NO_2 and TMA is helpful [53]. A number of semiconductor nanorods including ZnO, GaN, GaP, and InP have been coated with ALD alumina [54, 55] from which alumina nanotubes can be obtained if the semiconductor core is dissolved away [56]. The use of templates is the common method to make nanotubes by ALD as mentioned earlier in connection with CNTs [52] and semiconductor nanorods [56]. Shin and coworkers [57] have used another approach to nanotubes as they employed a polycarbonate membrane as a template. They coated the inner walls by ALD TiO_2 or ZrO_2 films and dissolved the membrane. Nevertheless, the use of templates in ALD is quite rare. The thin nanofilm usually deposited on nanosize objects by ALD is Al_2O_3, but other oxides and even Al_2O_3/W multilayers have also been made [51]. Even biomolecules have been used as templates in ALD for decorating biomaterials or making nanotubes [58]. Porous materials and the modification of pore size and composition of inner walls of pores have also attracted interest in ALD. It is considered that the membranes have been aimed for catalytic separation or detection purposes [59, 60]. The deposited materials have again been mostly oxides while nitrides have also been tried [61]. ALD allows the subtle tuning of both the pore size in angstrom scale and the chemical property. The synthesis of inverse opal structures is by infiltrating thin nanofilm into colloidal crystals using ALD and then dissolving the template particles to form an interesting potential functional nanofilm. TiO_2 and ZnS, as two typical thin films, have been investigated [62, 63]. ALD possesses several merits in the fabrication, including high degree of step coverage, conformality, dense films, and possibility to tune the chemical composition of the film. In the following, the making of nanolaminate structures, the use of fiber and pore templates in making nanorods and nanotubes, and the deposition of thin films inside nanopores are exemplified. It can be seen that ALD has been applied widely for the modification of nanomaterials and as such making the nanosize objects.

6.3.1 Nanolaminates and Nanofilms

Nanolaminates, nanofilms, and superlattice materials, as well as mixtures or solid solutions of variable or arbitrary composition, can be obtained by ALD. Nanolaminate is a special concept utilized in ALD since early 1980s, which is indexed to the relatively thick (from 50 to 200 nm) dielectric nanofilms suitable for applying in dielectric devices or electroluminescent displays [64, 65]. Recently, the nanolaminate has been extensively investigated for considerably thinner (<50 nm) high-k dielectrics in transistors and memory capacitors [65, 66]. The goal is to improve the electrical properties by fine-tuning the composition and layer thicknesses.

Nanolaminate of TiO_2 and Ho_2O_3 were grown on Si (001) substrates by ALD at 300 °C from alkoxide and β-diketonate-based metal precursors and ozone [67]. Unilaminar thicknesses were 2 nm for TiO_2 and 4.5 nm for Ho_2O_3, respectively. As-prepared films were uniform and amorphous as being detected by X-ray. After annealing at 800 °C or higher temperatures, the nanolaminate structure was destroyed by solid-state reaction to form $Ho_2Ti_2O_7$. The cross-sectional TEM, HRTEM image, and EDX map of the $Ho_2O_3-TiO_2$ nanolaminate in the as-deposited state are displayed in Figure 6.5 [67]. The films demonstrated diamagnetic or paramagnetic behavior in the as-deposited state. After annealing, the films possessed net magnetic moment, allowing one to record saturation magnetization and weak coercivity.

Dr Buchalska exploited a TiO_2 nanofilm preparation method via ALD routine, where the TiO_2 was deposited on the native oxide covered Si (100) [68]. The nanofilms were grown by applying various numbers of deposition cycles at different temperatures. FESEM images of ALD TiO_2 nanofilms deposited at 350 °C with an increasing number of deposition cycles are displayed in Figure 6.6 [68]. The resulting coatings possessed different thickness, crystallinity, morphology, and roughness. The photocatalytic activity of the films has been studied comparatively. Anatase–rutile composites (coatings deposited at 350 °C) indicate the highest activity in the process of TA oxidation (the highest efficiency of hydroxyl radical production).

6.3.2 Nanostructures as Templates

Due to the interesting electrical properties and even potential applications in microelectronics, oxide nanotubes have been paid plenty of attention. But, only a few techniques exist for making oxide nanotubes, and hence the related literature is limited. The usage of fiber templates combining with the ALD techniques offers a versatile and subtle method to obtain inorganic nanotubes. Many different types

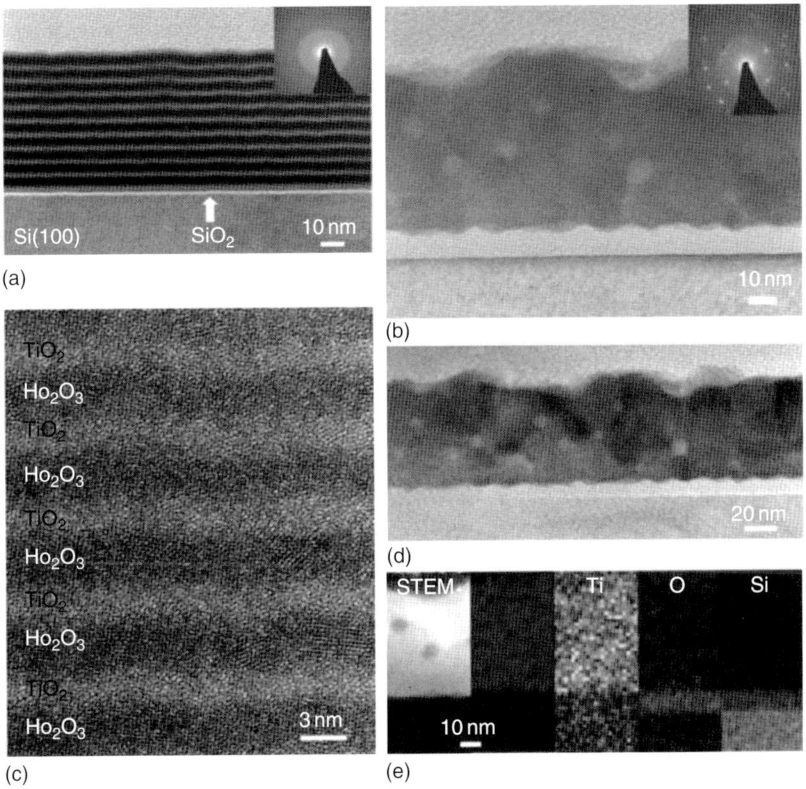

Figure 6.5 (a) Cross-sectional TEM, (b–d) HRTEM images of a Ho_2O_3–TiO_2 nanolaminate in the as-deposited state, and (e) compositional EDX map from the nanolaminate. (Kukli *et al.* [67]. Reproduced with permission of Elsevier.)

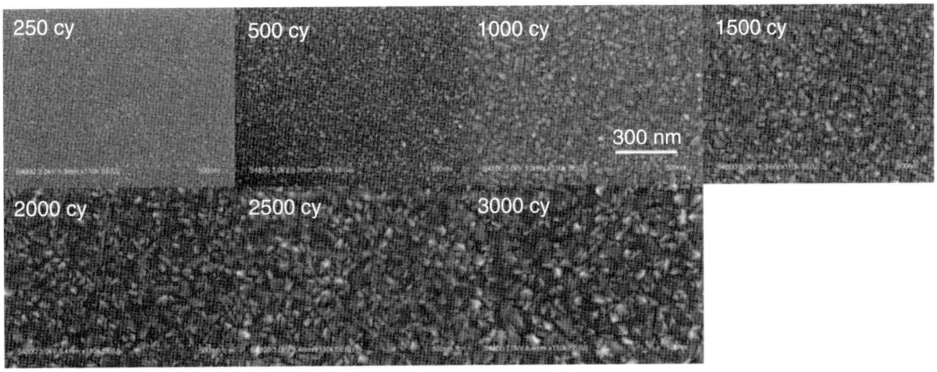

Figure 6.6 FESEM images of ALD TiO_2 films deposited at 350 °C with an increasing number of deposition cycles. The scale bar applies to all images. (Buchalska *et al.* [68]. Reproduced with permission of Elsevier.)

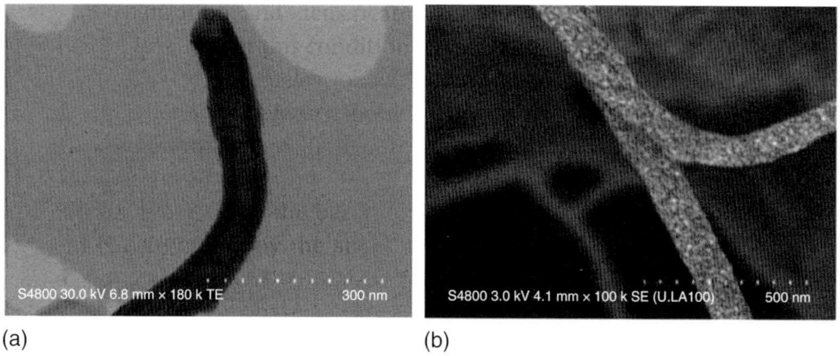

Figure 6.7 (a) A STEM image of a TiO_2 nanotube formed by replication of a cellulose nanofiber. (b) A BSE image of an Ir/TiO_2/cellulose composite formed by ALD of TiO_2 and Ir on filter paper at 250 °C. (Leskelä *et al.* [69]. Reproduced with permission of Elsevier.)

of nanofibers can be used as substrates or templates in making of nanocomposites or nanotubes. In ALD, till now, cellulose and some synthetic polymers have been utilized as templates for oxide materials [69–71]. If the fiber core is not removed, a flexible composite material is obtained, but if the material is annealed in air, oxide nanotubes are formed as depicted in Figure 6.7 [69] where cellulose fibers have been used as a template. Conformal deposition of TiO_2 by ALD was supplied on the cellulose fibers of filter paper at 150–250 °C to form TiO_2/cellulose composites. The TiO_2 deposited at 250 °C was crystalline with anatase phase, whereas the TiO_2 deposited at 150 °C was amorphous. The filter paper was removed by annealing at 450 °C in air to form a TiO_2 replica of the cellulose nanofibers. The process has been continued by depositing Ir by ALD on top of anatase TiO_2 using the method reported in literature [72]. The anatase TiO_2/cellulose composites have indicated to be photocatalytically active and the additive Ir improves the catalysis [70, 73]. The limitation in the ALD on polymers and especially on natural fibers is often the process temperature. Luckily, there are processes such as trimethylaluminum plus water and titanium chloride plus water, which are suitable at a low temperature or even room temperature. If higher process temperatures are needed, an oxide barrier layer can be deposited first to protect the underlying substrate material. The use of radical-enhanced ALD facilitates low-temperature depositions, but the high conformality requirements often faced by nanomaterials may be difficult to achieve [74]. Electrospinning provides a facile and versatile method for synthesizing thin polymer or ceramic fibers [75]. The direct electrospinning of polymers from a solution is straightforward. In the case of inorganic fibers, the solution contains polymer and metal precursor, and a composite fiber is first formed and the pure inorganic fiber material is obtained after annealing. ALD can be employed to coat the electrospun nanofibers. When polymer fibers are covered by an oxide layer and annealed, oxide nanotubes are obtained as exemplified in Figure 6.8 where alumina tubes have been made using polymer fibers as templates [76].

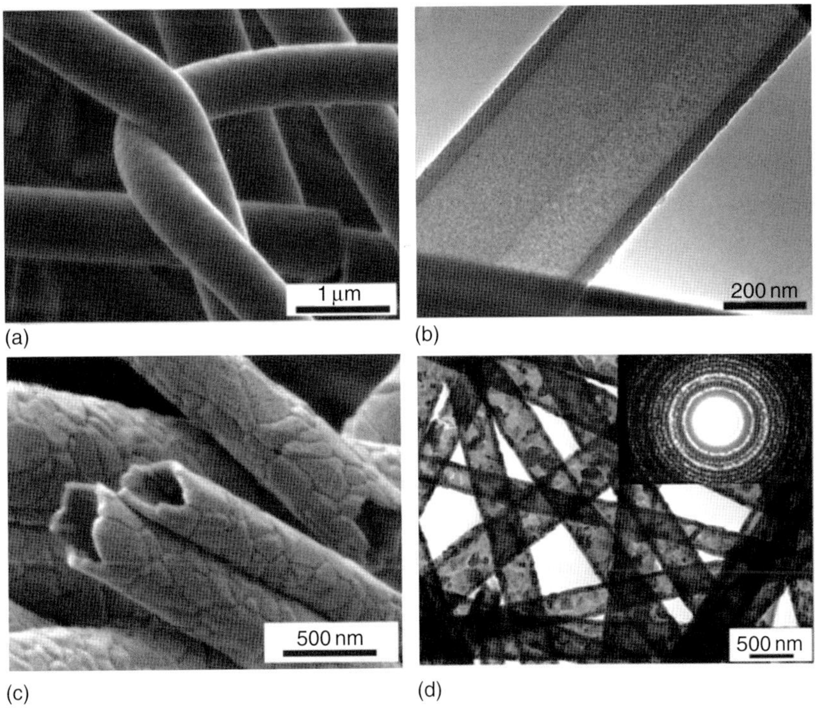

Figure 6.8 (a) SEM image and (b) TEM image of as-TiO_2 deposited electrospinning fibers. (c) A close-up SEM image and (d) TEM image of several titania tubes after calcination. (Kim *et al.* [76]. Reproduced with permission of American Chemical Society.)

6.3.3 Nanostructured Modification

Porous nanostructure, as one of the important nanomaterials, provides many useful and interesting properties for catalysis, filtering, and so on. As mentioned earlier, ALD has been utilized in the deposition of thin films or modifying of pore structures. However, the encountered bottleneck is that the process is very slow and noneffective [59–61, 77, 78]. The gaseous precursor molecules have to diffuse into the pores, and at the same time the reaction products have to diffuse out. The timescale in one cycle is in minutes rather than in seconds. The fill-in-hold-pump-down reactor concept proposed by Gordon *et al.* [79] is advantageous for porous substrates because it allows time for the precursors to penetrate into the pores, and the waste of chemical is minimized. Regretfully, the designed reactor does not overcome the slowness.

As illuminated in Figure 6.9 [69], an ALD nanofilm, as an example, was deposited in the pores of macroporous silicon with high aspect ratio. The conformal Ru nanofilm can be seen clearly all the way from the top to the bottom of the pores, even though some pores are bottleneck shaped. The obstacles are due to the fact that the macroporous Si was formed by random nucleation.

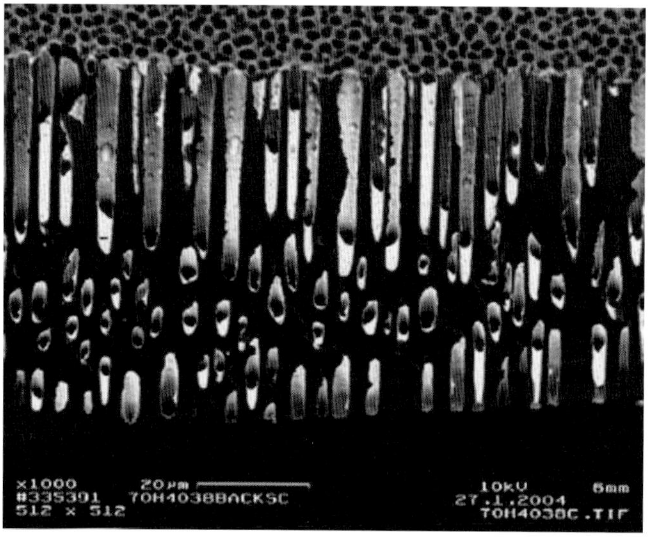

Figure 6.9 A BSE image of a conformal Ru thin film deposited on the pore walls of macroporous Si by ALD. (Leskelä *et al.* [69]. Reproduced with permission of Elsevier.)

The macroporous silicon has been tested as a high-density capacitor support using metal–oxide–semiconductor (MOS) structure made by ALD. A new reactor design version has been recently developed, which is very usable for such porous substrates where the pores extend through the whole substrate. This configuration allows growing films as short pulse and purge times as is used for planar substrates. Metallization of high aspect ratio and through substrate holes represents one interesting application with great application potential also in microelectronics [80]. Aerogels are nanoporous solids arising from highly cross-linked wet gels. Their surface area is typically more than $50\,m^2/g$. For applications as catalysts or sensors, the silica, alumina, or carbon aerogels have to be modified, and ALD has been explored as a potential modification method. It have been demonstrated that ZnO and some noble metal nanofilms can be successfully grown on aerogels via ALD routine [81].

6.4 Summary

The merits of ALD are precise thickness control at the angstrom or monolayer level. The self-limiting aspect of ALD results in an excellent step coverage and conformal deposition on high aspect ratio structures. Some surface areas will react before other surface areas because of different precursor gas fluxes. However, the precursors will adsorb and subsequently move out from the surface areas where the reaction has reached completion. The precursors will then proceed to react with other rest surface areas and produce a very even deposition. ALD processing is also extendible to very large substrates and to parallel processing of multiple substrates. The ALD species are gas-phase molecules, and they fill all space independent of substrate geometry and do not require line of sight to the substrate. ALD is only limited by the size of the reaction chamber. The ALD process is also dominated by surface reactions. Because the surface reactions are carried out sequentially, the two gas-phase reagents are not in contact in the gas phase. This separation of the two reactions limits possible gas-phase reactions that can form particles that could reduce the quality of nanofilms.

ALD is a useful process for the fabrication in microelectronics due to its capability to produce accurate thicknesses and uniform surfaces for high-quality film production using various different materials. In microelectronics, ALD is widely studied as a potential technique to deposit high-k (high permittivity) gate oxides, high-k memory capacitor dielectrics, ferroelectrics, metals, and nitrides. In high-k gate oxides, where the control of ultrathin nanofilms is essential, ALD is only likely to come into wider usage at the 45 nm technology. In metallizations, where conformal nanofilms are necessary, ALD may be used in mainstream production. In DRAMs, the conformality requirements are even higher and ALD is the only method that can be used when feature sizes become smaller than 100 nm. Several products that use ALD include magnetic recording heads, MOSFET gate stacks, DRAM capacitors, nonvolatile ferroelectric memories, and many others.

Understanding and being able to denote the surface properties on biomedical devices is critical in the biomedical industry, especially regarding devices that are implanted in the body. A material interacts with the environment at its surface, so the surface properties largely direct the interactions of the material with its environment. Surface chemistry and surface topography affect protein adsorption, cellular interactions, and the immune response. Some current applications of ALD in biomedical applications include creating flexible sensors; modifying nanoporous membranes, polymer ALD; and creating thin biocompatible coatings. ALD has been used to deposit TiO_2 films to create optical waveguide sensors as diagnostic tools. Furthermore, ALD is beneficial in creating flexible sensing devices that can be used in the clothing of athletes to detect movement or heart rate, for example. ALD is a possible manufacturing approach for flexible organic field-effect transistors (OFETs) because it is a low-temperature deposition method.

Nanoporous materials are emerging throughout the biomedical industry in drug delivery, implants, and tissue engineering. Unlike many other methods, the superiority of using ALD to modify the surfaces of nanoporous materials is that the saturation and self-limiting nature of the reactions means that even deeply embedded surfaces and interfaces are coated with a uniform nanofilm. Nanoporous surfaces can have their pore size reduced further in the ALD process because the conformal coating will completely coat the insides of the pores. This reduction in pore size may be advantageous in some certain applications.

Acknowledgments

This work is supported by the National Basic Research Program of China (973 Program) under Grant No. 2015CB351905, the National Natural Science

Foundation of China (No. 61306015), and Guangdong Innovative Research Team Program (No. 201001D0104713329).

References

1. George, S.M. (2010) *Chem. Rev.*, **110**, 111.
2. George, S.M., Ott, A.W., and Klaus, J.W. (1996) *J. Phys. Chem.*, **100**, 13121.
3. Puurunen, R. (2005) *J. Appl. Phys.*, **97**, 121301.
4. Sneh, O., Clark-Phelps, R.B., Londergan, A.R., Winkler, J., and Seidel, T.E. (2002) *Thin Solid Films*, **402**, 248.
5. Aleskovskii, V.B. and Khim, Z.P. (1974) *J. Appl. Chem. USSR*, **47**, 2207.
6. Malygin, A.A. (2006) *J. Ind. Eng. Chem.*, **12**, 1.
7. Suntola, T. and Antson, J. Growth of thin films of compounds on base materials — by alternate deposition of mono-atomic layers of the individual elements (1977) US Patent 4,058,430.
8. Suntola, T., Pakkala, A., and Lindfors, S. (1983) Growing compound thin films using multi-vapour source — and carrier gas which acts as diffusion barrier between separately supplied vapours. US Patent 4,389,973.
9. Parsons, G.N., Elam, J.W., George, S.M., Haukka, S., Jeon, H., Kessels, W.M.M., Leskelä, M., Poodt, P., Ritala, M., and Rossnagel, S.M. (2013) *J. Vac. Sci. Technol., A*, **31**, 050818.
10. Suntola, T. (1992) *Thin Solid Films*, **216**, 84.
11. Liu, R., Han, L., Huang, Z., Ferrer, I.M., Smets, A.H.M., Zeman, M., Brunschwig, B.S., and Lewis, N.S. (2015) *Thin Solid Films*, **586**, 28.
12. Leskela, M. and Ritala, M. (2002) *Thin Solid Films*, **409**, 138.
13. Ritala, M. and Leskela, M. (2001) Atomic layer deposition, in *Handbook of Thin Film Materials*, (ed. H.S. Nalwa), Academic Press, San Diego, CA.
14. Kim, H. (2003) *J. Vac. Sci. Technol., B*, **21**, 2231.
15. Ritala, M. and Leskela, M. (1999) *Nanotechnology*, **10**, 19.
16. Higashi, G.S. and Fleming, C.G. (1989) *Appl. Phys. Lett.*, **55**, 1963.
17. Soto, C. and Tysoe, W.T. (1991) *J. Vac. Sci. Technol., A*, **9**, 2686.
18. Goldstein, D.N., McCormick, J.A., and George, S.M. (2008) *J. Phys. Chem. C*, **112**, 19530.
19. Kim, J.B., Kwon, D.R., Chakrabarti, K., Lee, C., Oh, K.Y., and Lee, J.H. (2002) *J. Appl. Phys.*, **92**, 6739.
20. Dillon, A.C., Ott, A.W., Way, J.D., and George, S.M. (1995) *Surf. Sci.*, **322**, 230.
21. Ferguson, J.D., Weimer, A.W., and George, S.M. (2000) *Thin Solid Films*, **371**, 95.
22. Ferguson, J.D., Weimer, A.W., and George, S.M. (2004) *Chem. Mater.*, **16**, 5602.
23. Juppo, M., Rahtu, A., Ritala, M., and Leskela, M. (2000) *Langmuir*, **16**, 4034.
24. Rahtu, A., Alaranta, T., and Ritala, M. (2001) *Langmuir*, **17**, 6506.
25. Elam, J.W., Groner, M.D., and George, S.M. (2002) *Rev. Sci. Instrum.*, **73**, 2981.
26. Hakim, L.F., Blackson, J., George, S.M., and Weimer, A.W. (2005) *Chem. Vap. Deposition*, **11**, 420.
27. Ritala, M., Leskela, M., Dekker, J.P., Mutsaers, C., Soininen, P.J., and Skarp, J. (1999) *Chem. Vap. Deposition*, **5**, 7.
28. Groner, M.D., Elam, J.W., Fabreguette, F.H., and George, S.M. (2002) *Thin Solid Films*, **413**, 186.
29. Knez, M., Niesch, K., and Niinisto, L. (2007) *Adv. Mater.*, **19**, 3425.
30. Klaus, J.W., Sneh, O., and George, S.M. (1997) *Science*, **278**, 1934.
31. Klaus, J.W. and George, S.M. (2000) *Surf. Sci.*, **447**, 81.
32. Ferguson, J.D., Smith, E.R., Weimer, A.W., and George, S.M. (2004) *J. Electrochem. Soc.*, **151**, G528.
33. Klaus, J.W., Ott, A.W., Johnson, J.M., and George, S.M. (1997) *Appl. Phys. Lett.*, **70**, 1092.
34. Parks, G.A. (1965) *Chem. Rev.*, **65**, 177.
35. Jung, S.H. and Kang, S.W. (2001) *Jpn. J. Appl. Phys., Part 1*, **40**, 3147.

36. Fabreguette, F.H., Sechrist, Z.A., Elam, J.W., and George, S.M. (2005) *Thin Solid Films*, **488**, 103.
37. Klaus, J.W., Ferro, S.J., and George, S.M. (2000) *Thin Solid Films*, **360**, 145.
38. Aaltonen, T., Ritala, M., Sajavaara, T., Keinonen, J., and Leskela, M. (2003) *Chem. Mater.*, **15**, 1924.
39. Martensson, P. and Carlsson, J.O. (1997) *Chem. Vap. Deposition*, **3**, 45.
40. Martensson, P. and Carlsson, J.O. (1998) *J. Electrochem. Soc.*, **145**, 2926.
41. Huo, J.S., Solanki, R., and McAndrew, J. (2002) *J. Mater. Res.*, **17**, 2394.
42. Solanki, R. and Pathangey, B. (2000) *Electrochem. Solid-State Lett.*, **3**, 479.
43. Elam, J.W., Zinovev, A., Han, C.Y., Wang, H.H., Welp, U., Hryn, J.N., and Pellin, M.J. (2006) *Thin Solid Films*, **515**, 1664.
44. Utriainen, M., Kroger-Laukkanen, M., Johansson, L.S., and Niinisto, L. (2000) *Appl. Surf. Sci.*, **157**, 151.
45. Chae, J., Park, H.S., and Kang, S.W. (2002) *Electrochem. Solid-State Lett.*, **5**, C64.
46. Lim, B.S., Rahtu, A., and Gordon, R.G. (2003) *Nat. Mater.*, **2**, 749.
47. Ritala, M. and Leskelä, M. (2001) in *Handbook of Thin Films*, vol. **1** (ed. H.S. Nalwa), Academic Press, San Diego, CA, p. 103.
48. Haukka, S., Lakomaa, E.-L., and Suntola, T. (1998) *Stud. Surf. Sci. Catal.*, **120**, 715.
49. Wank, J.R., George, S.M., and Weimer, A.W. (2004) *Powder Technol.*, **142**, 59.
50. Hakim, L.F., George, S.M., and Weimer, A.W. (2005) *Nanotechnology*, **16**, S375.
51. Herrmann, C.F., Fabreguette, F.H., Finch, D.S., Geiss, R., and George, S.M. (2005) *Appl. Phys. Lett.*, **87**, 123110.
52. Lee, J.S., Min, B., Cho, K., Kim, S., Park, J., Lee, Y.T., Kim, N.S., Lee, M.S., Park, S.O., and Moon, J.T. (2003) *J. Cryst. Growth*, **254**, 443.
53. Farmer, D.B. and Gordon, R.G. (2004) *Nano Lett.*, **6**, 699.
54. Min, B., Lee, J.S., Cho, K., Hwang, J.W., Kim, H., Sung, M.Y., Kim, S., Park, J., Seo, H.W., Bae, S.Y., Lee, M.S., Park, S.O., and Moon, J.T. (2003) *J. Electron. Mater.*, **32**, 1344.
55. Min, B., Lee, J.S., Hwang, J.W., Keem, K.H., Kang, M.I., Cho, K., Sung, M.Y., Kim, S., Lee, M.S., Park, S.O., and Moon, J.T. (2003) *J. Cryst. Growth*, **252**, 565.
56. Hwang, J., Min, B., Lee, J.S., Keem, K., Sung, M.Y., Lee, M.S., and Kim, S. (2004) *Adv. Mater.*, **16**, 422.
57. Shin, H., Jeong, D.-K., Lee, J., Sung, A.M., and Kim, J. (2004) *Adv. Mater.*, **16**, 1197.
58. Knez, M., Kadri, A., Wege, C., Gösele, U., Jeske, H., and Nielsch, K. (2006) *Nano Lett.*, **6**, 1172.
59. Elam, J.M., Routkevitch, D., Mardilovich, P.P., and George, S.M. (2003) *Chem. Mater.*, **15**, 3507.
60. Pellin, M.J., Stair, P.C., Xiong, G., Elam, J.M., Birrell, J., Curtiss, L., George, S.M., Han, C.Y., Iton, L., Kung, H., Kung, M., and Wang, H.-H. (2005) *Catal. Lett.*, **102**, 127.
61. Chen, P., Mitsui, T., Farmer, D.B., Golovchenko, J., Gordon, R.G., and Branton, D. (2004) *Nano Lett.*, **4**, 1333.
62. Rugge, A., Becker, J.S., Gordon, R.G., and Tolbert, S.H. (2003) *Nano Lett.*, **3**, 1293.
63. King, J.S., Heineman, D., Graugnard, F., and Summers, C.J. (2005) *Appl. Surf. Sci.*, **244**, 511.
64. Kukli, K., Ihanus, J., Ritala, M., and Leskelä, M. (1996) *Appl. Phys. Lett.*, **68**, 3737.
65. Kukli, K., Ritala, M., and Leskelä, M. (1997) *J. Electrochem. Soc.*, **144**, 300.
66. Kukli, K., Ritala, M., Leskelä, M., Sajavaara, T., Keinonen, J., Gilmer, D., Bagchi, S., and Prabhu, L. (2002) *J. Non-Cryst. Solids*, **303**, 35.
67. Kukli, K., Lu, J., Link, J., Kemell, M., Puukilainen, E., Heikkila, M., Hoxha, R., Tamm, A., Hultman, L., Stern, R., Ritala, M., and Leskela, M. (2014) *Thin Solid Films*, **565**, 165.
68. Buchalska, M., Surowka, M., Hamalainen, J., Livonen, T., Leskela, M., and Macyk, W. (2015) *Catal. Today*, **252**, 14.
69. Leskelä, M., Kemell, M., Kukli, K., Pore, V., Santala, E., Ritala, M., and Lu, J. (2007) *Mater. Sci. Eng., C*, **27**, 1504.
70. Kemell, M., Pore, V., Ritala, M., Leskelä, M., and Linden, M. (2005) *J. Am. Chem. Soc.*, **127**, 14178.

71. Ras, R.H.A., Kemell, M., de Wit, J., Ritala, M., ten Brinke, G., Leskelä, M., and Ikkala, O. (2007) *Adv. Mater.*, **19**, 102.
72. Aaltonen, T., Ritala, M., Sammelselg, V., and Leskelä, M. (2004) *J. Electrochem. Soc.*, **151**, G489.
73. Kemell, M., Pore, V., Ritala, M., and Leskelä, M. (2006) *Chem. Vap. Deposition*, **12**, 419.
74. Niskanen, A., Arstila, K., Ritala, M., and Leskelä, M. (2005) *J. Electrochem. Soc.*, **152**, F90.
75. Li, D. and Xia, Y. (2004) *Adv. Mater.*, **16**, 1151.
76. Kim, G.M., Lee, S.M., Michler, G.H., Roggendorf, H., Gosele, U., and Knez, M. (2008) *Chem. Mater.*, **20**, 3085.
77. McCool, B.A. and DeSisto, W.J. (2004) *Chem. Vap. Deposition*, **10**, 190.
78. Cameron, M.A., Gartland, I.P., Smith, J.A., Diaz, S.F., and George, S.M. (2000) *Langmuir*, **16**, 7435.
79. Hausmann, D.M., Kim, E., Becker, J., and Gordon, R.G. (2002) *Chem. Mater.*, **14**, 4350.
80. Aaltonen, T., Ritala, M., Arstila, K., Keinonen, J., and Leskela, M. (2004) *Chem. Vap. Deposition*, **10**, 215.
81. Kucheyev, S.O., Biener, J., Wang, Y.M., Baumann, T.F., Wu, K.J., Van Buuren, T., Hamza, A.V., Satcher, J.H. Jr.,, Elam, J.W., and Pellin, M.J. (2005) *Appl. Phys. Lett.*, **86**, 083108.

7
Nanomaterial Development with Liquid-Phase Epitaxy

Weiqing Yang

7.1
Introduction

Liquid-phase epitaxy (LPE) is a method to grow semiconductor crystal layers isostructural with the substrate crystal in various solvents including water, polyols, ionic liquids, and mixed solvents. This happens at temperatures well below the melting point of the deposited semiconductor. The semiconductor is dissolved in the melt of another material. When the corresponding growing environments are close to the equilibrium between dissolution and deposition, the deposition of the semiconductor crystal on the substrate is relatively fast and uniform. In order to facilitate nucleation and to avoid tension in the grown layer, the thermal expansion coefficient of substrate and grown layer should be similar. Compared with other epitaxy methods, LPE has the outstanding features such as the lower temperatures, the better control of the amount of the crystalline phase grown or removed by dissolution, the lower densities of crystallization defects, and the more considerable reduction of the vapor pressure of volatile components of the compounds [1, 2].

Hydrothermal technique is one of the most important LPE techniques. This technique can facilitate the fabrication of most materials with desired physical and chemical properties. Compared with the other conventional processes, there are several advantages such as energy saving, simplicity, cost-effectiveness, good nucleation control, pollution free, high dispersion, high rate of reaction, good shape control, low temperature of operation in the presence of an appropriate solvent, and so on. In addition, the hydrothermal technique has a lot of other advantages. For example, it accelerates interactions between solid and fluid species. Phase pure and homogeneous materials can be achieved. Reaction kinetics can be enhanced. The hydrothermal fluids offer high diffusivity and low viscosity, facilitating mass transport and high dissolving power. Most importantly, the chemical environment can be suitably tailored. Especially, the microwave-assisted hydrothermal/solvothermal method has attracted more attentions due to the effective overcoming of the long reaction time of traditional

Advanced Nano Deposition Methods, First Edition. Edited by Yuan Lin and Xin Chen.

hydrothermal/solvothermal method. Recently, a great variety of nanomaterials including metals, semimetals, alloys, metal oxides, metal sulfides, metal selenides, metal tellurides, calcium-based salts and other oxysalts, inorganic/inorganic nanocomposites, and inorganic/organic nanocomposites have been synthesized by using this technique. Therefore, in this chapter, we mainly introduced the newly functional nanostructures by the LPE hydrothermal method.

7.2 Hydrothermal Method

7.2.1 Development of Hydrothermal Method

Water is the most commonly used solvent for the synthesis of inorganic nanostructures. About 70% of the earth's surface is composed of water, making water the most abundant liquid. In fact, water was the only solvent available to chemists to carry out chemical reactions for hundreds of years. In this part, we briefly introduce the evolution of hydrothermal technique. In the early 1845, E. T. Schafthual's group firstly applied hydrothermal method to prepare fine particles of quartz in a Papin's digester [3]. After that, more than 150 mineral species including various silicates, clays, hydroxides–oxides minerals, and diamond [4] were synthesized one after another until the early twentieth century. In 1908, K.J. Bayer's group firstly employed the hydrothermal technique to prepare the commercial aluminum by using raw bauxite mineral [5], which opened up a new window for hydrothermal research in the area of metallurgy. In the 1970s and 1980s, the multienergy hydrothermal research was mainly conducted in Japan and the United States.

There is a great difference between the hydrothermal research carried out during the previous century and the early twenty-first century. During the mid-twentieth century, hydrothermal technology mainly focused on the high temperature and pressure regime of materials processing due to the lack of knowledge on the solubility of several compounds and also on the selection of an appropriate solvent. The First International Hydrothermal Symposium (1982) held at the Tokyo Institute of Technology brought together specialists from interdisciplinary branches of science [6]. After that, the knowledge on the physical chemistry and pressure volume temperature (PVT) relationship in the hydrothermal systems were greatly improved, which helped in drastically reducing the temperature and pressure conditions of the processing. Similarly, the solvothermal and supercritical processing, which used a variety of other solvents such as organic and organometallic complexes in materials processing, was developed, thereby taking this technology toward green chemistry. The hydrothermal technology has attended toward green technology for sustained human development [7, 8]. Many of the hydrothermal researches based on the intelligent modeling of the hydrothermal reactions prior to the actual experiments greatly help predict the experimental conditions to obtain a desired phase with a controlled shape and size [9, 10].

7.2.2 Microwave-Assisted Hydrothermal Method

Recently, a relatively new and green chemical method, that is, the microwave-assisted hydrothermal synthesis (MAHT), is used to facilitate reactions and could be another avenue for green synthesis of nanomaterials [11]. Compared with the convectional hydrothermal methods, this method has lots of advantages such as the greener syntheses, the shorter reaction time, the lower energy consumption, and the higher product yield. This microwave-assisted technology can be combined with other green chemistry strategies to make it more appealing to green synthesis. For example, ionic liquids, solvent-free reactions, and nontoxic precursors can be easily utilized in conjunction with microwave heating.

In the middle of the 1980s, the pioneering work on the microwave-assisted synthesis was carried out. The research work of Komarneni and Roy [12] in 1985 was the beginning of microwave-assisted inorganic synthesis in liquid phase. The as-prepared titania gel spheres were shown in Figure 7.1. Gedye *et al.* [14] and Giguere *et al.* [15] first reported the use of microwaves for the organic synthesis in 1986. In their pioneering work [12], Komarneni and Roy reported the microwave-assisted synthesis of titania gel spheres by the dispersal of the polymerized titanium ethoxide solution into a two-phase liquid suspension in kerosene followed by heating in a microwave oven. Since these reports on the microwave-assisted synthesis, the microwave heating technology has received increasing attention as a promising heating method [15]. The microwave heating technology is emerging as an alternative popular heat source for rapid chemical reactions, and materials can be synthesized in minutes, instead of hours or even days usually required by the conventional heating methods. Over the past two decades, the microwave chemistry has rapidly developed from a laboratory curiosity to a well-established synthetic technique used in many academic and industrial laboratories around the world. This technology will open new synthetic pathways and allow the use of more environmentally friendly solvents, and it will also yield cleaner products that will not require much purification [16]. The microwave-assisted preparation of inorganic nanostructured materials in liquid

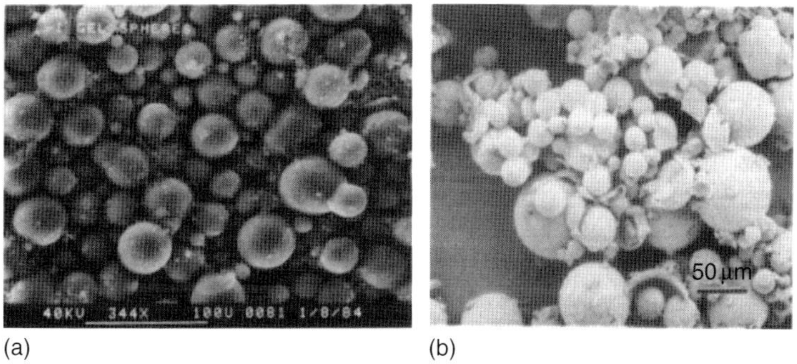

(a) (b)

Figure 7.1 Scanning electron micrograph of (a) titania gel spheres and (b) titania gel sphere and flakes from broken spheres. (Riman *et al.* [13]. Reproduced with permission of Elsevier.)

phase is currently a fast-growing area of research. It is expected that the related research activities will continue to grow rapidly in the coming years. Although most research on microwave-assisted synthesis is still performed on a laboratory (milliliter) scale, it is expected that this novel technology will soon be used on a larger, perhaps even production, scale [17].

Nowadays, many laboratories of materials science as well as organic and pharmaceutical chemical laboratories have been equipped with microwave reactors. Most early publications on microwave-assisted synthesis were carried out in household microwave ovens. Experimental parameters such as microwave power, reaction temperature, and pressure inside the vessel were not precisely known in household microwave ovens. These uncertainties led to poor control over the synthesis and a lack of reproducibility of experiments. Therefore, the use of commercially available specially designed microwave reactors with built-in magnetic stirrers and direct monitoring of temperature and pressure by various probes and sensors is strongly recommended. It is possible to combine microwave chemistry with the well-established liquid-phase synthesis techniques, and the possible combinations open unique and exciting opportunities for the preparation of inorganic nanostructures in liquid phase [11].

In recent years, some reports have been published on the microwave-assisted synthesis of nanostructured materials [17–24]. In this part, we mainly focus on microwave-assisted preparation of inorganic nanostructured materials in liquid phase, which are of great importance and significance for the rapid development of the related research field.

7.2.2.1 Microwave-Assisted Preparation of Nanostructures in Aqueous Solution

Water is cheap, widely available, and environment friendly, and water is also a polar solvent and a good microwave absorber [25, 26]. A simplified schematic illustration of the microwave heating mechanism of water molecules is shown in Figure 7.2. During the microwave heating, polarized water molecules try to orientate with the rapidly changing alternating electric field; thus heat is generated by the rotation, friction, and collision of water molecules. The loss tangents of the solvents, which are related to the ability of the solvent to absorb microwave energy, depend on the relaxation times of the molecules. These relaxation times depend critically on the nature of the functional groups and

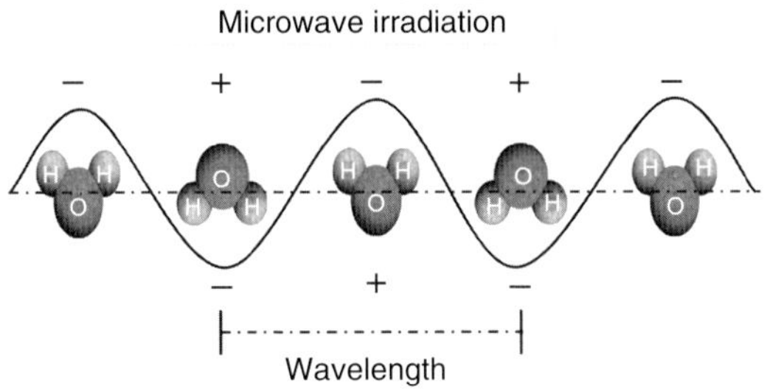

Figure 7.2 Water molecules in an alternating electrical field under microwave irradiation [25]. (Zhu *et al.* [11]. Reproduced with permission of the American Chemical Society.)

the volume of the molecule. Functional groups capable of hydrogen bonding have a particularly strong influence on the relaxation time. The relaxation times of solvents decrease as the temperature of the solvent increases. Water has a relaxation time of 9×10^{-12} s at 20 °C; that is, it has a relaxation frequency of 18 GHz. The most effective conversion of microwave energy into thermal energy will occur in this frequency region [25]. However, the majority of commercial and household microwave appliances use a frequency of 2.45 GHz, which is displaced from this maximum.

It is fortunate that even when the relaxation time is one or two orders of magnitude different from the one corresponding to the microwave frequency operating in the cavity, the solvent is still capable of acting as an effective medium for microwave dielectric heating because its loss tangent is sufficiently large. The loss tangent of water at 2.45 GHz and room temperature is 0.123, which is much higher than those at lower frequencies (0.045 at 900 MHz, 0.001 at 27.12 MHz), and the heating rates are sufficiently high at this frequency for the food processing industry [27].

Hu *et al.* [28] prepared monodisperse Ag nanoparticles via the rapid microwave-assisted LPE method in a $AgNO_3$ aqueous solution, using basic amino acids such as L-lysine or L-arginine as a reducing agent and soluble starch as a protecting agent, at 150 °C for a very short period of time (only 10 s), and the transmission electron microscope (TEM) image of the product is shown in Figure 7.3. The as-prepared Ag nanoparticles had a spherical shape and a mean particle diameter of 26.3 nm with a standard deviation of 2.1 nm and were readily redispersed

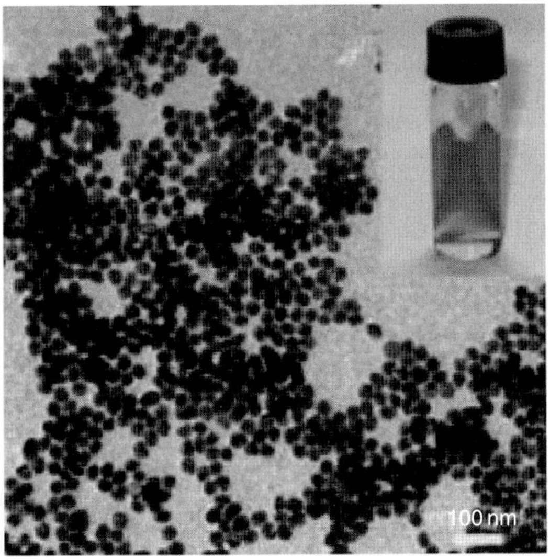

Figure 7.3 Typical TEM image of Ag nanoparticles prepared in an aqueous system using $AgNO_3$ as the silver source, L-lysine as a reducing agent, and soluble starch as a capping agent via microwave irradiation at 150 °C for only 10 s. (Hu *et al.* [28]. Reproduced with permission of the American Chemical Society.)

in water, exhibiting a transparent light brown suspension of well-dispersed Ag nanoparticles (the inset of Figure 7.2). The aqueous dispersion of Ag nanoparticles was stable without precipitate for more than 2 months at room temperature.

Using continuous flow-type microwave reactors is a promising strategy, and exciting progress has been made in recent years. Horikoshi *et al.* [29] reported a continuous flow reactor system consisting of a Pyrex pipe placed horizontally in the microwave waveguide through which the carboxymethyl cellulose (CMC)/$[Ag\text{-}(NH_3)_2]^+$ aqueous solution was circulated by a peristaltic pump, as shown in Figure 7.4. The as-grown Ag nanoparticles are with a narrow size distribution (0.7–2.8 nm, average size 1.5 nm).

The continuous flow-type microwave reactor with a pressurized function allowing homogeneous heating of the reaction solution flow is desirable for rapid and scaling up synthesis of nanostructured materials. Nishioka *et al.* [30] reported a kind of innovative continuous flow-type single-mode microwave reactor equipped with an elevated-pressure attachment. This microwave reactor could form a uniform electromagnetic field along a tubular reactor located in the center of a cylindrical microwave cavity, as shown in Figure 7.5. This microwave reactor system was characterized by a variable frequency microwave generator, pressure pump, and a single-mode cavity in which a quartz tubular reactor was mounted along the central axis. The key feature of the microwave reactor system was the automatic detection and compensation of the resonance frequency shift. The temperature

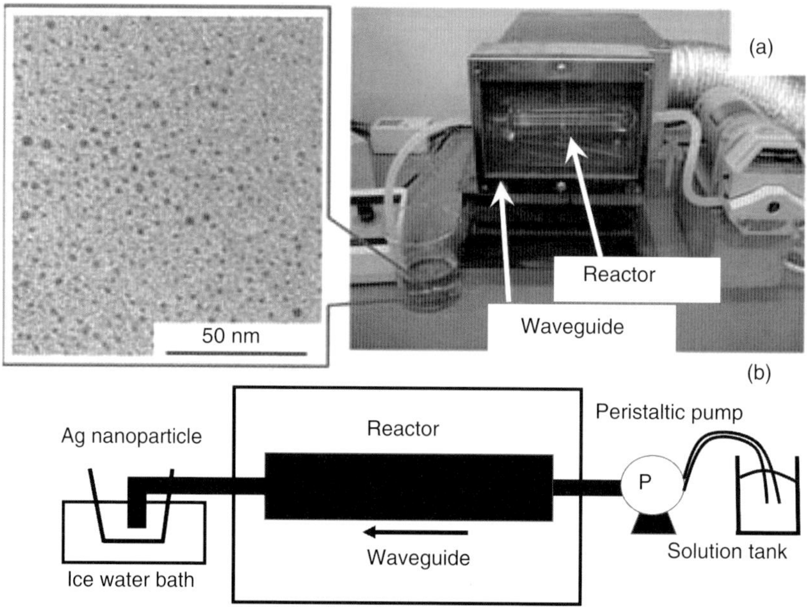

Figure 7.4 (a) Photograph of the continuous flow reactor system for the microwave-assisted synthesis of Ag nanoparticles; the TEM image of the as-prepared Ag colloid is also displayed. (b) Schematic illustration of the overall experimental setup. (Horikoshi *et al.* [29]. Reproduced with permission of the Royal Society of Chemistry.)

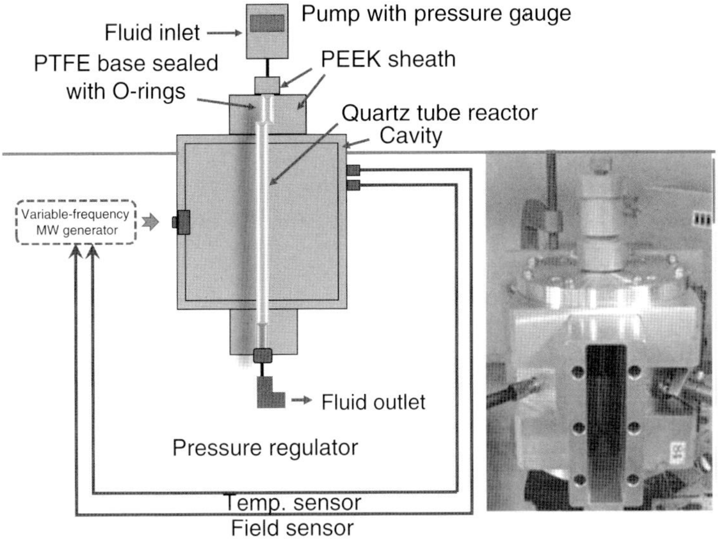

Figure 7.5 Experimental setup of the microwave-assisted continuous flow reactor system and a photograph of the microwave cavity part. (Nishioka *et al.* [30]. Reproduced with permission of the American Chemical Society.)

of the reaction solution was measured using a radiation thermometer through the open slit of the cavity. The reaction fluid was pressurized by a pump and was guided to the quartz reactor tube inlet. The pressure was controlled manually using a pressure regulator, and it was measured using the pressure gauge. The quartz reactor tube was connected with the PEEK sheath as designed to tolerate up to 10 MPa and was sealed with either a Viton or a Kalrez O-ring. Continuous flows of polar solvents including water, ethylene glycol (EG), and ethanol were heated instantaneously beyond their boiling points by the application of pressure. Acceleration of the reaction was exemplified in continuous synthesis of Cu nanoparticles at the reaction temperature beyond the boiling point of the solvent of EG at 2 MPa.

7.3 Nanostructures Fabricated Using LPE

7.3.1 Core–Shell Structures

The core–shell nanostructures are generally synthesized through a layer-by-layer coating process that involves the deposition of shell layers on the surface of preformed core nanoparticles [31]. Hierarchical design of the core–shell structures is shown in Figure 7.6. A dense surface coverage with crystalline layers matching

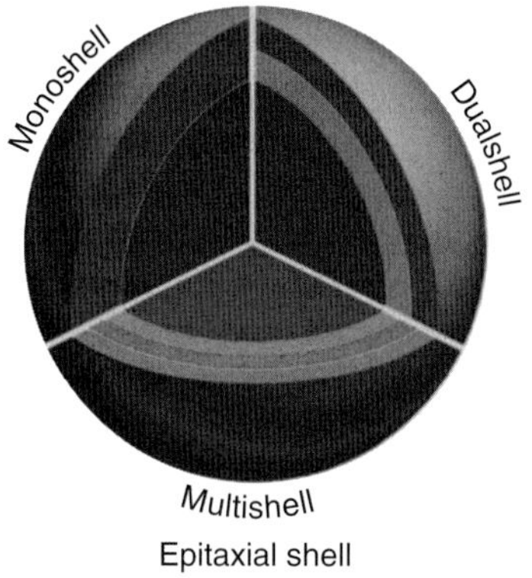

Figure 7.6 Schematic architectures of core–shell nanoparticles. (Chen *et al.* [31]. Reproduced with permission of the Royal Society of Chemistry.)

the core nanoparticle lattice, that is, epitaxial shell, can also be attached to the surface of the core nanoparticle by forming an assembly layer. Epitaxial coatings offer high chemical stability and facile formation of multishelled structures. With the ability of nanochemistry to produce a shell of stoichiometric composition and a shell thickness with monolayer precision, the electronic structure as well as the physiochemical structure of the core–shell system can be purposely aligned for the especially functional nanomaterials, such as upconverted nanoparticles (UCNPs) core–shell quantum dots. For this kind of UCNPs, there are lots of advantages of the core–shell structure [31]. (i) Compared to the conventional UCNPs, it has a higher upconversion efficiency due to suppression of the surface-related quenching effect produced by an epitaxial growth of a shell layer. (ii) It can achieve nanoscopic control of ion–ion interactions through spatial confinement of different types of lanthanide dopants in separated shell layers or the core domain. (iii) It induces coupling with new excitation electronic states of dyes or localized surface plasmon resonance (LSPR) for enhanced or broadband upconversion. (iv) It generates new functionalities and multifunctionalities for theranostic applications.

The core–shell nanostructures utilized in recent studies can be broadly classified into two categories: (i) a core–shell nanoparticle containing epitaxial layers with a matched or nearly matched crystalline lattice and (ii) a core–shell nanoparticle containing nonepitaxial layers. The core nanoparticle here can be either conventional UCNPs without any epitaxial shell or core–shell UCNPs with epitaxial shell layers. Figure 7.7 gives the examples of six types of core–shell UCNPs from these two categories, which constitute most of the current core–shell design to produce selective UC emissions as well as to deliver multifunctionalities to reach comprehensive imaging and therapeutic modalities [31–36].

7.3.2
The Epitaxial Preparation Methods of Core–Shell Structures

Use of nanophotonic approaches to control the excitation dynamics of upconversion in a core–shell structure significantly relies on the ability of nanochemistry

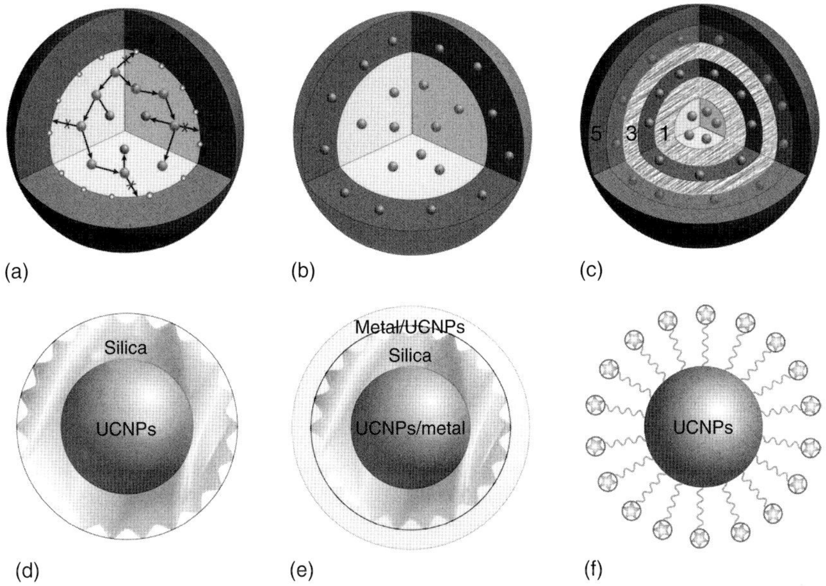

Figure 7.7 Different types of core–shell structures. The upper panel illustrates core–shell UCNPs with epitaxial layers: (a) core–shell UCNPs with an "inert" shell layer, (b) core–shell UCNPs with an active shell layer, (c) core–shell UCNPs with multiple shell layers, (c) core–multishell type, (d) UCNPs@silica core–shell nanoparticles, (e) core–shell–shell UCNPs@silica@metal nanoparticles, and (f) UCNPs coated with organic dye molecules. (Chen *et al.* [31]. Reproduced with permission of the Royal Society of Chemistry.)

to produce shells in a precisely defined manner. This is of particular importance for epitaxial core–multishell UCNPs, as each shell may play a specific distinct role. This means that the host lattice, thickness, crystal structure, as well as concentration and type of dopants might vary in different shell layers. Yet the entire multishell coating has to be grown in a well-defined manner. This section provides an account of nanochemical approaches to prepare these epitaxial core–shell structures with single or multiple shell layers with monolayer deposition precision, as well as to synthesize nonepitaxial mesoporous silica-coated UCNPs and plasmonic core–shell nanoparticles. The core-only UCNPs have been typically prepared using solution-based colloidal nanochemistry via three main types of synthetic approaches: (i) thermal decomposition, (ii) Ostwald ripening, and (iii) hydrothermal methods. By carefully adjusting experimental parameters, such as concentrations of precursors, type of solvents, reaction temperature, and time, the core-only UCNPs with a high crystallinity, narrow size distribution, and exceptional optical properties can be prepared [31, 37–39].

7.3.2.1 General Nanochemical Approaches to Prepare Epitaxial Core–Shell UCNPs with a Single Shell Layer

Typically, there are three techniques to produce a single shell layer on the core nanoparticle surface.

1. The most widely used technique is the seed-mediated liquid-phase epitaxial growth, which can prepare high-quality epitaxial core–shell structure nanoparticles with a uniform morphology and size. The most important tuning process of the high-quality inorganic nanoparticles with a narrow size distribution is to separate well the nucleation stage and the growth stage in time [39–41]. For seed-mediated epitaxial growth, the core nanoparticles are utilized as nuclei for seeding the growth of the shell layer. In Figure 7.8, Mai *et al.* [41] successfully applied the seed-mediated growth method to prepare both homogeneous and heterogeneous core–shell nanostructures,

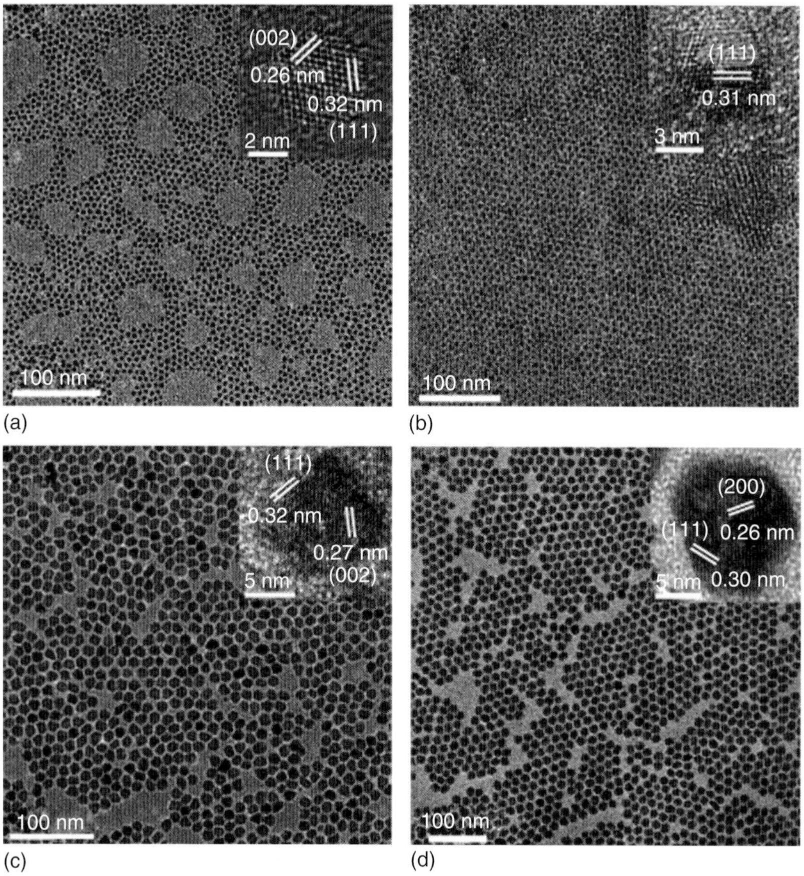

Figure 7.8 TEM and HRTEM (inset) images of α-$NaNdF_4$ (a), α-$NaEuF_4$ (b), α-$NaYF_4$ (c), and α-$NaYbF_4$ (d) nanopolyhedra. (Mai *et al.* [41]. Reproduced with permission of the American Chemical Society.)

in which the thickness of the shell can be controlled via adjusting the shell precursor concentration. Additionally, the shell precursors can be either directly injected into the growing solution of the core nanoparticles to induce a further growth of the shell, or the core nanoparticles can be centrifuged out, and the shell layer is grown in a separate step [41].

2. Seed-mediated Ostwald ripening. Ostwald ripening refers to the process that favors larger particles with smaller surface-to-volume ratios over the energetically less stable smaller particles, resulting in the growth of larger particles at the expense of smaller ones. As a result, the Ostwald ripening process is able to dissolve energetically less stable small sacrificial nanoparticles into molecules of shell materials, which can then deposit on the larger stable core nanoparticles. Veggel and coworkers recently exploited this process to perform layer-by-layer shell growth onto larger core nanoparticles [42]. As shown in Figure 7.9, small sacrificial alpha-$NaYF_4$ nanoparticles were firstly synthesized and used as the shell precursor. Subsequent injection

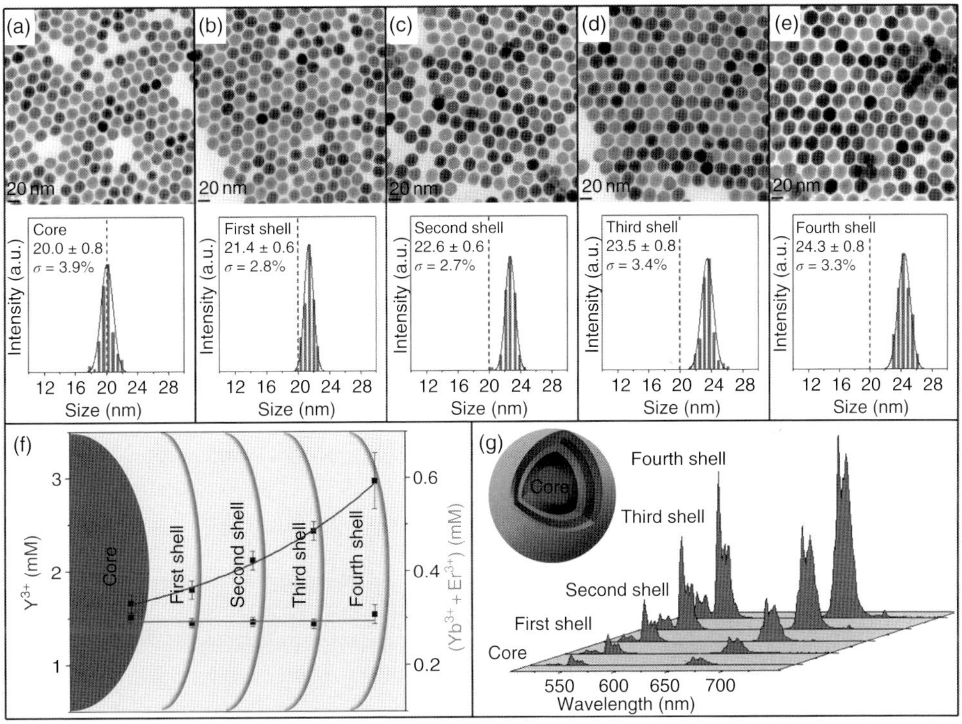

Figure 7.9 (a–e) TEM images and size distribution of $NaYF_4$:Yb^{3+}/Er^{3+} (15/2%) core NCs ($t = 0$), $NaYF_4$:Yb^{3+}/Er^{3+} (15/2%) core/$NaYF_4$ shell NCs after successive layer-by-layer epitaxial growth at $t = 5$, 10, 15, and 20 min, respectively; (f) ICP-MS elemental analysis of the core and core–shell NCs with same number concentration of NCs; and (g) upconversion emission spectra of the hexane dispersions of core and core–shell NCs with same number concentration of NCs under 980 nm excitation. (Johnson *et al.* [42]. Reproduced with permission of the American Chemical Society.)

of these sacrificial nanoparticles into a hot solution of $NaYF_4$: Ln^{3+} core nanoparticles induces a rapid dissolution of the sacrificial nanoparticles and their deposition onto the larger core NPs (a self-focusing Ostwald ripening process), which resulted in the formation of the core–shell structured nanoparticles with a well-defined shape, narrow size distribution, and a control to provide a tunable shell growth. Thus, the shell thickness can be precisely controlled by manipulating the amount of the injected sacrificial nanoparticles. This method is flexible and holds promise to prepare other high-quality homogeneous and heterogeneous core–shell nanoparticles.

3. Cation exchange reaction. This strategy involves a reversible chemical reaction between an insoluble solid and a solution, during which cationic ions are interchanged. As a result, cationic ions from the surface of UCNPs are replaced by other cations present in solution, thus producing a unique heterogeneous core–shell structure. Typically, John A. Capobianco group used the cation exchange approach to prepare multifunctional $NaYF_4:Yb^{3+}/Tm^{3+}$@$NaGdF_4$ core–shell NPs with tunable shell thickness, enhanced UCPL, and outstanding paramagnetic performance (see Figure 7.10) [43], which open up a versatile alternative to impart new imaging modality to UCNPs through the formation of a core–shell structure.

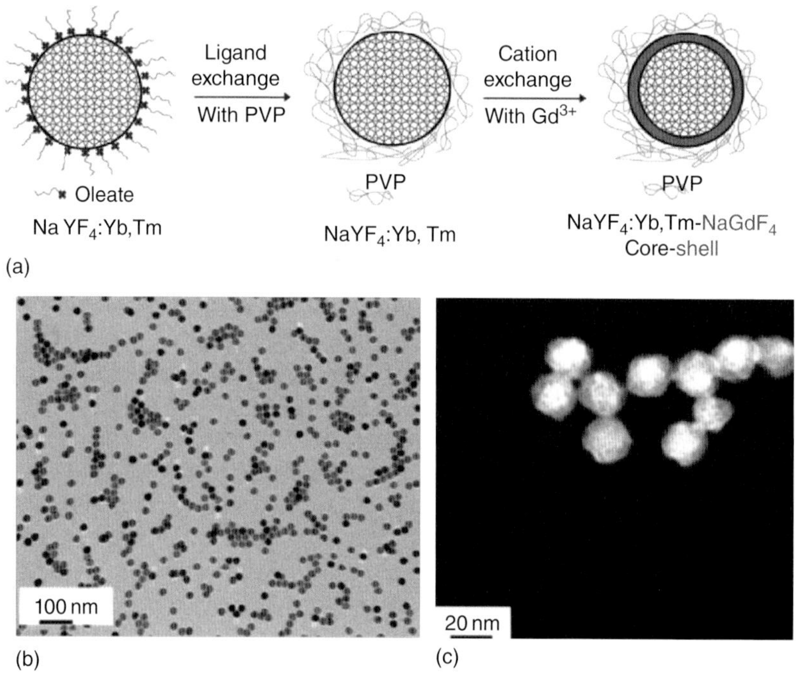

Figure 7.10 (a) Sketch of $NaYF_4$: Yb, Tm-NaGdF4 core–shell nanoparticles via cation exchange and (b,c) TEM images of $NaYF_4:Yb^{3+}/NaGdF_4:Tm^{3+}$ core–shell nanoparticles. (Dong *et al.* [43]. Reproduced with permission of the American Chemical Society.)

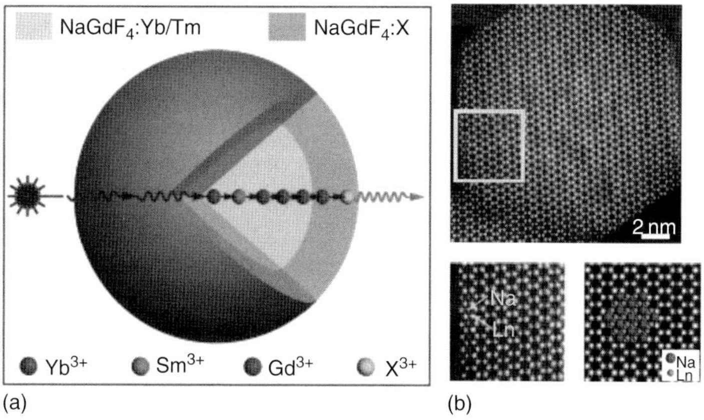

Figure 7.11 Tuning upconversion through energy migration in core–shell nanoparticles. (a) Schematic design of a lanthanide-doped $NaGdF_4$@$NaGdF_4$ core–shell nanoparticle for EMU (X: activator ion). (b) An enlarged view of the selected area indicated by a light gray box, showing lanthanide (Ln) and Na atomic columns. Digitally processed STEM image by imposing projection symmetry to enhance the signal-to-noise ratio. (Wang *et al.* [45]. Reproduced with permission of Nature Publishing Group.)

7.3.2.2 Layer-by-Layer Approach to Prepare Core–Multishell UCNPs with Monolayer Thickness Precision

The key to produce core–multishell UCNPs is the deposition of shells in a layer-by-layer way. A straightforward approach is to use a multistep approach to prepare each shell using seed-mediated liquid-phase epitaxial growth. The thickness of the shell could be precisely tuned by controlling the shell precursor dosage. Recently, many researchers had employed the multistep method to prepare core–multishell nanostructures, such as $NaYbF_4$@Na(Yb, Gd)F_4@$NaGdF_4$ [44], $NaGdF_4$:Yb^{3+}/Tm^{3+}@$NaGdF_4$@$NaGdF_4$:Tb^{3+} [45] (Figure 7.11), $NaYF_4$:Yb^{3+}/Er^{3+}@$NaYF_4$:Yb^{3+}@$NaYF_4$:Nd^{3+} [46] (Figure 7.12), and $NaYF_4$:Yb^{3+}/Er^{3+}@$NaGdF_4$ [47]. This investigation opened up a new milestone for nanophotonic control of core–shell nanostructure with a subnanometer precision.

7.3.2.3 Mesoporous Silica Coating

Silica coating is a strategy for inorganic surface treatment to make nanoparticles water dispersible and biocompatible as well as to introduce new therapeutic functionalities when using mesoporous silica to entrap a therapeutic agent and deliver it to the disease site [31]. Generally, there are two types of conventional chemistry to coat nanoparticles with silica, that is, Stöber method and reverse micelle microemulsion. Recently, three methods were developed to prepare the mesoporous silica shell, and their sketches of preparation process as well as their examples were shown in Figure 7.13. It obviously illustrated three methods to prepare a mesoporous silica shell. They were (i) the calcination method

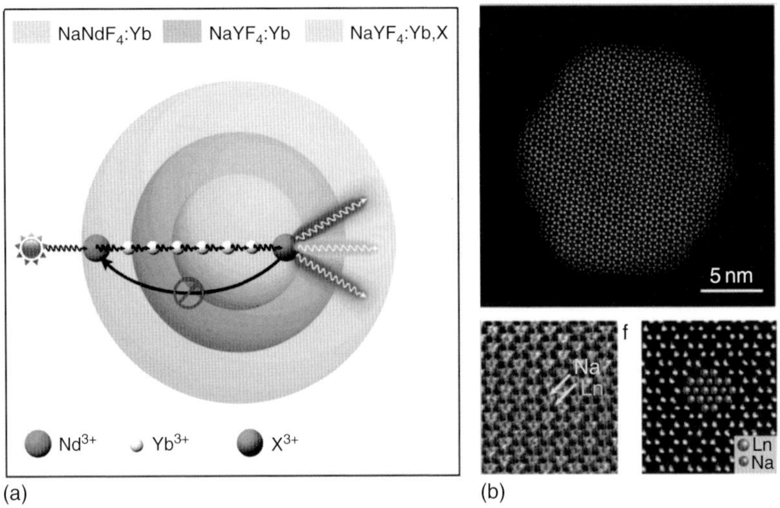

Figure 7.12 (a) Schematic illustration of the proposed energy-transfer mechanisms in the quenching-shield sandwich-structured UCNPs upon 800 nm excitation. (b) Aberration-corrected high-angular annular dark-field (HAADF) STEM image of a single quenching-shield sandwich nanoparticle projected from the [001] direction. Corresponding enlarged annular bright field (ABF) STEM image reveals the distribution of Na^+ ions and Ln^{3+} ions. Simulated HAADF image based on the fast Fourier transformation (FFT) multislice method to enhance the signal-to-noise ratio. (Zhong *et al.* [46]. Reproduced with permission of Wiley.)

(Figure 7.13a), (ii) the organic template method (Figure 7.13b), and (iii) the etching method (Figure 7.13c).

7.3.2.4 Coupling of UCNPs with Plasmonics Using Core–Shell Architecture

Coupling of plasmonic properties with UCNPs can be of great importance for obtaining plasmonically enhanced UCNPs. We simply introduce a kind of metallic@silica@UCNPs nanostructure. Figure 7.14a illustrates a typical procedure for creating metallic@silica@UCNPs core–shell–shell nanostructure, where a plasmonic metallic core of silver nanospheres is utilized as an example [51]. The first step is to coat silica onto metallic nanoparticles by using either the Stöber method or the reverse micelle microemulsion method (Figure 7.14b). By controlling the coating process, the resulting metallic@SiO_2 (e.g., Ag@SiO_2) can have a uniform diameter and contain one metallic nanoparticle positioned at the center. The thickness of the silica can also be precisely controlled via adjusting the amount of the utilized TEOS as well as by varying other parameters. In the second step, the metallic@SiO_2 can then be coated with a layer of Y:Ln(OH)CO_3·H_2O through the LPE method from yttrium nitrate, lanthanide nitrate, and urea (Figure 7.8c). Finally, heat treatment at high temperature can convert the Y:Ln(OH)CO_3·H_2O layer into a crystallized lanthanide-doped Y_2O_3 upconverting layer (Figure 7.8d). The thickness of this layer is determined by the

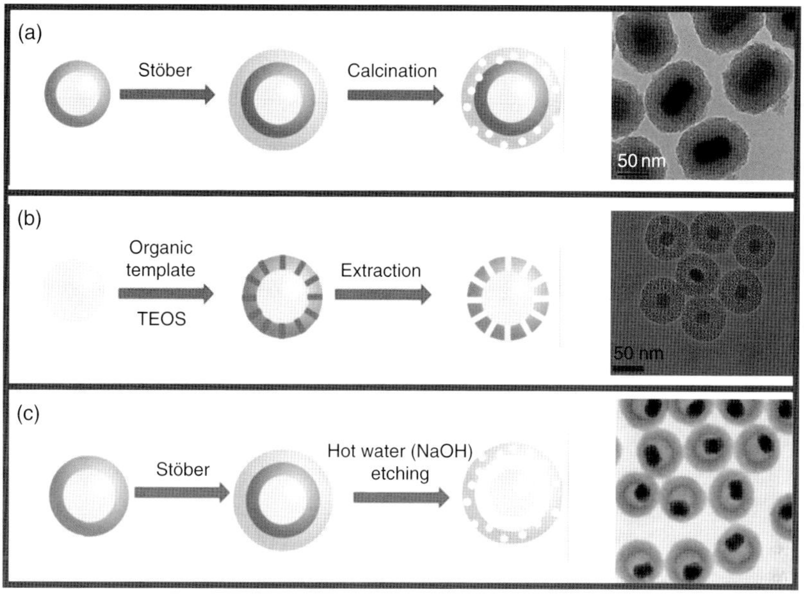

Figure 7.13 Coating of UCNPs with a mesoporous silica shell. (a) Schematic of the calcination method and corresponding TEM images of mesoporous silica-coated $NaYF_4$:Yb^{3+}/Er^{3+}@silica nanoparticles. (b) Schematic of the template method and dark-field STEM image of $NaYF_4$:Tm^{3+}/Yb^{3+}@$NaYF_4$@$mSiO_2$. (c) Schematic of the etching method and HAADF STEM image of $NaYF_4$:Yb/Tm@NaYF4@$hmSiO_2$ [48–50]. (Chen 2014. Reproduced with permission of the Royal Society of Chemistry.)

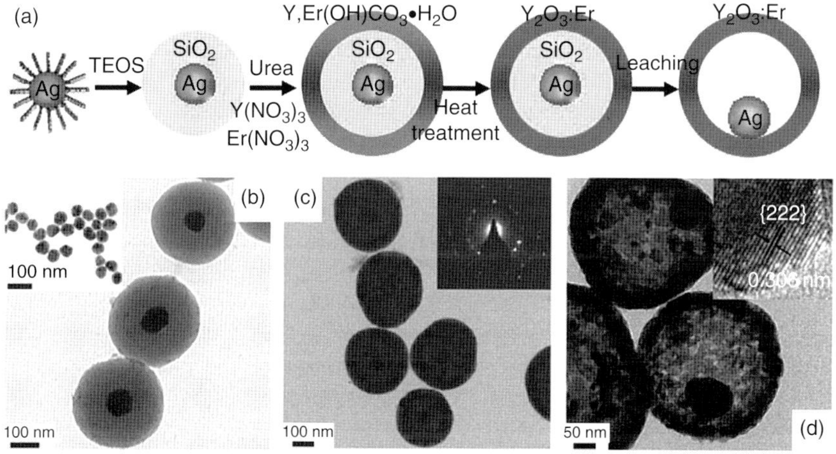

Figure 7.14 (a) Schematic illustration for preparing core–shell–shell structures of metallic@silica@UCNPs, along with (b–d) TEM images of the nanostructures obtained in each step. (Zhang *et al.* [51]. Reproduced with permission of the American Chemical Society.)

thickness of the layer of Y:Ln(OH)CO_3·H_2O. This procedure was employed by Zhang *et al.* to prepare Ag@SiO_2@Y_2O_3: Er^{3+} nanoparticles [51].

7.4 Summary

In this chapter, we have introduced the LPE technique, especially the hydrothermal method to grow nanostructures. The development history and advantages of hydrothermal methods are demonstrated in detail. Moreover, the growth mechanism and the corresponding results of microwave-assisted preparation for nanostructures in aqueous solution are discussed. Another important example of LPE method was given for nanophotonic approaches to control the excitation dynamics in core–shell UCNPs and various synthetic nanochemistries to prepare epitaxial core–shell UCNPs.

Acknowledgments

The authors acknowledge the National Natural Science Foundation of China (No. 51202023) and the scientific and technological projects for distinguished young scholars of Sichuan Province (No. 2015JQ0013).

References

1. Herman, M.A., Richter, W., and Sitter, H. (2004) *Epitaxy-Physical Principle and Technical Implementation*, Springer-Verlag, Berlin, Heidelberg, New York.
2. Yoshimura, M. and Byrappa, K. (2008) *J. Mater. Sci.*, **43**, 2085.
3. Schafthaul, K.F.E. (1845) *Gelehrte Anzeigen Bayer Akal*, **20**, 557.
4. Morey, G.W. and Niggli, P. (1913) *J. Am. Chem. Soc.*, **35**, 1086.
5. Bayer, K.J. (1887) *A Textbook of Hydrometallurgy*, Libraire Universitaire du Quebec, Quebec.
6. Somiya, S. (1983) *Proceedings of the First International Symposium on Hydrothermal Reactions, 1982*, Gakujutsu Bunken Fukyukai Publications, Tokyo, p. 965.
7. Byrappa, K. (2005) In: *Kirk-Othmer Encyclopedia of Chemical Technology*, John Wiley & Sons, Inc., London.
8. Jelinski, L.W., Graedal, T.E., Laudise, R.A., McCall, D.W., and Patel, C.K.N. (1992) *Proc. Natl. Acad. Sci. U.S.A.*, **89**, 793.
9. Lencka, M.M., Andreko, A., and Riman, R.E. (1995) *J. Am. Ceram. Soc.*, **78**, 2609.
10. Hao, Y. and Teja, A.S. (2003) *J. Mater. Res.*, **18**, 415.
11. Zhu, Y.J. and Chen, F. (2014) *Chem. Rev.*, **114**, 6462.
12. Komarneni, S. and Roy, R. (1985) *Mater. Lett.*, **3**, 165.
13. Riman, R.E., Suchanek, W.L., Byrappa, K., Chen, C.W., Shuk, P., and Oakes, C.S. (2002) *Solid State Ionics*, **151**, 393.
14. Gedye, R., Smith, F., Westaway, K., Ali, H., Baldisera, L., Laberge, L., and Rousell, J. (1986) *Tetrahedron Lett.*, **27**, 279.
15. Giguere, R.J., Bray, T.L., Duncan, S.M., and Majetich, G. (1986) *Tetrahedron Lett.*, **27**, 4945.
16. Rao, K.J., Vaidhyanathan, B., Ganguli, M., and Ramakrishnan, P.A. (1999) *Chem. Mater.*, **11**, 882.
17. Collins, M.J. Jr., (2010) *Future Med. Chem.*, **2**, 151.

18. Nadagouda, M.N., Speth, T.F., and Varma, R.S. (2011) *Acc. Chem. Res.*, **44**, 469.
19. Tsuji, M., Hashimoto, M., Nishizawa, Y., Kubokawa, M., and Tsuji, T. (2005) *Chem. Eur. J.*, **11**, 440.
20. Zhang, X.Y. and Liu, Z. (2012) *Nanoscale*, **4**, 707.
21. Park, S.E., Chang, J.S., Hwang, Y.K., Kim, D.S., Jhung, S.H., and Hwang, J.S. (2004) *Catal. Surv. Asia*, **8**, 91.
22. Baghbanzadeh, M., Carbone, L., Cozzoli, P.D., and Kappe, C.O. (2011) *Angew. Chem. Int. Ed.*, **50**, 11312.
23. Bogdal, D., Prociak, A., and Michalowski, S. (2011) *Curr. Org. Chem.*, **15**, 178.
24. Bilecka, I. and Niederberger, M. (2010) *Nanoscale*, **2**, 1358.
25. Tompsett, G.A., Conner, W.C., and Yngvesson, K.S. (2006) *ChemPhysChem*, **7**, 296.
26. Polshettiwar, V. and Varma, R.S. (2010) *Green Chem.*, **12**, 743.
27. Gabriel, C., Gabriel, S., Grant, E.H., Halstead, B.S.J., and Mingos, D.M.P. (1998) *Chem. Soc. Rev.*, **27**, 213.
28. Hu, B., Wang, S.B., Wang, K., Zhang, M., and Yu, S.H. (2008) *J. Phys. Chem. C*, **112**, 11169.
29. Horikoshi, S., Abe, H., Torigoe, K., Abe, M., and Serpone, N. (2010) *Nanoscale*, **2**, 1441.
30. Nishioka, M., Miyakawa, M., Daino, Y., Kataoka, H., Koda, H., Sato, K., and Suzuki, T.M. (2013) *Ind. Eng. Chem. Res.*, **52**, 4683.
31. Chen, X., Peng, D.F., Ju, Q., and Wang, F. (2015) *Chem. Soc. Rev.*, **44**, 1318.
32. Zhang, J., Shade, C.M., Chengelis, D.A., and Petoud, S. (2007) *J. Am. Chem. Soc.*, **129**, 14834.
33. Huang, X., Han, S., Huang, W., and Liu, X. (2013) *Chem. Soc. Rev.*, **42**, 173.
34. Chen, G., Shen, J., Ohulchansk, T.Y., Patel, N.J., Kutikov, A., Li, Z., Song, J., Pandey, R.K., Ågren, H., Prasad, P.N., and Han, G. (2012) *ACS Nano*, **6**, 8280.
35. Park, Y.I., Kim, J.H., Lee, K.T., Jeon, K.-S., Na, H.B., Yu, J.H., Kim, H.M., Lee, N., Choi, S.H., Baik, S.-I., Kim, H., Park, S.P., Park, B.-J., Kim, Y.W., Lee, S.H., Yoon, S.-Y., Song, I.C., Moon, W.K., Suh, Y.D., and Hyeon, T. (2009) *Adv. Mater.*, **21**, 4467.
36. Ye, X., Collins, J.E., Kang, Y., Chen, J., Chen, D.T.N., Yodh, A.G., and Murray, C.B. (2010) *Proc. Natl. Acad. Sci. U.S.A.*, **107**, 22431.
37. Wang, F. and Liu, X.G. (2009) *Chem. Soc. Rev.*, **38**, 976–989.
38. Liu, Y.S., Tu, D.T., Zhu, H.M., and Chen, X.Y. (2013) *Chem. Soc. Rev.*, **42**, 6924–6958.
39. Zhou, J., Liu, Z., and Li, F.Y. (2012) *Chem. Soc. Rev.*, **41**, 1323–1349.
40. Boyer, J.C., Vetrone, F., Cuccia, L.A., and Capobianco, J.A. (2006) *J. Am. Chem. Soc.*, **128**, 7444–7445.
41. Mai, H.X., Zhang, Y.W., Si, R., Yan, Z.G., Sun, L.D., You, L.P., and Yan, C.H. (2006) *J. Am. Chem. Soc.*, **128**, 6426.
42. Johnson, N.J.J., Korinek, A., Dong, C.H., and van Veggel, F. (2012) *J. Am. Chem. Soc.*, **134**, 11068.
43. Dong, C.H., Korinek, A., Blasiak, B., Tomanek, B., and van Veggel, F. (2012) *Chem. Mater.*, **24**, 1297.
44. Wen, H.L., Zhu, H., Chen, X., Hung, T.F., Wang, B.L., Zhu, G.Y., Yu, S.F., and Wang, F. (2013) *Angew. Chem. Int. Ed.*, **52**, 13419.
45. Wang, F., Deng, R.R., Wang, J., Wang, Q.X., Han, Y., Zhu, H.M., Chen, X.Y., and Liu, X.G. (2011) *Nat. Mater.*, **10**, 968.
46. Zhong, Y.T., Tian, G., Gu, Z.J., Yang, Y.J., Gu, L., Zhao, Y.L., Ma, Y., and Yao, J.N. (2014) *Adv. Mater.*, **26**, 2831.
47. Zhang, F., Che, R.C., Li, X.M., Yao, C., Yang, J.P., Shen, D.K., Hu, P., Li, W., and Zhao, D.Y. (2012) *Nano Lett.*, **12**, 2852.
48. Qian, H.S., Guo, H.C., Ho, P.C.L., Mahendran, R., and Zhang, Y. (2009) *Small*, **5**, 2285.
49. Li, C.X., Yang, D.M., Ma, P.A., Chen, Y.Y., Wu, Y., Hou, Z.Y., Dai, Y.L., Zhao, J.H., Sui, C.P., and Lin, J. (2013) *Small*, **9**, 4150.
50. Zhang, F., Braun, G.B., Pallaoro, A., Zhang, Y.C., Shi, Y.F., Cui, D.X., Moskovits, M., Zhao, D.Y., and Stucky, G.D. (2012) *Nano Lett.*, **12**, 61.
51. Zhang, F., Braun, G.B., Shi, Y.F., Zhang, Y.C., Sun, X.H., Reich, N.O., Zhao, D.Y., and Stucky, G. (2010) *J. Am. Chem. Soc.*, **132**, 2850.

8
Nanostructural Thin Film Development with Chemical Solution Deposition

Yanda Ji and Yuan Lin

8.1
Introduction

Among the well-developed deposition techniques, chemical solution deposition (CSD) is noted for its easy setup, low cost, and the ability of coating on irregular surfaces. In principle, a CSD technique is based on the hydrothermal or solvothermal mechanism, and the structural and physical properties of the final products can be optimized by controlling the parameters of the process from liquid phase to solid phase [1–7].

Usually, a typical CSD process for thin film deposition includes three steps: precursor solution preparation, surface coating, and thermal treatment. We will describe the three steps in the following.

8.2
Precursor Solution Preparation

The CSD process starts from the precursor solution preparation. Some requirements on the solvent, solute, substrates, and processing conditions should be carefully taken into consideration:

1. The solubility of the solute should be as high as possible to form the precursor solution with a high concentration.
2. The chemical composition of precursor should be as simple as possible to make sure that all of the chemical elements except the desired cations can be released completely during the annealing.
3. The precursor solution should be uniform, stable, and no phase separation.
4. The wetting between the precursor and the substrate should be as good as possible and the interdiffusion during the thin film preparing should be well controlled.
5. The viscosity of the precursor solution should be properly adjusted to avoid thickness variations and crack formation.

Advanced Nano Deposition Methods, First Edition. Edited by Yuan Lin and Xin Chen.

8.2.1
Chemical Strategies for Precursor Solutions

To meet these requirements, several types of precursor systems have been developed. The first type is the carboxylates. In chemistry, metal carboxylate salts are easy to dissolve in their own carboxylic acids. For example, the Ba source of $BaTiO_3$ ferroelectric thin films can be induced by Ba acetate. Although this chemical can be well dissolved in both water and acetate acids, acetate acids are generally chosen as the solvent. The reason is that the processing stability would be negatively affected by the water-induced hydrolysis of the precursor solution.

The second precursor system is alkoxide. Alkoxides are the metal salts of alcohols. The hydrolysis process of alkoxides is very sensitive. The following chemical reaction would undergo when the hydrolysis process begins in the alkoxides:

$$\text{M-OR} + \text{H}_2\text{O} \rightarrow \text{M-OH} + \text{ROH}$$

To develop thin films from this kind of precursor solution, the hydrolysis process should be carefully manipulated to form the short-chain oligomers, which are only reacted by a few monomer molecules.

The third common system for CSD precursor is β-diketonate. β-Diketonates can exhibit strong chelate effects to metal alkoxides. Moreover, it is well dissolved in alcohols, ketones, and ethers to form the uniform solution. For some special solutes, it is too active in the single solvent, such as titanium tetraisopropoxide, which is usually employed as the Ti source. In this case, the mixed solvent strategy of acetylacetone and acetic acid is used. The role of acetylacetone is to stabilize the titanium tetraisopropoxide in acid.

As mentioned earlier, after the precursor preparation, some chemical process may happen in the solution, which would affect the stability of the precursor solution. According to the different routes to form the precursor solutions, the CSD techniques can be divided into sol–gel method, metal-organic deposition (MOD), and some other techniques such as newly developed polymer-assisted deposition (PAD).

8.2.2
Sol–Gel Method

Sol–gel method is a well-known CSD fabrication technique [8–10]. The nature for sol–gel method is the hydrolysis of ester. For instance, the silica can be prepared from the tetraethyl orthosilicate. The chemical route can be described as [11]

$$\text{M-(OR)}_x + \text{H}_2\text{O} \rightarrow \text{M(OR)}_{x-1}(\text{OH}) + \text{ROH} \tag{8.1}$$

$$2\text{M(OR)}_{x-1}(\text{OH}) \rightarrow \text{M}_2\text{O(OR)}_{2x-3}(\text{OH}) + \text{ROH} \tag{8.2}$$

$$2\text{M(OR)}_{x-1}(\text{OH}) \rightarrow \text{M}_2\text{O(OR)}_{2x-2}(\text{OH}) + \text{H}_2\text{O} \tag{8.3}$$

It should be noted that the aforementioned chemical reaction cannot be stopped once it starts. This reaction will happen in both water and alcohol. As a result, it is rather difficult to control the formation of precursor gel. In practice, the aging effect for the gel must be taken into consideration. The difference in hydrolysis time may induce variation in the structural and physical properties of the final production. Apart from the hydrolysis of ester, another synthesis approach is based on the chemical reaction between carboxylate and alkoxides. The same aging problem would occur when the chemical reaction starts.

8.2.3 Metal-Organic Deposition

The MOD method can overcome the disadvantage of the aging effects in hydrolysis [12–14]. The normal MOD uses the carboxylate–carboxylic acid system. However, the reactivity of original chemical reagents is low, which may cause some limitations. For instance, cracking may take place in the films if the overshrinkage happens during the thermal process. Many attempts have been tried to improve it. By using the carboxylate and β-diketonate compounds, the risk of cracking problem can be decreased due to the monomeric compounds formed by the chelating effect in β-diketonate, which can reduce the shrinkage during the thermal treatment. In addition, the solvent for this process is generally organic solvent without water. Another organic solvent-based CSD route is the so-called Pechini method [15]. In this method, metal nitrates are first dissolved in the water. Then, the citric acid is added to the solution to chelate the metal cations. After forming a polybasic acid, the ethylene is added to form organic ester compounds. In this case, the metal ions can be uniformly dispersed in the organic ester compounds, and the organic can be removed at a rather low temperature. Clearly, the organic solvent will greatly increase the cost for the whole fabrication process and may cause possible environmental pollution. So developing a novel CSD technique with water-soluble route and avoiding the aging effects are quite necessary.

8.2.4 Polymer-Assisted Deposition

PAD, which was invented in 2004, was developed to solve the aforementioned issues [16–19]. This CSD technique is very similar to MOD. In this method, inorganic metal ions are chelated into a water-soluble polymer, such as polyethyleneimine (PEI), to form the precursor water solution. Ethylenediaminetetraacetic acid (EDTA) is usually added to enhance the chelating ability of polymer. In this technique, there is no hydrolysis process since the metal ions are chelated into the polymer. Thus, the problems in previous CSD method such as aging effect and use of the organic solvent can be avoided. Unlike traditional CSD routes, the precursor solution should be filtrated by a filtration equipment as shown in Figure 8.1. Thus, the noncoordinated cation and anion species can be removed. The concentration of effective ions can be measured exactly by the inductively coupled plasma atomic emission spectroscopy (ICP-AES). In this way, the accurate stoichiometry for each kind of ions in the complex compound thin

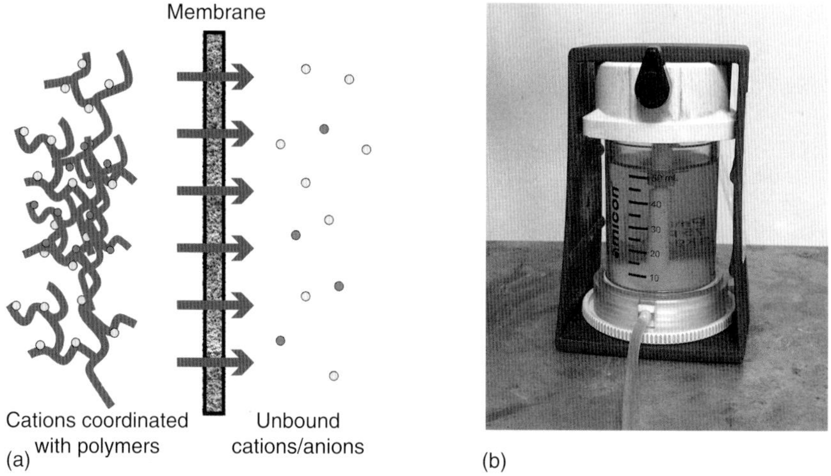

Figure 8.1 (a) Schematic illustration of filtration used in the PAD process and (b) photograph of an Amicon ultrafiltration cell used in the experiment. (McCleskey *et al.* [19]. Reproduced with permission from the Royal Society of Chemistry.)

films can be achieved, which would be critical for the growth of the high-quality thin films. The solution preparation of PAD is unique, and the stability of the solution has been greatly improved. The majority of the functional materials have been prepared using the PAD technique, including oxides, nitrides, carbides, and some metals such as Ge.

8.3 Coating

After solution preparation, coating should be performed to form the wet films on the substrates. Usually, the coating techniques include spin coating, dip coating, and spray coating. The most commonly used technique in the laboratory is spin coating [20, 21]. The substrate is first placed on the holder of the spinner by a vacuum at the backside. After the precursor solution is dropped on the substrate, the substrate is rotated at the speed of 1000–8000 rotations per minute. The thickness of the as-prepared wet film can be controlled by rotation speed, spinning time, and the solution viscosity. Dip coating is a more straightforward way for the precursor wet film deposition [22]. The general steps for dip coating are as follows: (i) keep the substrate immersed in the precursor solution and (ii) take the substrate out of the solution. The factors impacting the thickness of the dip-coated film include gravitational force, viscosity, and the surface tension. As discussed earlier, it is necessary that the precursor solution should be with a proper viscosity for both spin-coating and dip-coating techniques. When viscosity is too low to meet the requirements of these two kinds of techniques, spray coating can be carried out to realize the film deposition. In this technique, the precursor solution is transformed to aerosol by nebulizer or pressure. Here, the solution viscosity should be very low for nebulizing or the formation of droplets. The droplets are transferred by a carrier gas. Finally, the coating on the substrate is formed by spraying of the nebulized droplets. In this case, the gas pressure and the spraying time are the key factors for the final products.

After coating of the precursor, the gel process may take place. Usually, this process should be chemical polymerization or polycondensation reactions. The films cannot be dissolved again when the reaction is finished. It is also possible that there is no gel process, especially in some MOD techniques in which the solvent is evaporated upon heating without any other reactions. It should be noted that the choice of the solvent is very important to control the behavior during the formation of the film [23]. Not only the properties of the aimed precursor solution (dissolution, stability, and solution aging behavior) but also other solvent characteristics such as evaporation rate and wetting behavior should be considered for the selected solvent [24].

8.4 Thermal Treatment

To get the desired prepared products, the organic solvent remaining in the film and the gel network should be removed. Moreover, the formation of the proper phase may also need heating. During the process of thermal treatment, the organic chemical bonds are broken, and the associated volatile species are removed to form M–O–M network. Usually, the amorphous structure is formed in this case. The general reactions in this transformation for oxide thin films include (i) thermolysis, (ii) pyrolysis, (iii) dehydration, and (iv) oxidation. In the thermolysis period, the volatile organic species are firstly formed. And then they will react with oxygen at high temperature to produce CO and CO_2. Then, the OH groups begin to form H_2O in the dehydration period. Finally, the amorphous thin film is synthesized.

Technically, the processes for the fabrication of the aimed films can be realized in two ways: two-step process and one-step process. In the two-step method, removing the organic species (pyrolysis/thermolysis) in the thin films is prior to the crystallization of the films [25]. In the first step, the wet film is normally placed at 200–400 °C for the organic species to volatilize. It has been proposed that this approach can remove organic content in the amorphous network; thus cracking and blistering can be avoided. And then the film is further heated to a higher temperature for crystallization. In the one-step process, the film is heated to a relatively high temperature (~700 °C) for crystallization directly. The removal of the organic materials and the crystallization of the film both take place at the same temperature [26, 27]. Usually, the rapid thermal annealing (RTA) furnaces are employed to anneal the samples. Because rapid heating can delay the removal of the organic materials to a higher temperature, it would greatly enhance the chemical kinetics of the pyrolysis and thermolysis [28–30]. In fact, the removal of the organic materials and the crystallization of the film can co-occur within a time of 15 s when using the RTA treatment. As a result, the rapid heating treatment can enhance thin film densification by delaying the crystallization to a higher temperature. This is an important advantage for the preparation of high-density film materials. In summary, it is feasible to control the densification behavior of the film by the chemistry in precursor solution and the heating rate.

8.5 Control of the Microstructures in Thin Films Prepared by CSD Techniques

Undoubtedly, the microstructure and orientation of the thin films can be significantly impacted by the solution and material chemistry, substrates, and annealing conditions in the CSD technique. The residual intermediate phase such as fluorite, pyrochlore, or carbonate may occur during the deposition, which would degrade the physical performance of the thin films. Thus, it is important to understand the underlying kinetic and thermodynamic aspects of the fabrication processes that affect the microstructures of thin films.

8.5.1 Thermodynamics for CSD-Delivered Thin Films

Usually, after the pyrolyzation of the wet precursor film, the amorphous structure of the thin film would form, especially in the two-step process in which the nucleation of the crystals would start at a higher temperature than that of the pyrolyzation. In nature, the interface nucleation of the final crystalline phase is columnar, and the film is polycrystalline with equiaxed grains. From the viewpoint of thermodynamics, transformation from the amorphous film into the crystalline one needs a driving force, and it plays a rather significant role in nucleation and the microstructure of the films. The driving force comes from the difference of the free energy between the amorphous phase and the crystalline phase (ΔG_v), as shown in Figure 8.2 [2, 23, 31]. This diagram is greatly helpful to figure out the fabrication condition of optimizing the nucleation and final microstructure of the thin film. In practice, the nucleation can take place at the locations of interface, surface of the film, or bulk in the film once the growth condition is proper.

The nucleation barriers can be overcome by thermal energy which is induced by a heat process during the structure crystallization. Due to the larger surface area in the amorphous phase, the free energy is greater than the supercooled equilibrium liquid, and the residual hydroxyl induces excess free volume to make contributions to free energy. From the theory of crystal nucleation and growth, the energy barriers for homogeneous and heterogeneous nucleation can be described by [2]

$$\Delta G^*_{\text{homo}} = \frac{16\pi\gamma^3}{3(\Delta G_v)^2}$$

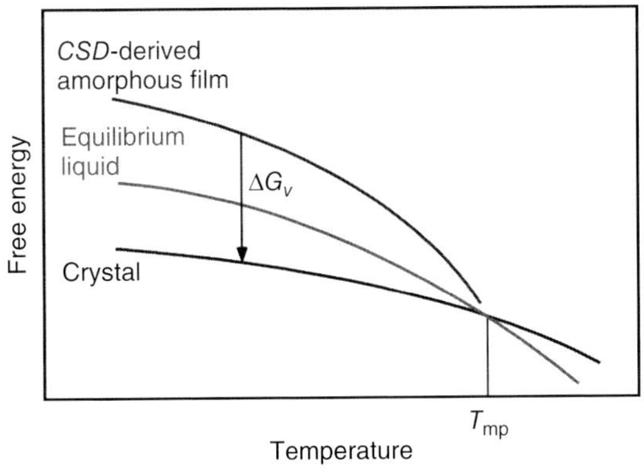

Figure 8.2 Schematic diagram of the free energies of a CSD-derived amorphous film.

and

$$\Delta G^*_{\text{hetero}} = \frac{16\pi\gamma^3}{3(\Delta G_v)^2} f(\theta)$$

where γ is the interfacial energy, ΔG_v is the free energy for a unit volume transformed from amorphous phase to crystalline phase, and $f(\theta)$ is the function of the contact angle, which can be described by the formula

$$f(\theta) = \frac{2 - 3\cos\theta + (\cos\theta)^3}{4}$$

From these equations, the barrier heights for nucleation of thin films are determined by the surface energy, the contact angle with the substrates, and the driving force (free energy difference) for crystallization.

To fully understand the effects of thermodynamic driving force, the factors of the free energy in amorphous/crystalline states, the annealing temperature, and the transformation pathway should be carefully taken into consideration. In practice, it is almost impossible to estimate the real free energy of the amorphous state. The conclusions for crystallization behavior based on the basic nucleation theory are revealed as follows:

1. Due to the $f(\theta)$ term in the equation, interface nucleation is with a lower energy barrier. So, the nucleation will take place at the interface at first. When the crystallization driving force gradually increases, bulk nucleation will occur in addition to the interface nucleation. When the crystallization driving force is very high, it would have sufficient energy to overcome all the energy barriers for nucleation.
2. When the driving force for forming films from amorphous to crystalline is rather low, the differences in energy barrier heights for nucleation events would be more obvious. In this case, the morphology of substrate would be of great influence on thin film microstructure.
3. It is necessary to clarify the roles of crystallization temperature on the nucleation and driving force. When the heating rate is high, the processes

of crystallization and densification of the film would occur at a higher temperature. Under this condition, crystallization needs lower driving forces, and heterogeneous nucleation events become dominated.

4. The crystalline temperature should be below the melting point of the thin film.
5. Usually, the film crystallization begins during heating process. When the temperature of the sample increases to a certain value, thermal energy is high enough to overcome the barriers for nucleation. The microstructures of the film is determined by the stage in the growth process which is with more nucleation points.
6. The intermediate phases also have a key effect on nucleation and growth behavior.

8.5.2 Epitaxial Thin Film Growth

To grow high-quality epitaxial thin films, control of the nucleation at the substrate/film interface is important. In practice, lattice matching with the substrate, thermal processing parameters, and material chemistry are key factors for the growth of textured or epitaxial films. In fact, to produce high-quality oriented or epitaxial films, crystallization of the thin film should be initiated from the heterogeneous nucleation at the interface. For example, rapid thermal processing can delay nucleation to a higher temperature, which is beneficial to prepare high-quality thin films [1, 32–36]. As shown in Figure 8.2, the nucleation at higher temperatures will decrease the driving force of crystallization process. As a result, the heterogeneous nucleation at the interface between the film and the substrate becomes the dominant role in defining the microstructure of the film. Specially, once the nucleation on the substrate interface occurs, the crystallization process will continue along a certain direction in the lattice with the lowest nucleation energy, which can result in oriented films. In this case, the phase transformation from amorphous to crystalline starts from the grains nucleated at the interface between the film and the substrate. In addition, epitaxial films could also be obtained by nucleation occurring at both the interface and the bulk of the film. The role of substrate crystal chemistry in the deposition of oriented films is critical. If the lattice of substrate highly matches with that of the thin film, the $f(\theta)$ term decreases, which makes the heterogeneous nucleation more preferred than other kinds of nucleation. Apart from the effects of the substrate, the role of the precursor chemistry on both microstructural evolution and orientation selection also needs to be investigated. In the case of PZT thin films grown on Pt/Si substrate, the fine FTIR analysis was performed to monitor the decomposition of the precursor wet thin film during the heating process [37]. From the results, it can be concluded that the pyrolyzed processes have obvious impacts on the growth orientation of the films: (i) if there are residual organics in the film, a highly (111) orientated thin film will be produced, and (ii) if there is a high density of OH bonds before the thin film crystallization, the film will be dominated by a mixture of (111)/(100) orientations.

In principle, the epitaxial grain growth should have identical structures and small lattice mismatch between the film and the substrate. Depending on structural differences and lattice matching between the thin film and substrate, the deposition of epitaxial films can be described as follows:

1. If the thin film is well lattice-matched to the substrate with a few percentages of lattice mismatch, the prepared epitaxial films may involve nanometer-sized grains at the film/substrate interface [32, 38–42]. This kind of grain grows along the interface, and they are always randomly oriented to destroy the epitaxial structure. And single-crystal islands may form instead of continuous epitaxial thin films in the case of preparing the ultrathin films due to the microstructural instability phenomenon [36, 43, 44].
2. If the thin film and the substrate have different crystal structures and large lattice mismatches, the epitaxial growth of the film can still be possible. Usually, it can be achieved by a two-step process. First, the polycrystalline film is fabricated, and then it is heated to form isolated grains with a low interfacial energy. Then, another layer is deposited onto the surface, and the isolated grains act as nucleation seeds for the growth of a highly oriented film [45, 46]. However, the role of isolated grains is still unclear. One of the roles is to improve the contact angle and make the nucleation following the seed layer to be preferred. Another is to produce lattice matching in a gradient [47–49].
3. If the processing temperature is higher than the melting point, epitaxial films may also be achieved in a liquid phase [50].
4. If the solute concentrations of the precursor solution are low, the material is likely to follow a layer growth model during the annealing. In this case, heterogeneous interface can be achieved, and epitaxial growth of the film would occur [51].

8.6 Examples of Nanostructural Thin Films Prepared by CSD Techniques

8.6.1 Sol–Gel-Delivered Nanostructured Materials

The sol–gel technique has been widely used in the preparation of nanostructured materials. The obvious advantage of this method includes its low cost, easy setup, and convenience in preparing composite materials. For example, nanostructural TiO_2, an important photocatalytic material, has been widely produced by the sol–gel method. The typical sol–gel route for TiO_2 is by the hydrolysis and condensation reaction of titanium tetraisopropoxide in acidic aqueous solution. Technically, the solution is formed by 5% of titanium tetraisopropoxide, 5% acetic acid, 1.4% hydrochloric acid, and water. The TiO_2 nanopowder can be formed by heating at 60 °C under vigorous stirring [52]. Usually, ultrasonic process is used to accelerate the hydrolysis and condensation reactions. Such extreme conditions in a small reactor would induce great changes in the morphology of TiO_2

nanostructures [53]. Alternately, by adding a second solution of $HAuCl_4$ into the alcoholic (isopropanol) solution of titanium isopropoxide, the Au/TiO_2 nanostructure can be delivered, which can be considered as the source of surface plasma for enhancing the optical properties [54]. Nano-B_2O_3/TiO_2 composites can also be prepared by adding boric acid into a solution which contains ethanol and titanium *tert*-butoxide. Then, the mixture of HCl, ethanol, and water was added to start the chemical reactions during the mixture refluxing. After being stirred and dried for 24 h, the material is heated at about 450 °C to get nano-B_2O_3/TiO_2 composite [55].

8.6.2 MOD of Nanostructured Materials

The MOD method has also been reported to grow various nanomaterials. For example, the solvothermal method can form pure anatase nano-TiO_2, which is considered as the most effective phase for light absorption. The route is as follows: 25 g titanium *n*-butoxide is added in 100 ml of toluene and then the solution is placed in an autoclave. The autoclave is filled with nitrogen and is heated up to 573 K for 2 h. The final obtained TiO_2 nanopowder needs to be washed by methanol and dried in air [56].

Another example is the nano-sized Bi/TiO_2 composites prepared from the mixture of titanium tetraisopropoxide and bismuth(III) chloride. They are mixed with ethyl alcohol and put in an autoclave heated at 200 °C. The molar ratio of Bi:Ti is 1 : 10. During the thermal treatment, Ti and Bi are hydrolyzed, and then the crystallized nano-sized Bi/TiO_2 composites occur. The resulting powder also needs to be washed by distilled water and dried. Finally, these samples are annealed at 500 °C for 3 h to remove impurities [57].

8.6.3 PAD-Delivered Nanostructured Materials

According to the previous reports, the PAD technique can be employed to grow nanoscale high-quality epitaxial thin films. Compared to other conventional CSD techniques, PAD uses water-soluble polymers in the solution. This route can prevent the metal ions from unwanted hydrolysis reactions by binding them to the main polymer chain. In addition, noncoordinated cations and anions are removed by ultrafiltration. Thus, the stoichiometry of each metal ion can be strictly controlled. In this way, the chemistry of the precursor solution during the thin film growth is very simple, and possible intermediate phase can be avoided. For example, the $BaTiO_3$ thin films were prepared by PAD. The X-ray diffraction (XRD) results for the films prepared from precursor solutions without and with filtration are compared, as shown in Figure 8.3. Both the films were annealed at the same temperature of 900 °C. The XRD pattern of the film prepared from the precursor solution without filtration shows weak diffraction peaks from unknown phases and orientations, which are possibly from rutile or anatase

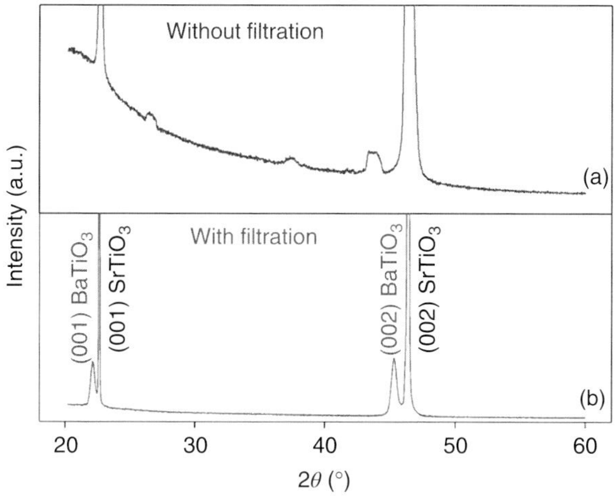

Figure 8.3 X-ray diffraction patterns of converted films on $SrTiO_3$ substrates. (a) Without and (b) with filtration of the solutions. (McCleskey *et al.* [19]. Reproduced with permission from the Royal Society of Chemistry.)

TiO_2, BaO_2, and so on. This is possible because the chemical reaction of unbind individual metal ions starts at the beginning of thermal treatment. However, the XRD pattern from the film produced from the precursor solution with filtration shows clearly only (00l) diffraction peaks. It is indicated that the film is not only *c*-axis oriented but also pure phase without any other intermediate phases.

The PAD technique has been widely used for the epitaxial thin film deposition. Lin Y *et al.* prepared binary Eu_2O_3 and ternary ferroelectric $Ba_{1-x}Sr_xTiO_3$ (BST) thin films on the $LaAlO_3$ substrate, respectively [58, 59]. The transmission electron microscope (TEM) images shown in Figure 8.4 reveal that the epitaxial relationship between the film and the substrate is Eu_2O_3(004)//LAO(001) and Eu_2O_3(440)//LAO(020). As for the BST thin films grown by PAD, the dielectric tunability can reach 69% at the frequency of 1 MHz, which is as good as the one prepared by pulsed laser deposition. Other magnetic functional thin films such as $La_{0.67}Sr_{0.33}MnO_3$ (LSMO) and $La_{0.67}Ca_{0.33}MnO_3$ (LCMO) thin films were fabricated by Jain *et al.* [60, 61]. The magnetic resistances for LSMO and LCMO are 50 and 88%, respectively. Recently, Y.D. Ji *et al.* employed the PAD technique to grow the vanadium oxides [62]. Due to the different chemical valences of V, the family of vanadium oxides is very complex for different kinds of their oxide phases. By carefully controlling the anneal temperature and atmosphere, high-quality V_2O_3, VO_2,V_4O_7, and V_2O_5 thin films were prepared as shown in Figure 8.5. H.M. Luo *et al.* also made great efforts to prepare nitride thin films using the PAD technique [63, 64]. The ultrathin epitaxial NbN films with lattice mismatch of 12.6% on the $SrTiO_3$ substrate were prepared, with the superconducting transition temperature of 14 K. The epitaxial complex ternary nitride thin film of $SrTiN_2$ with good physical properties was also grown on the $LaAlO_3$ substrate [65]. The epitaxial relationships can be described as $SrTiN_2$ (001)//$LaAlO_3$ (001) and $SrTiN_2$

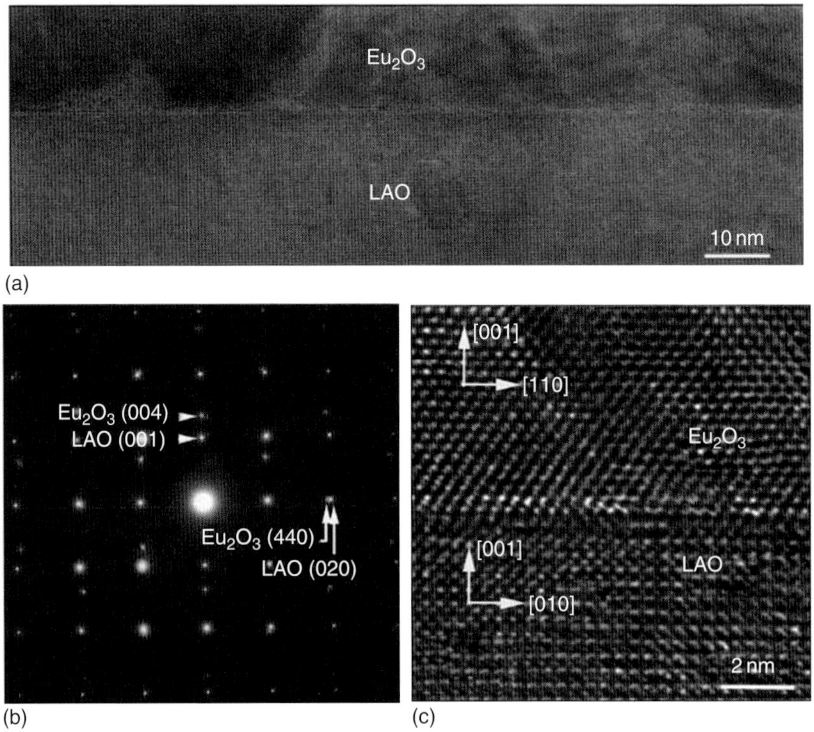

Figure 8.4 TEM (a), SAED (b), and the corresponding HRTEM image (c) of the Eu_2O_3 thin film on the $LaAlO_3$ substrate. (Zou *et al.* [18]. Reproduced with permission from the Royal Society of Chemistry.)

[102]//$LaAlO_3$ [101]. G.F. Zou *et al.* fabricated the Ge epitaxial thin film on Si, which opens a new route for the fabrication of Ge/Si heterojunctions [66]. They also developed the epitaxial carbide thin films by PAD [67], such as NbC thin film with the superconductivity transition temperature of 10 K and the hardness of 19.54 GPa [68]. These results proved that the PAD technique is a feasible CSD technique to fabricate high-quality epitaxial thin films.

The PAD also takes advantage in producing the nanostructural thin films by its good controllability of the thermal dynamics in the growth stages as demonstrated earlier. One example is the composite thin film with embedded amorphous SiO_2 nanoparticles in epitaxial $SrTiO_3$ and $CoFe_2O_4$ matrices prepared by the PAD technique, as shown in Figure 8.6. SiO_2 nanoparticles bind to PEI by a surfactant-like interaction of the polymer with the surface of the nanoparticles. Thus, it is easy to form the final precursor solution of the composite by simply mixing the separate polymeric solutions of metal ions and SiO_2 nanoparticles. Specifically, the Sr binds to PEI polymer as an EDTA complex using $Sr(NO_3)_2$ as the metal precursor. Ti binds to carboxylated polyethyleneimine (PEIC) with $TiCl_4$ as the precursor. Co and Fe bind to PEI using $CoCl_2$ and $FeCl_3$ as precursors. Ultrafiltration and concentration were carried out using Amicon stirred cells.

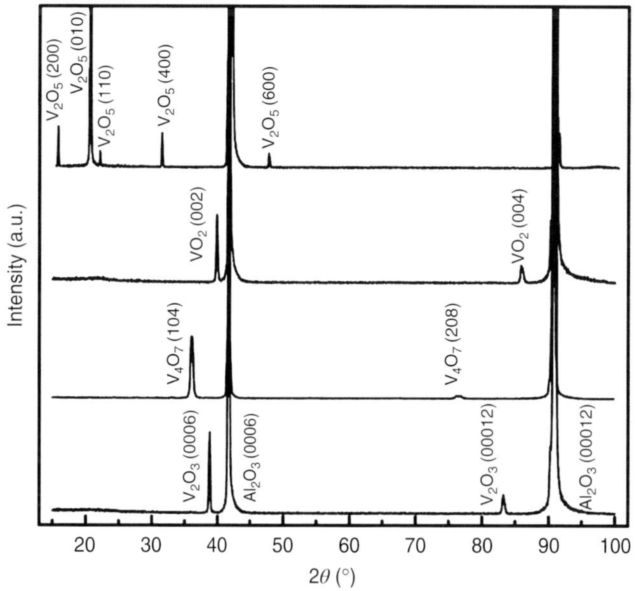

Figure 8.5 XRD patterns from θ to 2θ scans of vanadium oxide films prepared under the optimal conditions for V_2O_3, V_4O_7, VO_2, and V_2O_5, respectively.

The concentrations of the metal ions were measured by ICP-AES. Then the final solution was mixed according to the stoichiometric molar ratios. The resulting precursor solutions were spin-coated onto the $LaAlO_3$ (LAO) substrates. Then, the precursor SiO_2–STO and SiO_2–CFO films were annealed in flowing oxygen for 2 h at 860 and 950 °C, respectively. It is demonstrated that the ferrimagnetic properties of the as-prepared SiO_2–CFO films can be tuned by the amount of embedded SiO_2 nanoparticles. Moreover, as is known, the silica can be removed by HF or NaOH. In this way, it is a reasonable technique for the fabrication of porous epitaxial oxide thin films [69].

The carbon nanotubes (CNTs) and NbC composite thin films have also been prepared by the PAD technique through thermal reaction under controlled environment. As-synthesized CNT forests were soaked by the Nb precursor solution. After the solution was permeated into CNT forests, the CNT forest was detached from the Si wafer simply by peeling off the CNT forest from the Si substrate using a tweezer. The CNT forests with Nb precursor solution were heated to 650 °C in the mixture gases of ethylene (10 sccm) and forming gas (10 sccm). Then, the sample was annealed in such an environment for 1 h. The mixture gases were switched to Ar (10 sccm) while the furnace was ramped to 1000 °C in 1 h. Then, the sample was annealed at this temperature in Ar for 3 h. Finally, the sample was naturally cooled to room temperature [70]. In this composite, the highly oriented CNTs embedded in NbC matrix effectively enhance the superconducting properties of the NbC.

Another example is the fabrication of self-assemble nanocomposite oxide thin films. By simply mixing the precursors of the two materials which cannot form

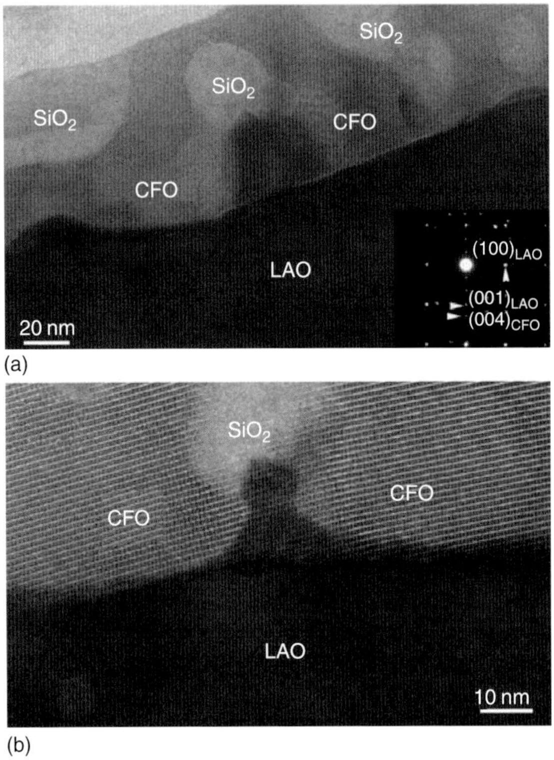

Figure 8.6 (a,b) Cross-sectional HRTEM images of an SiO_2–CFO (30 : 70) composite film on an LAO substrate and the corresponding SAED pattern (inset) taken from the interface between the film and the substrate. Luo *et al.* [69]. Reproduced with permission from Wiley.

solid solutions, such as $BaTiO_3$ and $NiFe_2O_4$, self-assembled nanocomposite can be formed. The surface energy of the materials is found to play a critical role in determining the microstructures of the composite. A typical example is the fabrication of the multiferroic composite thin films. The self-assembled two-phase nanocomposite films, which combine perovskite ferroelectrics (such as $BaTiO_3$, $Pb(Zr,Ti)O_3$, and $BiFeO_3$) with spinel ferromagnetic materials (such as $CoFe_2O_4$ and $NiFe_2O_4$), show enormous magnetoelectric coefficient at room temperature. This may have a great potential of applications due to the unique magnetoelectric coupling effect. As reported by Yan *et al.*, to grow the $BaTiO_3$–$CoFe_2O_4$ composite films using the PAD technique, four separate solutions with Ba, Ti, Co, and Fe bound to polymers were prepared, respectively [71]. Ba and Fe were bound as an EDTA complex to PEI. Co was bound to PEI directly. The Ti solution was made by adding a mixture of 30% peroxide and titanium chloride in deionized water to the solution mixed with PEI and EDTA. The solutions were all purified and concentrated in an Amicon filtration unit, respectively. Inductively coupled plasma optical emission spectrometer (ICP-OES) was used to check the final concentrations of Ba, Ti, Co, and Fe ions in the four solutions, respectively. These solutions were then mixed to the final solution with the molar ratio of Ba:Ti:Co:Fe = 1 : 1 : 1 : 2. The resulting solution was spin-coated on the LAO substrates. The thermal treatment was then performed with oxygen. The polymer was pyrolyzed by initially slowly heating it to 550 °C at the rate of 1 – 5 °C/min. Samples were then annealed by heating at 900 °C for 1 h to induce crystallization. XRD characterizations suggested that the $BaTiO_3$ thin film is epitaxial while $CoFe_2O_4$ is *c*-axis oriented. Magnetic force microscopy (MFM) was used to check the magnetic response from $CoFe_2O_4$. The magnetic domains can be observed, and it reveals that $CoFe_2O_4$ phase is dispersed pretty uniformly in the films. The cross-sectional TEM has been used to check the microstructure of the composite. It is observed that the $BaTiO_3$ phase is epitaxial to the substrate and form the matrix in the composite. $CoFe_2O_4$ nanoparticles are embedded in the $BaTiO_3$ matrix. Similarly, $BaTiO_3-NiFe_2O_4$ nanocomposite was fabricated using the PAD technique by Luo *et al.* [72]. However, in contrast to the $BaTiO_3-CoFe_2O_4$ system, the $BaTiO_3$ is embedded in the spinel $NiFe_2O_4$ matrix. By carefully comparing the properties of $CoFe_2O_4$, $BaTiO_3$, and $NiFe_2O_4$, the surface energy is believed to be the main reason to explain the differences. It is found that the surface energy of (001) $BaTiO_3$ ($1.26\,J/m^2$) is larger than (001) $NiFe_2O_4$ ($1.161\,J/m^2$) but smaller than (001) $CoFe_2O_4$ ($1.486\,J/m^2$). The material which has a lower surface energy would like to wet the one with a higher surface energy, forming a matrix surrounding the latter one. As a result, $BaTiO_3$ is the matrix in the $BaTiO_3-CoFe_2O_4$ system, while $NiFe_2O_4$ is the matrix in the $BaTiO_3-NiFe_2O_4$ system.

Nanostructural thin films can also be induced by the surface pattern of the substrates. For example, the epitaxial anatase TiO_2 film with nanometer-scaled periodic surface strip patterns can be produced on the (001) LAO substrate when the sample is annealed at 900 °C and with a proper temperature ramping rate, as shown in Figure 8.7. It is found that the ramping rate of the temperature during annealing is an important factor to determine the morphology of the surface pattern. The formation of the regular surface strip pattern of the anatase TiO_2 thin films can be considered to be somewhat associated with the diffusion limit growth dynamics and the surface step terrace features of the LAO substrates. There are two different types of surface terminations, namely, La–O and Al–O step terraces, on a (001) $LaAlO_3$ substrate. The existing potentials at the edges of the step terraces, usually referring to the Schwoebel potentials or barriers, will generate bias in the diffusion process during the nucleation of the films on the step terraces. When the diffusing species (either atoms or nanocrystal grains) diffuse to the ascending steps, they will stick and bond with the steps during the annealing processes. On the other hand, the diffusing species will be reflected back by the Schwoebel potentials when the species are diffusing toward the edges of the descending steps. As a result, as the sample grows with a high temperature ramping rate, the diffusing species either directly diffuse to the surface step terraces or are bounded by the Schwoebel potential at the step terraces to diffuse and be caught by the ascending step edges. However, when the sample grows with a lower temperature ramping rate, the sample will stay at the temperature lower than the

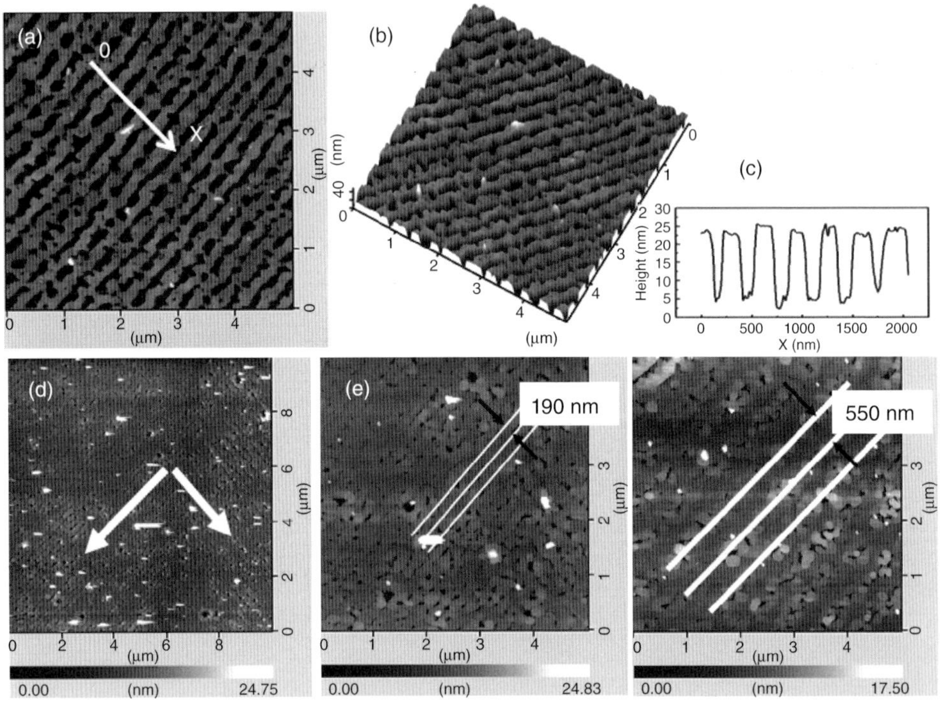

Figure 8.7 (a) The AFM image of sample TiO_2 thin films prepared at 900 °C; (b) 3D image of (a); (c) the profile along the OX line in (a); (d) perpendicular strips on the surface of the film; and (e,f) strips with different widths on the same sample.

final annealing temperature (900 °C) for a longer time. Thus, during the ramping process, small diffusing species have a short diffusing path and easily coalesce and form into large-sized islands on the surface step terraces before they encounter the step edges. These large-sized islands are randomly distributed on the step terraces and need high energy to diffuse to the edge areas. Hence, the high ramping rate will result in the clearly defined mesa–valley structure, whereas the low ramping rate will make such a structure blurred. Moreover, since there are plenty of twinning structures in the $LaAlO_3$ crystals, the mesa–valley strips will also follow the twinning features to form the near perpendicular strips. The different strip widths observed on the same sample may result from the different step widths on the twinning surfaces of the LAO substrate. As proved by the experiments, such a surface pattern structure will strongly affect the change of hydrophilic properties of the anatase thin film before and after UV illumination [73].

8.7 Summary

In summary, the CSD techniques have been widely used for the fabrication of nanostructural thin films. Especially, some recently developed CSD methods such as PAD have shown great feasibility in the preparation of high-quality nanostructural epitaxial thin films.

References

1. Lange, F.F. (1996) *Science*, **273**, 903.
2. Schwartz, R.W. (1997) *Chem. Mater.*, **9**, 2325.
3. Schwartz, R.W., Schneller, T., and Waser, R. (2004) *C.R. Chim.*, **7**, 433.
4. Bhuiyan, M.S., Paranthaman, M., and Salama, K. (2006) *Supercond. Sci. Technol.*, **19**, R1.
5. Wang, D., Xie, T., and Li, Y. (2009) *Nano Res.*, **2**, 30.
6. Lavalle, P., Voegel, J.-C., Vautier, D., Senger, B., Schaaf, P., and Ball, V. (2011) *Adv. Mater.*, **23**, 1191.
7. Pawar, S.M., Pawar, B.S., Kim, J.H., Joo, O.-S., and Lokhande, C.D. (2011) *Curr. Appl Phys.*, **11**, 117.
8. Maiwa, H., Iizawa, N., Togawa, D., Hayashi, T., Sakamoto, W., Yamada, M., and Hirano, S. (2003) *Appl. Phys. Lett.*, **82**, 1760.
9. Hayashi, T., Ohji, N., Hirohara, K., Fukunaga, T., and Maiwa, H. (1993) *Jpn. J. Appl. Phys.*, **32**, 4092.
10. Kakihana, M. (1996) *J. Sol-Gel Sci. Technol.*, **6**, 7.
11. Khelifa, F., Druart, M.-E., Habibi, Y., Benard, F., Leclere, P., Olivier, M., and Dubois, P. (2013) *Prog. Org. Coat.*, **76**, 900.
12. Klee, M., Eusemann, R., Waser, R., Brand, W., and Vanhal, H. (1992) *J. Appl. Phys.*, **72**, 1566.
13. Ousibenomar, W., Xue, S.S., Lessard, R.A., Singh, A., Wu, Z.L., and Kuo, P.K. (1994) *J. Mater. Res.*, **9**, 970.
14. Joshi, V., Dacruz, C.P., Cuchiaro, J.D., Araujo, C.A., and Zuleeg, R. (1997) *Integr. Ferroelectr.*, **14**, 133.
15. Liu, M.L. and Wang, D.S. (1995) *J. Mater. Res.*, **10**, 3210.
16. Jia, Q.X., McCleskey, T.M., Burrell, A.K., Lin, Y., Collis, G.E., Wang, H., Li, A.D.Q., and Foltyn, S.R. (2004) *Nat. Mater.*, **3**, 529.
17. Burrell, A.K., McCleskey, T.M., and Jia, Q.X. (2008) *Chem. Commun.*, 1271.
18. Zou, G.F., Zhao, J., Luo, H.M., McCleskey, T.M., Burrell, A.K., and Jia, Q.X. (2013) *Chem. Soc. Rev.*, **42**, 439.
19. McCleskey, T.M., Shi, P., Bauer, E., Highland, M.J., Eastman, J.A., Bi, Z.X., Fuoss, P.H., Baldo, P.M., Ren, W., Scott, B.L., Burrell, A.K., and Jia, Q.X. (2014) *Chem. Soc. Rev.*, **43**, 2141.
20. Birnie, D.P. (2001) *J. Mater. Res.*, **16**, 1145.
21. Haas, D.E., Birnie, D.P., Zecchino, M.J., and Figueroa, J.T. (2001) *J. Mater. Sci. Lett.*, **20**, 1763.
22. Hungerford, G., Pereira, M.R., Ferreira, J.A., Viseu, T.M.R., Coelho, A.F., Isabel, M., Ferreira, C., and Suhling, K. (2002) *J. Fluoresc.*, **12**, 397.
23. Schwartz, R.W., Voigt, J.A., Tuttle, B.A., Payne, D.A., Reichert, T.L., and DaSalla, R.S. (1997) *J. Mater. Res.*, **12**, 444.
24. Hoffmann, S., Klee, M., and Waser, R. (1995) *Integr. Ferroelectr.*, **10**, 155.
25. Garino, T.J. and Harrington, M. (1991) *MRS Online Proc. Lib.*, **243**.
26. Reaney, I.M., Brooks, K., Klissurska, R., Pawlaczyk, C., and Setter, N. (1994) *J. Am. Ceram. Soc.*, **77**, 1209.
27. Griswold, E.M., Weaver, L., Sayer, M., and Calder, I.D. (1995) *J. Mater. Res.*, **10**, 3149.
28. Wilkinson, A.P., Speck, J.S., Cheetham, A.K., Natarajan, S., and Thomas, J.M. (1994) *Chem. Mater.*, **6**, 750.
29. Keddie, J.L. and Giannelis, E.P. (1991) *J. Am. Ceram. Soc.*, **74**, 2669.
30. Brinker, C.J., Drotning, W.D., and Scherer, G.W. (1984) *MRS Online Proc. Lib.*, **32**.
31. Roy, R. (1969) *J. Am. Ceram. Soc.*, **52**, 344.
32. Seifert, A., Lange, F.F., and Speck, J.S. (1995) *J. Mater. Res.*, **10**, 680.
33. Miller, K.T. and Lange, F.F. (1991) *J. Mater. Res.*, **6**, 2387.
34. Seifert, A., Lange, F.F., and Speck, J.S. (1993) *J. Am. Ceram. Soc.*, **76**, 443.
35. Waser, R.M. (1996) *Curr. Opin. Solid State Mater. Sci.*, **1**, 706.

36. Langjahr, P.A., Wagner, T., RÜhle, M., and Lange, F.F. (1995) *MRS Online Proc. Lib.*, **401**.
37. Fè, L., Norga, G.J., Wouters, D.J., Maes, H.E., and Maes, G. (2001) *J. Mater. Res.*, **16**, 2499.
38. Lakeman, C.D.E., Xia, Y., Kim, J.-H., Wu, X., Eckert, H.G., and Lange, F.F. (1998) *J. Mater. Res.*, **13**, 1596.
39. Kodaira, T., Nishio, K., Yamaguchi, I., Suzuki, S., Tsukada, K., and Tsuchiya, T. (2003) *J. Sol-Gel Sci. Technol.*, **26**, 1049.
40. Bae, S.Y. and Wang, S.X. (1996) *Appl. Phys. Lett.*, **69**, 121.
41. Manabe, T., Yamaguchi, I., Kondo, W., Mizuta, S., and Kumagai, T. (1997) *J. Mater. Res.*, **12**, 541.
42. Hasenkox, U., Mitze, C., Waser, R., Arons, R.R., Pommer, J., and Güntherodt, G. (1999) *J. Electroceram.*, **3**, 255.
43. Seifert, A., Vojta, A., Speck, J.S., and Lange, F.F. (1996) *J. Mater. Res.*, **11**, 1470.
44. Langjahr, P.A., Wagner, T., Rühle, M., and Lange, F.F. (1999) *J. Mater. Res.*, **14**, 2945.
45. Shimizu, M., Okaniwa, M., Fujisawa, H., and Niu, H. (2002) *Jpn. J. Appl. Phys.*, **41**, 6686.
46. Kwok, C.K. and Desu, S.B. (1993) *J. Mater. Res.*, **8**, 339.
47. Neumayer, D.A., Duncombe, P.R., Laibowitz, R.B., Saenger, K.L., Purtell, R., Ott, J.A., Shaw, T.M., and Grill, A. (1997) *Integr. Ferroelectr.*, **18**, 319.
48. Schwartz, R.W., Clem, P.G., Voigt, J.A., Byhoff, E.R., Van Stry, M., Headley, T.J., and Missert, N.A. (1999) *J. Am. Ceram. Soc.*, **82**, 2359.
49. Muralt, P., Maeder, T., Sagalowicz, L., Hiboux, S., Scalese, S., Naumovic, D., Agostino, R.G., Xanthopoulos, N., Mathieu, H.J., Patthey, L., and Bullock, E.L. (1998) *J. Appl. Phys.*, **83**, 3835.
50. Golden, S.J., Lange, F.F., Clarke, D.R., Chang, L.D., and Necker, C.T. (1992) *Appl. Phys. Lett.*, **61**, 351.
51. Hoffmann, S., Hasenkox, U., Waser, R., Jia, C.L., and Urban, K. (1997) *MRS Online Proc. Lib.*, **474**.
52. Tung, W.S. and Daoud, W.A. (2008) *J. Colloid Interface Sci.*, **326**, 283.
53. Suslick, K.S. (1990) *Science*, **247**, 1439.
54. Uddin, M.J., Cesano, F., Scarano, D., Bonino, F., Agostini, G., Spoto, G., Bordiga, S., and Zecchina, A. (2008) *J. Photochem. Photobiol., A*, **199**, 64.
55. Kalfa, O.M., Yalçınkaya, Ö., and Türker, A.R. (2009) *J. Hazard. Mater.*, **166**, 455.
56. Supphasrirongjaroen, P., Praserthdam, P., Panpranot, J., Na-Ranong, D., and Mekasuwandumrong, O. (2008) *Chem. Eng. J.*, **138**, 622.
57. Kang, M., Ko, Y.-R., Jeon, M.-K., Lee, S.-C., Choung, S.-J., Park, J.-Y., Kim, S., and Choi, S.-H. (2005) *J. Photochem. Photobiol., A*, **173**, 128.
58. Lin, Y., Lee, J.S., Wang, H., Li, Y., Foltyn, S.R., Jia, Q.X., Collis, G.E., Burrell, A.K., and McCleskey, T.M. (2004) *Appl. Phys. Lett.*, **85**, 5007.
59. Lin, Y., Wang, H., Hawley, M.E., Foltyn, S.R., Jia, Q.X., Collis, G.E., Burrell, A.K., and McCleskey, T.M. (2004) *Appl. Phys. Lett.*, **85**, 3426.
60. Jain, M., Shukla, P., Li, Y., Hundley, M.F., Wang, H., Foltyn, S.R., Burrell, A.K., McCleskey, T.M., and Jia, Q. (2006) *Adv. Mater.*, **18**, 2695.
61. Jain, M., Li, Y., Hundley, M.F., Hawley, M., Maiorov, B., Campbell, I.H., Civale, L., and Jia, Q.X. (2006) *Appl. Phys. Lett.*, **88**, 232510.
62. Ji, Y.D., Pan, T.S., Bi, Z., Liang, W.Z., Zhang, Y., Zeng, H.Z., Wen, Q.Y., Zhang, H.W., Chen, C.L., Jia, Q.X., and Lin, Y. (2012) *Appl. Phys. Lett.*, **101**, 071902.
63. Luo, H., Lin, Y., Wang, H., Chou, C.-Y., Suvorova, N.A., Hawley, M.E., Mueller, A.H., Ronning, F., Bauer, E., Burrell, A.K., McCleskey, T.M., and Jia, Q.X. (2008) *J. Phys. Chem. C*, **112**, 20535.
64. Luo, H., Zou, G., Wang, H., Lee, J.H., Lin, Y., Peng, H., Lin, Q., Deng, S., Bauer, E., McCleskey, T.M., Burrell, A.K., and Jia, Q. (2011) *J. Phys. Chem. C*, **115**, 17880.
65. Luo, H., Wang, H., Bi, Z., Feldmann, D.M., Wang, Y., Burrell, A.K., McCleskey, T.M., Bauer, E., Hawley, M.E., and Jia, Q. (2008) *J. Am. Chem. Soc.*, **130**, 15224.
66. Zou, G., Luo, H., Ronning, F., Sun, B., McCleskey, T.M., Burrell, A.K., Bauer, E., and Jia, Q.X. (2010) *Angew. Chem. Int. Ed.*, **49**, 1782.

67. Zou, G., Wang, H., Mara, N., Luo, H., Li, N., Di, Z., Bauer, E., Wang, Y., McCleskey, T., Burrell, A., Zhang, X., Nastasi, M., and Jia, Q. (2010) *J. Am. Chem. Soc.*, **132**, 2516.
68. Zou, G., Luo, H., Zhang, Y., Xiong, J., Wei, Q., Zhuo, M., Zhai, J., Wang, H., Williams, D., Li, N., Bauer, E., Zhang, X., McCleskey, T.M., Li, Y., Burrell, A.K., and Jia, Q.X. (2010) *Chem. Commun.*, **46**, 7837.
69. Luo, H., Lin, Y., Wang, H., Baily, S.A., Lee, J.H., Hawley, M.E., McCleskey, T.M., Burrell, A.K., Bauer, E., Civale, L., and Jia, Q.X. (2008) *Angew. Chem. Int. Ed.*, **47**, 5768.
70. Zou, G.F., Luo, H.M., Baily, S., Zhang, Y.Y., Haberkorn, N.F., Xiong, J., Bauer, E., McCleskey, T.M., Burrell, A.K., Civale, L., Zhu, Y.T., MacManus-Driscoll, J.L., and Jia, Q.X. (2011) *Nat. Commun.*, **2**, 428.
71. Yan, L.H., Liang, W.Z., Liu, S.H., Huang, W., and Lin, Y. (2011) *Integr. Ferroelectr.*, **131**, 82.
72. Luo, H., Yang, H., Bally, S.A., Ugurlu, O., Jain, M., Hawley, M.E., McCleskey, T.M., Burrell, A.K., Bauer, E., Civale, L., Holesinger, T.G., and Jia, Q. (2007) *J. Am. Chem. Soc.*, **129**, 14132.
73. Lin, Y., Zeng, B., Ji, Y.D., Liang, W.Z., Feng, D.Y., Gao, M., Zhang, Y., Chen, X.X., Chen, B., and Chen, C.L. (2014) *Nanotechnology*, **25**, 014014.

9
Nanomaterial Development Using *In Situ* Liquid Cell Transmission Electron Microscopy

Xin Chen, Wangfan Zhou, Debiao Xie, and Hongliang Cao

9.1
Introduction

Nanomaterial synthesis and deposition usually involve liquid environments. Performing high-resolution *in situ* analysis during the nanomaterial growth procedure is not only important for understanding the fundamental material growth mechanisms but also attractive for developing new nanomaterial fabrication technologies. Electron microscopies, especially transmission electron microscopy (TEM), are promising due to the ultrahigh spatial resolutions and comprehensive characterizing capabilities [1]; however, they must solve the compatibility issue between the liquid samples and the vacuum environments in the electron microscopes.

The recent development of *in situ* liquid cell TEM has made it possible for high-resolution observation of nanomaterial growth under ambient pressure in liquid in real time and has drawn much attention internationally [2]. The number of publications in this field has been increased drastically, and there are already a couple of review papers addressing different aspects of this new technology [3–9], including the review on *in situ* observation of the growth of metal nanoparticles [10]. However, the review of this topic in research books is still relatively limited, and there is still the need to systematically introduce the technological development along with the application in the nanomaterial development.

In this chapter, the technological development of *in situ* liquid cell TEM will be presented, the main nanomaterial development results obtained with each type of *in situ* liquid cell TEM setup will be reviewed, and the future trend of *in situ* liquid cell TEM research will be addressed.

9.2
The Technological Development of *In Situ* Liquid Cell TEM

Viewing liquid samples as easily as solid samples has been a goal since TEM was invented; however, it is nontrivial to solve the incompatibility problem of a liquid

Advanced Nano Deposition Methods, First Edition. Edited by Yuan Lin and Xin Chen.

sample with the TEM vacuum environment [4]. Efforts have been made to study liquids encapsulated in another structure or liquids with low vapor pressure or to develop differentially pumped environmental chambers to study wet samples under a certain pressure [4, 11]. These efforts have achieved certain successfulness but have limitations in sample preparation, sample selection, and environmental pressures. The recently developed *in situ* liquid cell TEM technology, which allows regular liquids to be used, sealed with ease, under ambient pressure, thus opens up broad opportunities for studying liquid samples with TEM [2].

9.2.1 The Advent of the Modern *In Situ* Liquid Cell

In 2003, Ross and her group published a paper in *Nature Materials*, in which a modern *in situ* liquid cell setup was reported [2]. Figure 9.1 shows the schematic illustration and the photograph of the liquid cell [2]. In the setup, two silicon wafers were glued face to face, with a thin layer of liquid introduced in between, and two gold electrodes integrated into the *in situ* cell. The wafers have been coated with Si_3N_4 thin films and selectively etched to form freestanding Si_3N_4 windows in the center of the wafers. The top of Figure 9.1b shows a photograph of the Si_3N_4 window region of the *in situ* liquid cell, with a gold thin film electrode deposited on the bottom Si_3N_4 window surface. After it is assembled, the setup was loaded into a Hitachi H9000 TEM for the *in situ* liquid cell experiment [2].

One major breakthrough of this work is the introduction of the Si_3N_4 windows, which are not only thin enough to allow the electron beam to penetrate through to generate clear images, with minimal scattering, but also robust enough to seal the liquid from leaking into the vacuum chamber. With the development of the thin film microfabrication technology, the Si_3N_4 window on silicon substrates can be routinely made and well commercialized, thus becoming readily available to the general researchers. Another key breakthrough is the integration of metal electrodes into the *in situ* liquid cells, which demonstrated that apart from simply viewing liquid samples with TEM, the technology can be made multifunctional, allowing a broader range of research needs to be addressed.

9.2.2 Recent Technological Development of Liquid Cells

Apart from the aforementioned pioneering design, people have developed various *in situ* liquid cell setups to improve its performance and functionality. Metal *in situ* cell cages have been made using O-rings instead of epoxy to seal the liquid cell [12–14]. This kind of design aims to improve the liquid cell assembly easiness, reliability, and cell reusability. Figure 9.2a shows a schematic of the experimental setup of such a liquid cell. Here, two SiN_x window substrates have been placed in the middle, facing each other, with the liquid sample sandwiched in between, and the metal enclosure and the O-rings complete the cell cage and provide the sealing.

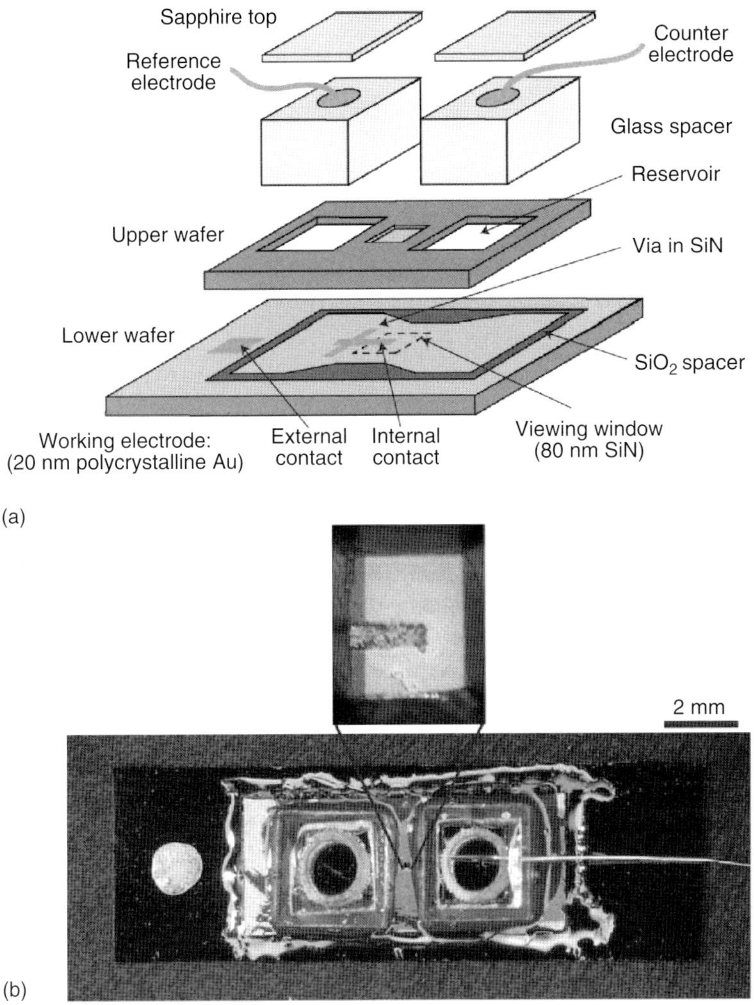

Figure 9.1 (a) Schematic and (b) photograph of an *in situ* liquid cell. The setup was used for observing electrochemical growth of metal nanoclusters in liquid with TEM. (Williamson *et al.* [2]. Reproduced with permission from Nature Publishing Group.)

In this example, the liquid cell was not only used to enclose a liquid sample in it but also had electrodes integrated with it to allow *in situ* electrochemical analysis.

The earlier *in situ* liquid cells only seal a limited amount of fluid samples in them, which results in the difficulty of adjusting the liquid sample composition during the *in situ* experiment, and the precursor depletion problem in many cases [2], so people developed flow cells to solve the problem [15]. Figure 9.2b shows an example of the *in situ* flow cell [15]. In this example, the *in situ* cell was fixed in place with a clamp inside a cartridge on the TEM sample stage, and two flow tubes were glued to the cell. These tubes pass through the center of the TEM stage and

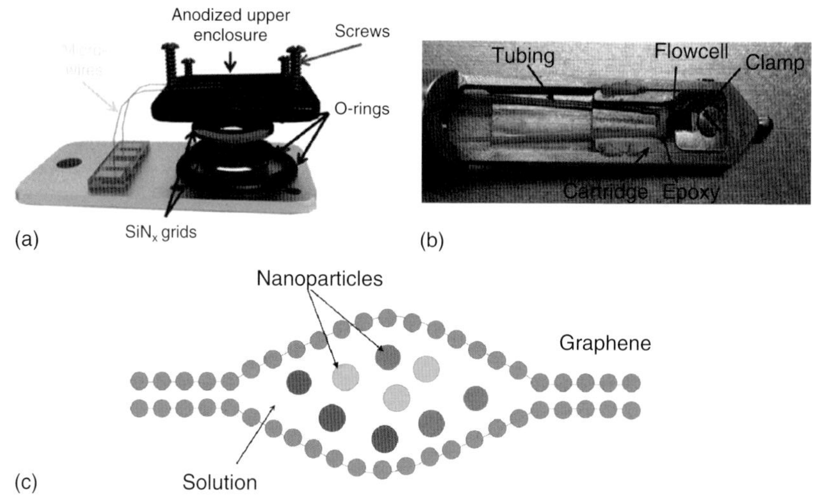

Figure 9.2 (a) Schematic of an O-ring-sealed *in situ* liquid setup, (b) an *in situ* fluid cell setup with flow channels, and (c) schematic of a graphene liquid cell. (Panel (a): Chen *et al.* [14]. Reproduced with permission from Elsevier. Panel (b): de Jonge *et al.* [15]. Reproduced with permission from American Chemical Society.)

connect to the outside equipment, allowing both liquid and gas to pass through the *in situ* cell. The picture shows the closeup of the TEM stage tip.

To study nanomaterial growths and nanomaterial properties in liquid, it is desired to be able to observe them not only at the room temperature but also under elevated or lowered temperatures. One way of doing this is to put mini-sized *in situ* liquid cells onto commercial TEM sample holders with temperature-regulating functions such as an *in situ* heating holder [16, 17]. Such an approach, however, puts limit on *in situ* liquid cell choices and may need to consider the cost issue of purchasing the *in situ* heating holder. Another approach is to design custom *in situ* liquid TEM setups with temperature-regulating functions [18]. Tai *et al.* designed the TEM sample stage which was used to observe liquid sample under elevated and lowered temperatures [18]. In the setup, a copper rod was introduced in the middle of the TEM stage, with the outer end connecting to a heat sink to lower the sample temperature. The inner end of the copper rod, which is located near the liquid cell, was wrapped with Pt metal wires for sample heating. Thermal isolation was made in this custom TEM sample stage to reduce heat leak from the setup to the TEM instrument. A thermal couple was placed near the liquid cell to monitor the sample temperature.

A further need is to push the limit to obtain atomic resolution images in liquid cells. The SiN_x window can scatter electrons and reduce the TEM image resolution; thus, it is important to find alternative window materials of smaller thicknesses and lower electron beam scattering. Graphene provides a material of only one single atomic layer thickness, which was used to develop such *in situ* liquid cells [19–21]. Figure 9.2c shows a schematic of the graphene-based liquid cell. Two layers of graphene were used to form the liquid cell, enclosing a tiny liquid drop in between. Such a graphene-based liquid cell was made on a TEM grid in order to be loaded in to the TEM system for the *in situ* analysis. Through *in situ* TEM, the process of nanocrystal growth has been detected with atomic resolution, and research using this type of liquid cells may become a hot topic in the field [9].

On the other hand, there are still challenges with the graphene-based liquid cells: to avoid electron beam damage to the graphene window, TEM systems that can be operated under acceleration voltages lower than 80 keV are needed, and there is the difficulty of introducing flow pipes and electrode connections with this type of liquid cells, so further works are still needed for the advanced *in situ* liquid TEM setup developments.

9.2.3
Commercial Liquid Cells

Homemade *in situ* liquid cell TEM setups are hardly reachable for the general TEM users; thus the appearance of the commercial products is a good news for the common TEM users who have interest in doing the *in situ* liquid cell TEM research. *In situ* liquid cell systems for TEM have now been produced and manufactured as commercial products, and currently there are two main manufacturers: Protochips Inc. and Hummingbird Scientific.

Protochips Inc. currently sells Poseidon 210 Liquid Flow Cell and Poseidon 510 Electrochemical Cell. Poseidon 210 has two-port and three-port system options, with the two-port system used to flow a liquid sample through the *in situ* liquid cell and the three-port system used for introducing additional chemistries for mixing. A Poseidon 500 series product [22] added electrodes to the *in situ* liquid cell and becomes an *in situ* microfluidic electrochemical STEM/TEM product. Two SiN_x window substrates of different sizes are used to form the liquid cell, and two O-rings are used to seal the *in situ* cell. Electrical contacts and microfluidic channels are integrated into the *in situ* TEM holder. The bigger (upper) SiN_x window substrate has metal electrode thin film layers deposited on it, forming an E-chip, and by pressing onto the electrode pads on the *in situ* TEM holder, the electrodes are connected from the *in situ* liquid cell to the outside testing circuits. There are three electrodes manufactured on the E-chip. Despite the working electrode, the other electrodes are made of polycrystalline platinum which is a stable precious metal. The bottom flowerlike cavity contains fluid channel openings that allow liquid sample to flow through the *in situ* liquid cell.

A set of the photographs of the Poseidon 510 product is shown in Figure 9.3a. Apart from the TEM holder and the E-chips, it also shows a syringe pump and a potentiostat. The syringe pump is used to push liquid sample into the *in situ* TEM holder and through the *in situ* liquid cell, while the potentiostat is used to supply bias and collect electrical signal for the *in situ* electrochemical experiments.

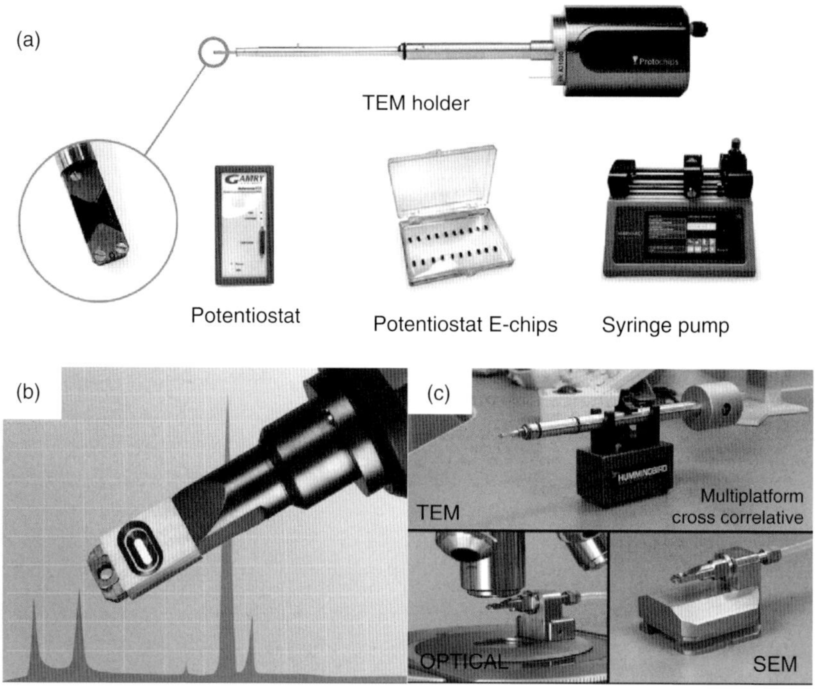

Figure 9.3 (a) Photographs of Poseidon 510 liquid cell with E-chip and accessory produced by Protochips Inc. (Reproduced with kind permission from Protochips) [23], (b) photograph of the Hummingbird Scientific *in situ* liquid spectroscopy holder tip and a liquid sample spectrum obtained with it [24] (Reproduced with kind permission from Hummingbird Scientific), and (c) Hummingbird cross-correlative holder with the tip being able to get removed and transferred to optical microscope and SEM [24] (Reproduced with kind permission from Hummingbird Scientific).

Hummingbird Scientific is another major company that sells *in situ* liquid cell TEM products. Apart from the liquid flow and electrochemical functions, it also provides further capabilities such as heating, spectroscopy, and cross-correlative analysis. A thin film coated on a SiN_x window substrate can be used as a Joule heater inside the liquid cell, which heats up the liquid to the boiling point (maximum 200 °C). For the spectroscopy option, the *in situ* liquid cell was designed with a large viewing angle, which ensures direct access from the sample to the energy-dispersive X-ray spectroscopy (EDS) detector. The chips are specially designed to ensure the thinnest possible liquid layer thickness to further allow electron energy loss spectroscopy (EELS) detection inside the liquid cell. Figure 9.3b shows the photograph of the *in situ* liquid spectroscopy holder tip and a liquid sample spectrum obtained with it [24]. The *in situ* TEM holder tip has also been made removable so that it can be transferred to optical, SEM, and X-ray characterization systems for cross-correlative analysis (Figure 9.3c).

In recent years, a bunch of research papers have been published using the commercial *in situ* liquid cell TEM holders, showing their great potential for the frontier *in situ* liquid cell TEM research [25–29]. On the other hand, there is one limitation with the commercial products though: the price level is still pretty high that reduced the number of people who could afford it.

9.3 Nanomaterial Development Using *In Situ* Liquid Cell TEM Technology

In situ liquid cell TEM has wide applications in basic science, nanoscience, new energy sources, and life sciences in liquids, which are hard to numerate one by one [9]. As this book is concentrated on advanced depositions, we will stress on the research on nanomaterial growth with the *in situ* liquid cell technology. Roughly correlated to the *in situ* liquid cell setup development, we will present this topic in the following parts: electrochemical bias-induced growth, irradiation-induced growth, heat-induced growth, and also some further nanomaterial development results from *in situ* liquid cell TEM.

9.3.1 Nanomaterial Growth Induced by Electrical Bias

In the breakthrough work of Williamson *et al.* [2], the electrodeposition of copper has been reported. High-resolution (~5 nm) real-time observation of liquid-phase cluster growth has been achieved. Under constant current densities, the nucleation and growth of individual copper nanoclusters on Au electrode have been observed. The dynamic growth behavior agrees well with electrodeposition simulations (Figure 9.4). The work is not only important for advanced copper film deposition for integrated circuit applications but also provides an experimental technique that can be applied to a broad range of solid–liquid system researches [2].

Radisic *et al.* [30] further studied *in situ* growth kinetics of electrodeposited copper on gold under constant voltage mode. The first direct measurements of growth exponents for individual islands during electrodeposition have been performed. At short times, the island growth exponent is 0.5 as expected for diffusion-limited growth of uncoupled hemispherical islands. While at longer times, the growth exponent is 1/6, corresponding to planar diffusion to the growing islands.

Chen *et al.* [14] demonstrated the electrodeposition and electropolishing procedures of nanograined nickel using an O-ring-based *in situ* electrochemical liquid cell (Figure 9.5). The liquid cell has two thin films: nickel electrodes and a 0.1 M aqueous $NiCl_2$ electrolyte. Biased at −1 V, an anisotropic electrodeposition of nickel film was observed, with growth speed on the substrate surface direction much faster than on the direction perpendicular to the substrate. During the electrodeposition, nanonickel grains have been observed to appear in front of the

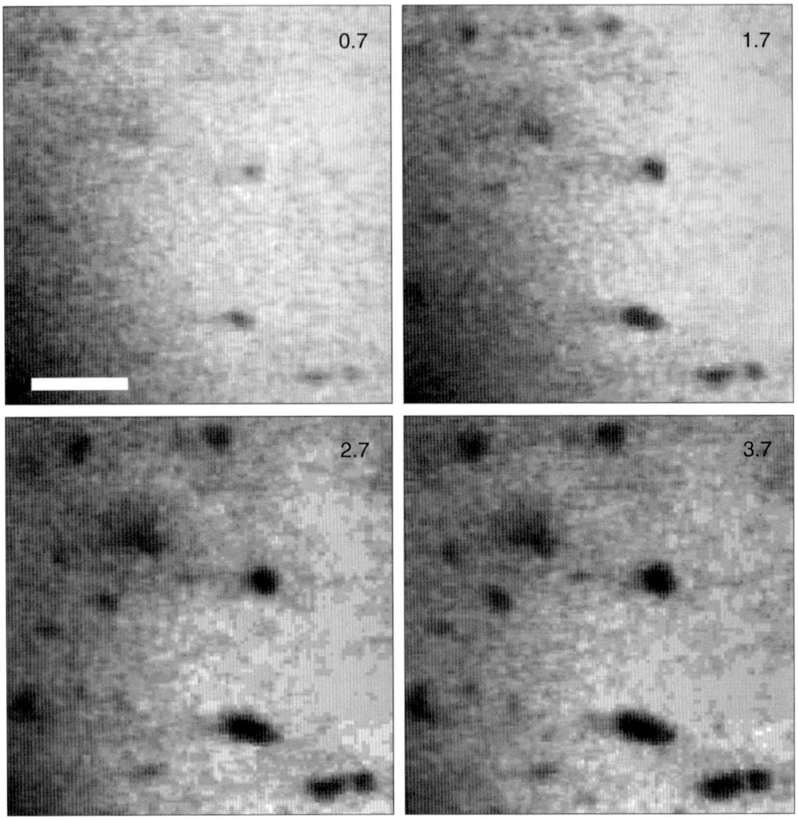

Figure 9.4 Copper cluster nucleation and growth observed *in situ*. Small regions extracted from a video recorded during deposition at 5 mA/cm^2. The images were acquired at the times shown (in seconds) after current flow began. Scale bar 500 nm. (Williamson *et al.* [2]. Reproduced with permission from Nature Publishing Group.)

existing film and then get merged into the growing film. The existence of a thin conducting adsorbate layer in front of the existing nickel film has been proposed to explain such a growth behavior.

Electrochemical growths of lead dendrites in a lead nitrate aqueous solution have been reported [31, 32]. White *et al.* [32] reported the electrochemical deposition of lead from an aqueous solution of lead(II) nitrate using the *in situ* TEM with a Hummingbird Scientific biasing sample holder. Both compact layer deposition and dendritic deposition have been achieved. Local Pb^{2+} concentration has been analyzed, and it shows excellent correlation with the deposition rate and the current passed by the *in situ* cell. Sun *et al.* [31] observed the dendrites developed by fast protrusion of lead branches and tip splitting. The fast-growing tip was composed of polycrystalline nanograins and eventually developed into single crystalline branch. It is also found that the lead concentration in the solution drastically influences the dendritic formation morphologies.

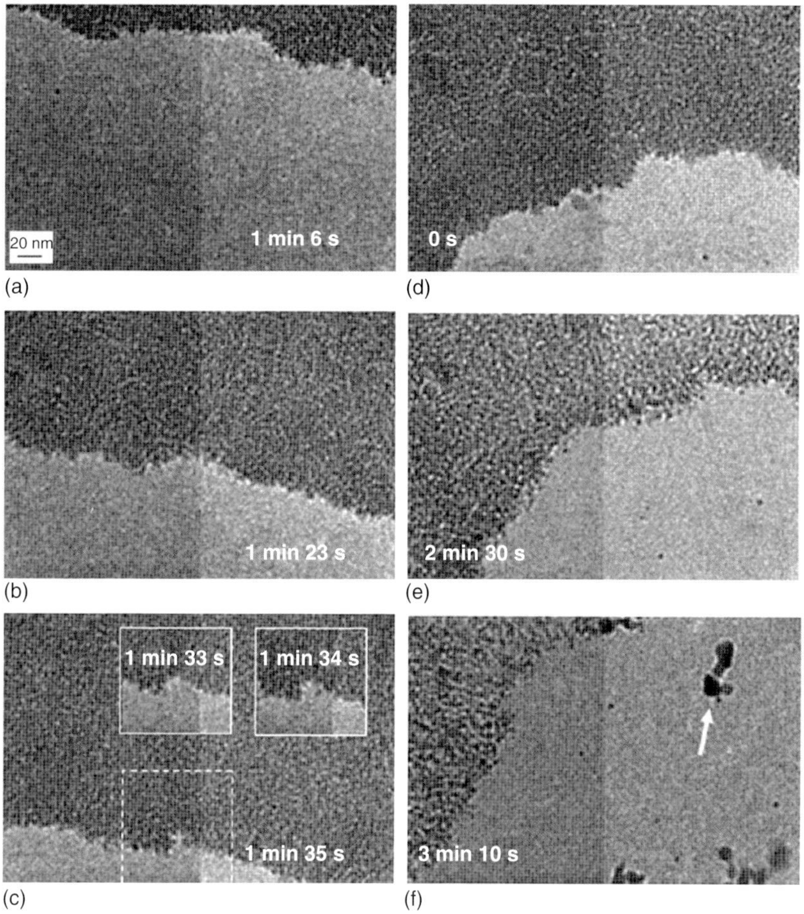

Figure 9.5 *In situ* TEM observations of nickel electrodeposition under −1 V bias at (a) 66 s, (b) 83 s, and (c) 95 s; and electropolishing under +1 V at (d) 0 s, (e) 2 min 30 s, and (f) 3 min 10 s. (Chen *et al.* [14]. Reproduced with permission from Elsevier.)

9.3.2
Nanomaterial Growth Induced by Irradiation

Electron beam has the ability to induce nucleation and growth of nanomaterials in liquid solutions [33]. Understanding nanomaterial growth mechanisms is essential for the syntheses of advanced materials with desired physical properties [33]. The electron beam source is readily available in the TEM system, and simple structured *in situ* liquid cells can be used for this work; thus there have been many papers published on beam-induced nanomaterial growth using *in situ* liquid cell TEM [10, 33–38].

Zheng *et al.* obtained subnanometer resolution images of growing Pt particles in a Pt(acetylacetonate)$_2$ solution. Complex particle growth trajectories have been

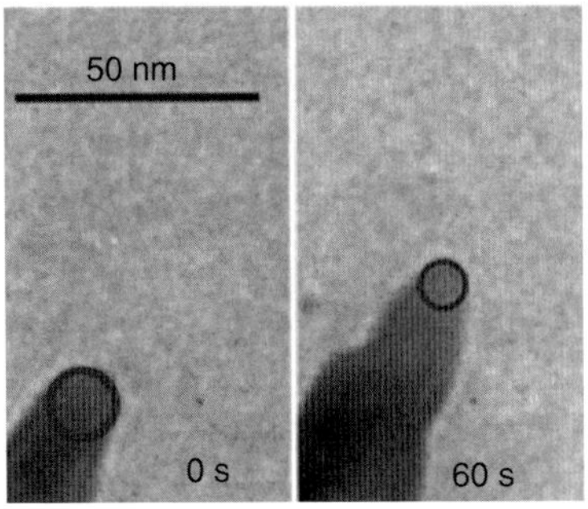

Figure 9.6 The picture shows the growth of the dendrite versus time *in situ* liquid TEM. (Kraus and de Jonge [39]. Reproduced with permission from American Chemical Society.)

observed upon beam irradiation, such as the monomer addition trajectory and the coalescence trajectory. Although the growing paths are different, the resulted final particles are quite uniform in size. The study not only is important for understanding the dynamic growth behavior of nanoparticles but also offers great potential for addressing many fundamental issues in materials science, chemistry, and other fields of science [33].

Kraus and de Jonge [39] reported the trajectories of gold nanoscale dendrite growth using *in situ* liquid cell TEM (Figure 9.6). Dendrite nucleation was induced by the electron beam, showing as an initial burst of growth. In the later growth process, the tip growth velocities can be observed which is between 0.1 and 2.0 nm/s for different dendrites. Those dendrites showing granularities in their structure experience the largest growth speed. Diffusion-limited growth mechanism has been shown to control the nanodendrite.

Liao *et al.* performed real-time imaging of Pt_3Fe nanogrowth in liquid solution [34]. $Pt(acetylacetonate)_2$ (20 mg/ml) and $Fe(acetylacetonate)_2$ (20 mg/ml) were dissolved in an organic solvent in which the solution was used as the precursor in the *in situ* liquid cell. Upon beam irradiation, Pt and Fe were reduced to form many small nanoparticles. These particles then grow larger by monomer attachment and coalescence trajectories, respectively. The particle building blocks then interact with each other, form nanochains, and got straightened to become nanorods.

Liu *et al.* [13] reported electron beam-induced deposition of silicon nanostructures from a $SiCl_4$ liquid precursor. The results demonstrate that the secondary electrons are the major reducing agent for the precursor that resulted in the nanodot growth. Thus, the primary electron beam, especially at high current densities, will suppress the growth rate of nanoparticles. Overall, the size and geometries of the silicon nanodots can be well controlled. A successful patterning of a bitmap image using this deposition method has been shown. This result demonstrates a new approach to fabricating silicon nanostructures and provides fundamental insights into the mechanism for liquid-phase electron beam-induced deposition.

Yuk *et al.* [19] reported a high-resolution TEM analysis of colloidal nanocrystal growth using the aforementioned graphene liquid cells. The unprecedented direct atomic resolution imaging allowed the visualization of critical steps in the nanocrystal growth process, including site-selective coalescence, structural reshaping, and surface faceting [19]. Further progresses include utilizing bubble/ultrathin liquid region to get high-resolution results on Pd nanostructural growth behavior [38], *in situ* TEM observation of Au–Pd core–shell nanostructure formation [36], and *in situ* liquid cell TEM observation of nanoparticle growth induced with both electron and light irradiation [40]. The field is moving very fast, and it is hard to address all the important progresses in a single book chapter.

9.3.3 Nanomaterial Formation Induced by Heating

Thermal heating is a very common method to induce the growth of nanoparticles. Xin and Zheng [16] reported the *in situ* observation of oscillatory growth of Bi nanoparticles at 180 °C. Figure 9.7a shows two particles performing pairwise Ostwald ripening and anti-Ostwald ripening collective oscillation. The experiment results suggested mass-transport zones around the particles that couple the growth kinetics of the particle pairs. Understanding such kinetics is important for controlling the sizes of synthesized nanoparticles [16].

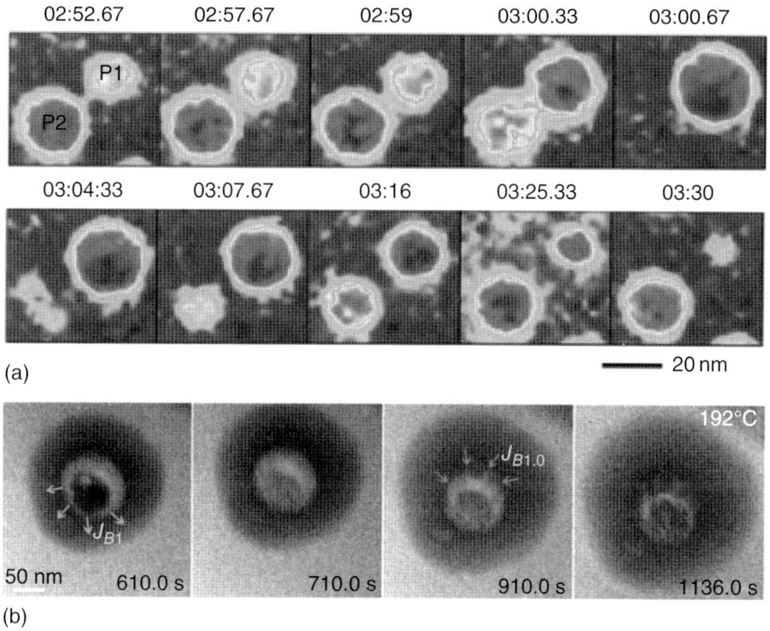

Figure 9.7 (a) Trajectories of two particles (P1 and P2) showing pairwise oscillation and (b) a bismuth oxide hollow nanoparticle formed at 190 °C and its collapse by inward mass diffusion. (Panel (a): Xin and Zheng [16]. Reproduced with permission from American Chemical Society. Panel (b): Niu *et al.* [17]. Reproduced with permission from American Chemical Society.)

Niu *et al.* [17] studied the bismuth oxide hollow nanoparticle formation through the Kirkendall effect, using *in situ* liquid cell TEM. The liquid precursor was prepared by dissolving Bi 2-ethylhexanoate in a mixture of oleylamine and dichlorobenzene. In the first stage, Bi nanoparticles were grown due to electron beam irradiation; in the second stage, the Bi particles were oxidized. Rich diffusion dynamics of Bi and the growth kinetics of bismuth oxide governed by the Kirkendall effect have been observed. Figure 9.7b shows one case, in which, under 190 °C, Bi diffused out of the particle to be oxidized and formed a relatively perfect hollow particle, but later on, due to inward diffusion, the void was gradually filled.

Liu *et al.* further performed *in situ* thermally activated precipitation of ZnO nanoparticles and studied the morphological evolution and growth kinetics [41]. Comparisons have been made between *in situ* and *ex situ* TEM results, and beam-induced precipitation effect has also been analyzed. The work provides insights into the rate-limiting mechanism and the reaction chemistry, which will be important for future hydrothermal nanomaterial synthesis.

9.3.4 Further Nanomaterial Development Results from *In Situ* Liquid Cell TEM

There are still further developments sprouting out in the field. For example, Evans *et al.* utilized a liquid cell with flow function in their beam-induced particle growth experiments [40]. Initiating the liquid flow can replenish the reactant and avoid a precursor depletion problem [2, 40]. Wu *et al.* observed the growth of Au on Pt icosahedral particles with liquid flow cell and low electron dose *in situ* TEM [42]. The liquid flow function allows real-time control of the fluid content in the liquid cell, thus enabling the observation of initial Au nucleation and growth events in TEM, while the low-dose TEM imaging minimized the electron beam interference with the growth procedure. The following growth steps have been observed in the experiment: (i) deposition of Au on the corner sites of the Pt nanoparticles, (ii) diffusion of Au from the corners to terraces and edges, and (iii) subsequent layer-by-layer growth of Au on Au surfaces to form Pt@Au core–shell nanoparticles [42].

Apart from simple electrodeposition, material growth in a working electrochemical device can also be observed. Mehdi *et al.* reported an observation of Li nanomaterial growth around the Pt working electrode in a lithium battery created in a commercial *in situ* STEM stage [25]. The electrodeposition and electrolyte breakdown processes were quantified directly on the nanoscale, which opens up possibilities for the future battery system developments [25].

Zheng *et al.* [43] used electron beam to manipulate and trap gold nanoparticles in the liquid cell. Multiple particles have been collected and assembled in to a nanocluster on the window surface. Figure 9.8 shows an example that by rapid changing of the electron beam size, scattered particles are collected together with the beam and formed a nanocluster.

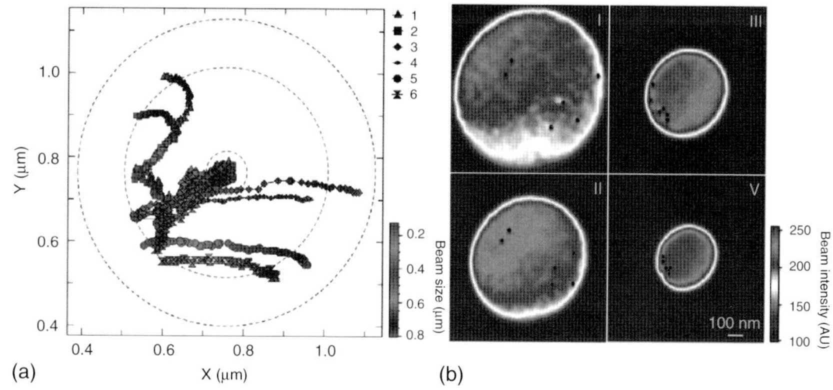

Figure 9.8 Assembly of gold nanoparticles on the window membrane surface by rapid changing of the electron beam size: (a) trajectories of the selected particles and (b) color maps show the electron beam density variations inside the beam region. The six selected particles in (a) are highlighted as black dots in (b). (Zheng *et al.* [43]. Reproduced with permission from American Chemical Society.)

9.4 Summary and Outlook

This chapter mainly refers to the technological advancement of *in situ* liquid cell TEM technology and nanomaterial development using this type of technology. There are different ways to construct a liquid cell, and various functions, such as electrical connections, flow channels, heating and cooling, and so on, can be integrated with the setup. The *in situ* liquid cell TEM technology has been used for studying electrochemical growth, irradiation-induced growth, and heating-induced growth of nanomaterials. It is further useful for studying chemical synthesis mechanism of advanced materials, nanostructured material development inside an operating electrochemical device, and electron beam-guided assembling of nanoclusters.

The *in situ* liquid cell TEM technology is moving forward quickly, and its capability for advanced nanomaterial development is far from getting fully utilized. The upsurge in activity, the achievements of recent years, and the broad range of application areas suggest that numerous exciting developments can be expected in the future [4].

Acknowledgments

The authors acknowledge the support from the Shanghai Leading Academic Discipline Project (B502), the Shanghai Key Laboratory Project (08DZ2230500), the Science and Technology Commission of Shanghai Municipality (11nm0507000), the State Key Laboratory of Functional Materials for Informatics Open Project (SKL201306), the Shanghai Pujiang Program (13PJ1401700), and the Scientific

Research Foundation for the Returned Overseas Chinese Scholars, State Education Ministry. They also acknowledge the support they got from Mr. C. Li and Ms. X. Q. Zhou.

References

1. Williams, D.B. and Carter, C.B. (eds) (2009) *Transmission Electron Microscopy: A Textbook for Materials Science*, Springer, New York.
2. Williamson, M.J., Tromp, R.M., Vereecken, P.M., Hull, R., and Ross, F.M. (2003) *Nat. Mater.*, **2**, 532.
3. Tao, F.F. and Salmeron, M. (2011) *Science*, **331**, 171.
4. de Jonge, N. and Ross, F.M. (2011) *Nat. Nanotechnol.*, **6**, 695.
5. Zaera, F. (2012) *Chem. Rev.*, **112**, 2920.
6. Ross, F.M. (2012) in *In-Situ Electron Microscopy: Applications in Physics, Chemistry and Materials Science*, vol. 7 (eds G. Dehm, J.M. Howe, and J. Zweck), Wiley-VCH Verlag GmbH & Co. KGaA, p. 171.
7. Abellan, T.J.W.P., Parent, L.R., Browning, N.D., Evansc, J.E., and Arsland, I. (2014) *Chem. Commun.*, **50**, 4873.
8. Wu, F. and Yao, N. (2015) *Nano Energy*, **11**, 196.
9. Chen, X., Li, C., and Cao, H.L. (2015) *Nanoscale*, **7**, 4811.
10. Liao, H.G., Niu, K.Y., and Zheng, H.M. (2013) *Chem. Commun.*, **49**, 11720.
11. Huang, J.Y., Zhang, L., Wang, C.M., Sullivan, J.P., Xu, W., Zhang, L.Q., Mao, S.X., Hudak, N.S., Liu, X.H., Subramanian, A., Fan, H.Y., Qi, L., Kushima, A., and Li, J. (2010) *Science*, **330**, 1515.
12. Franks, R., Morefield, S., Wen, J.G., Liao, D.X., Alvarado, J., Strano, M., and Marsh, C. (2008) *J. Nanosci. Nanotechnol.*, **8**, 4404.
13. Liu, Y., Chen, X., Noh, K.W., and Dillon, S.J. (2012) *Nanotechnology*, **23**, 385302.
14. Chen, X., Noh, K.W., Wen, J.G., and Dillon, S.J. (2012) *Acta Mater.*, **60**, 192.
15. de Jonge, N., Bigelow, W.C., and Veith, G.M. (2010) *Nano Lett.*, **10**, 1028.
16. Xin, H.L. and Zheng, H.M. (2012) *Nano Lett.*, **12**, 1470.
17. Niu, K.Y., Park, J., Zheng, H.M., and Alivisatos, A.P. (2013) *Nano Lett.*, **13**, 5715.
18. Tai, K., Liu, Y., and Dillon, S.J. (2014) *Microsc. Microanal.*, **20**, 330.
19. Yuk, J.M., Park, J., Ercius, P., Kim, K., Hellebusch, D.J., Crommie, M.F., Lee, J.Y., Zettl, A., and Alivisatos, A.P. (2012) *Science*, **336**, 61.
20. Yuk, J.M., Seo, H.K., Choi, J.W., and Lee, J.Y. (2014) *ACS Nano*, **8**, 7478.
21. Chen, Q., Smith, J.M., Park, J., Kim, K., Ho, D., Rasool, H.I., Zettl, A., and Alivisatos, A.P. (2013) *Nano Lett.*, **13**, 4556.
22. Unocic, R.R. *et al.* (2014) *Microsc. Microanal.*, **20**, 452.
23. Protochips http://www.protochips.com/products/poseidon‐500.html (accessed 11 April 2016).
24. Hummingbird http://hummingbirdscientific.com/products/liquid/ (accessed 11 April 2016).
25. Mehdi, B.L. *et al.* (2015) *Nano Lett.*, **15**, 2168.
26. van de Put, M.W.P., Carcouët, C.C.M.C., Bomans, P.H.H., Friedrich, H., de Jonge, N., and Sommerdijk, N.A.J.M. (2015) *Small*, **11**, 585.
27. Woehl, T.J., Jungjohann, K.L., Evans, J.E., Arslan, I., Ristenpart, W.D., and Browning, N.D. (2013) *Ultramicroscopy*, **127**, 53.
28. Parent, L.R., Robinson, D.B., Woehl, T.J., Ristenpart, W.D., Evans, J.E., Browning, N.D., and Arslan, I. (2012) *ACS Nano*, **6**, 3589.
29. Fischer, D.A., Alsem, D.H., Simon, B., Prozorov, T., and Salmon, N. (2013) *Microsc. Microanal.*, **19** (Suppl. 2), 476.
30. Radisic, A., Ross, F.M., and Searson, P.C. (2006) *J. Phys. Chem. B*, **110**, 7862.
31. Sun, M.H., Liao, H.-G., Niu, K.Y., and Zheng, H.M. (2013) *Sci. Rep.*, **3**, 3227.
32. White, E.R., Singe, S.B., Augustyn, V., Hubbard, W.A., Mecklenburg, M.,

Dunn, B., and Regan, B.C. (2012) *ACS Nano*, **6**, 6308.

33. Zheng, H.M., Smith, R.K., Jun, Y.W., Jun, Y., Kisielowski, C., Dahmen, U., and Alivisatos, A.P. (2009) *Science*, **324**, 1309.
34. Liao, H.-G., Cui, L.K., Whitelam, S., and Zheng, H.M. (2012) *Science*, **336**, 1011.
35. Li, D.S., Nielsen, M.H., Lee, J.R.I., Frandsen, C., Banfield, J.F., and De Yoreo, J.J. (2012) *Science*, **336**, 1014.
36. Jungjhann, K.L., Bliznakov, S., Sutter, P.W., Stach, E.A., and Sutter, E.A. (2013) *Nano Lett.*, **13**, 2964.
37. Chen, X., Zhou, L.H., Wang, P., Cao, H.L., Miao, X.L., Wei, F.F., and Chen, X. (2015) *Nano-Micro Lett.*, **7**, 385.
38. Zhu, G.M., Jiang, Y.Y., Lin, F., Zhang, H., Jin, C.H., Yuan, J., Yang, D., and Zhang, Z. (2014) *Chem. Commun.*, **50**, 9419.
39. Kraus, T. and de Jonge, N. (2013) *Langmuir*, **29**, 8427.
40. Evans, J.E., Jungjohann, K.L., Browning, N.D., and Arslan, L. (2011) *Nano Lett.*, **11**, 2809.
41. Liu, Y., Tai, K.P., and Dillon, S.J. (2013) *Chem. Mater.*, **25**, 2927.
42. Wu, J.B., Gao, W.P., Wen, J.G., Miller, D.J., Lu, P., Zuo, J.-M., and Yang, H. (2015) *Nano Lett.*, **15**, 2711.
43. Zheng, H.M., Mirsaidov, U.M., Wang, L.-W., and Matsudaira, P. (2012) *Nano Lett.*, **12**, 5644.

10
Direct-Writing Nanolithography

Min Gao

10.1
Introduction

The photolithographic processes use a mask to fabricate nanostructures, but the following direct-writing nanolithography processes discussed avoid the masks to build nanoscale patterns. Electron beams, ion beams, laser beams, and probes are used to write on the surface of samples. These beams or probes have a narrow radius, which can write a pattern pixel by pixel. The minimal size of the nanoscale structure is based on the properties of the beams or probes. In the next part, we will introduce several common methods which can directly fabricate nanopatterns.

10.2
Electron Beam Lithography

Electron beam (e-beam) lithography (EBL) uses electron beams to pattern nanostructures. Like photolithography, EBL also uses resist layer. The chemical properties of a resist layer can be changed after exposing to the electron beam. By focusing the electron beam on the desired position, the resist is patterned. EBL uses software to design the pattern and control the movement and blocking of the electron beams. In EBL, polymethylmethacrylate (PMMA) is the most commonly used photoresist. PMMA is a positive resist. The basic idea of EBL is shown in Figure 10.1. First, the substrate is coated with a layer of PMMA. Then the desired areas are exposed to the electron beam, as shown in Figure 10.1a. After exposure, the exposed PMMA is dissolved in a specific solvent, as shown in Figure 10.1b. The next step is depositing a metallic layer on the substrate as shown in Figure 10.1c. It shows that the metal layer is directly deposited on the substrate for the exposed areas, while the metal layer is deposited on the PMMA layer for the unexposed areas. The final step is dissolving the PMMA layer in an aggressive solvent, such as pyrrolidone and acetone. In Figure 10.1d, the metal

Advanced Nano Deposition Methods, First Edition. Edited by Yuan Lin and Xin Chen.

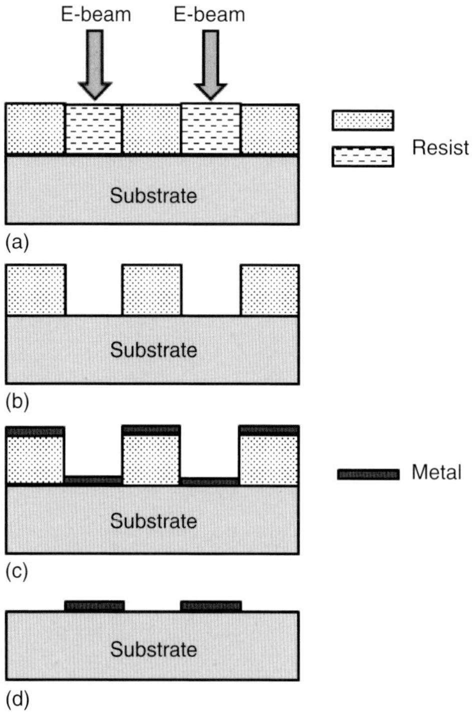

Figure 10.1 The basic idea of e-beam lithography.

layer deposited on the exposed area has remained on the substrate, whereas those on the other areas have been removed [1].

EBL has a resolution of 10 nm or less, which is higher than photolithography. This is because the electron de Broglie wavelength with a typical EBL operating condition, 50 keV, is less than 10 pm or 0.01 nm, which is far below typical atomic sizes. But compared with photolithography, EBL has a very low exposure speed. So it cannot replace photolithography in industry. However, EBL is used for mask fabrication due to its high resolution. Meanwhile, EBL is widely used in research because it can pattern precise nanostructure without mask.

To build an EBL machine, the most simple way is adding a beam blanker on a scanning electron microscopy (SEM). The most important part of a typical SEM is a column, which contains an electron gun, magnetic lenses, and a specimen chamber. The electron gun is the source of the beam, which is focused with several magnetic lenses and finally hits the surface of sample. To change a standard SEM into an EBL writer, the necessary modification is guiding the electron beam to scan the surface to draw a designed pattern pix by pix instead of scanning a whole square area. The diagram of an EBL system is shown in Figure 10.2. In Figure 10.2, the scan coils are the key part to guide electron beam.

The substrate for EBL must be relatively conductive, but the base should obviously be insulating. If the substrate was not conductive, the accumulated

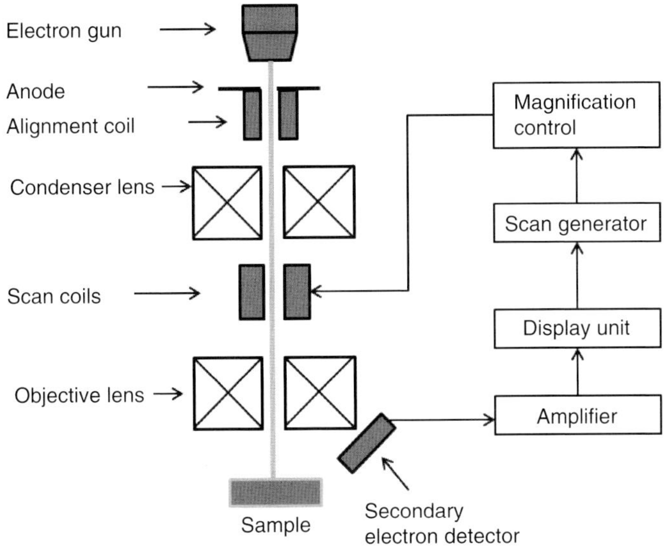

Figure 10.2 The diagram of an e-beam lithography system.

electrons would build up an electric field and deflect the electron beam, which would disturb the scanning of the electron beam. On the other hand, the entire circuit would be short-circuited if the base was not insulating. So the most widely used substrate in EBL is silicon with silicon dioxide on top. Glass plates coated with metal are also possible substrates. But the metal layer should be grounded before writing.

The diameters of e-beam are on the order of 1 nm. Although EBL is able to write extremely fine patterns, it needs correction methods to reach its best resolution. As the electrons hit the resist, the process becomes complex. Firstly, many of the electrons experience small-angle forward scattering, which could broaden the initial beam diameter. And as the electrons penetrate through the resist and reach the substrate, many of them experience backscattering, which would further broaden the beam diameter. This causes additional exposure in the resist and is also known as the e-beam proximity effect [2]. In a basic SEM conversion system, diffraction is not a limiting factor of the resolution, but the proximity effect limits the resolution to about 100 nm. The simplest correction method is using a double-layer e-resist. Other correction method usually used is adapting the beam current, the scanning speed, and the shape of the drawn features. Another factor that affects the resolution of EBL is resist. Although PMMA has extremely high resolution and its ultimate resolution has been demonstrated to be less than 10 nm [3], its major problems are its relatively poor sensitivity, poor dry etch resistance, and moderate thermal stability [2]. Unfortunately, developing other resists which have higher sensitivity and resolution compared to PMMA has not been realized. Figure 10.3 presented the SEM image of sub-10-nm-wide

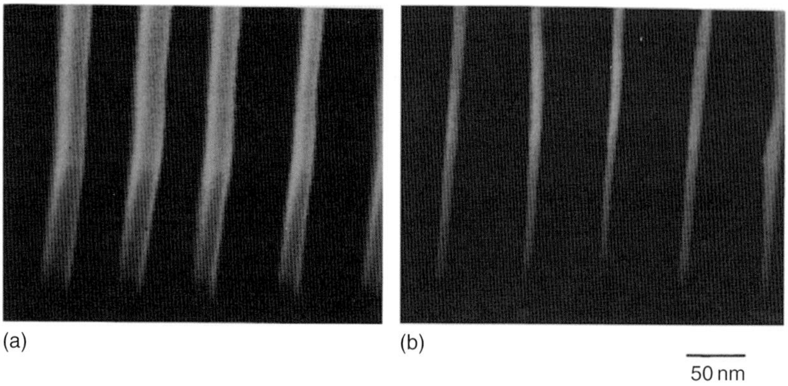

Figure 10.3 SEM image of sub-10-nm-wide Si patterns fabricated by e-beam lithography after etching in propanol-added KOH solution: (a) original Si patterns and (b) additionally etched Si. (Namatsu *et al.* 1995 [4]. Reproduced with permission from American Institute of Physics.)

Si patterns by EBL after etching in propanol-added KOH solution: (a) original Si patterns and (b) additionally etched Si patterns [4].

10.3 Focused Ion Beam Lithography

The focused ion beam (FIB) technique was developed in the late 1970s [5]. FIB uses a direct-write focused beam of slow heavy ions to sputter atoms, which can create patterns directly on the surface without photoresist. The diagram of FIB is shown in Figure 10.4. An FIB system consists of an ion source, ion optics, beam deflectors, and a substrate stage [6, 7]. The machine is similar to an SEM. And the

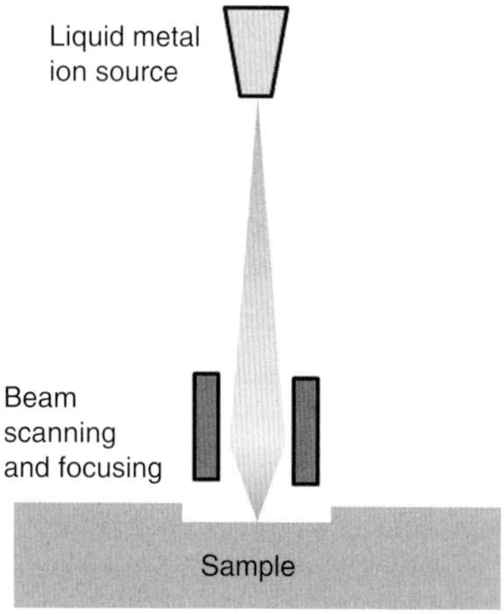

Figure 10.4 The diagram of FIB system.

system is also installed in a high vacuum chamber to minimize scattering of ions from the beam. The range of ion beam energies is 1–50 keV, and ions are sufficiently heavy to remove hard materials such as metals, ceramics, and inorganic semiconductors. This is a purely mechanical process. Typical spot size of the ion beam is about 50 nm. The ion beam is scanned across the surface in the desired points and blanked off at points where no material removal is desired. FIB can produce 3D patterns, which is unlike EBL. However, the speed of the FIB process to produce patterns is relatively slow.

In an FIB system, ion is a charged particle which is at least three orders of magnitude heavier than the electron. Different ions have different characteristics. Popular ion sources include As, Be, Ga, Si, Ar, B, and P [8]. Ga ions, the most commonly used ions in the FIB technique, are slow heavy ions. The interaction process between Ga ions and surface is that of altering the surface structure, modifying surface chemistry, and removing atoms from the surface through sputtering. On the other hand, the lighter ions, such as 100 keV to 3 MeV protons, can penetrate into the material much deeper with a minimal amount of surface disruption. Because the wavelength of a 100 keV proton is around 10^{-4} nm, the diffraction effects can be neglected for both light and heavy ions. Considering the properties of different ions, the fast light ion beams are a good choice to fabricate high aspect ratio 3D structures since the ions travel in a straight line. The penetration depth of the ion beam can be adjusted by changing the ion energy. In Figure 10.5, it shows the SEM image of an array of FIB-etched lines on a multilayered $AlF_3/GaAs_4$ sample.

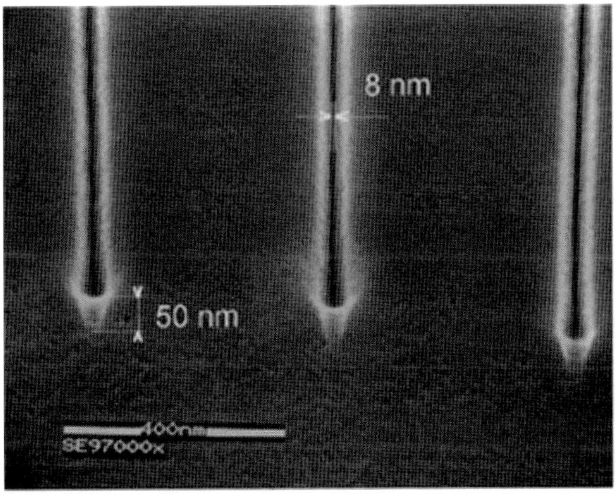

Figure 10.5 SEM image of an array of FIB-etched lines on a multilayered sample AlF_3 (50 nm thick)/$GaAs_4$. (Gierak *et al.* [9]. Reproduced with permission from Elsevier.)

10.4
Gas-Assisted Electron and Ion Beam Lithography

By utilizing gas-assisted etching, the scanned area is simultaneously exposed to reactive gas molecules, which can significantly enhance the etching rate of e-beam and ion beam lithography. For gas-assisted induced etching, the electron beam and ion beam can excite or dissociate the precursor gas, in which the resulting gas molecules chemically react with atoms on the surface of the substrate. This reaction forms volatile molecules containing substrate atoms, which can be easily pumped to the outside of the vacuum chamber.

In addition to gas-assisted etching, electron and ion beams can also be used to perform gas-assisted deposition of materials. In this process, the ambient gaseous molecules adsorbed on the surface are fragmented by the incident electron and ion beam and then deposited on the position focused by the beam. The first demonstration of this technique was that a polymer film was formed on a substrate in an SEM chamber when silicone vacuum pump oil was delivered into the SEM chamber. Recently, this e-beam-induced deposition process has been developed to deposit other desired materials. Dorp and Hagen list most materials and precursor gases that have been used for gas-assisted e-beam deposition [10]. Here, some commonly used materials and precursor gases for gas-assisted e-beam deposition are listed in Table 10.1. In Figure 10.6, it shows the gold nanoparticle patterning of silicon wafer by both chemical EBL and molecular recognition [11].

Table 10.1 Precursor names.

Material	Precursor
Al	Trimethylaluminum (TMA)
Au	$AuCl_3$
Au	Dimethylacetylacetonate gold (Me_2-Au-acac)
Co	Co_2CO_8
Co	$Co(CO_3)NO$
Cu	Bis-hexafluoroacetylacetonate copper ($Cu(hfac)_2$)
Cu	Hexafluoroacetylacetonate copper vinyltrimethylsilane (hfac-Cu-VTMS)
Cr	Chromyl chloride (CrO_2Cl_2)
Fe	Iron pentacarbonyl ($Fe(CO)_5$)
Ir	Trifluorophosphine diiridium ($(IrCl(PF3)_2)_2$)
Mo	Molybdenum hexacarbonyl ($Mo(CO)_6$)
Ni	Nickelocene ($Ni(C_5H_5)_2$)
Pt	Trimethyl-platinum-cyclopentadienyl ($CpPtMe_3$)
Si	Dichlorosilane (SiH_2Cl_2)
Si*Ox*	Tetraethoxysilane (TEOS)
W	Tungsten hexacarbonyl ($W(CO)_6$)

The data is from van Dorp and Hagen's report [10]. Reproduced with permission from American Institute of Physics.

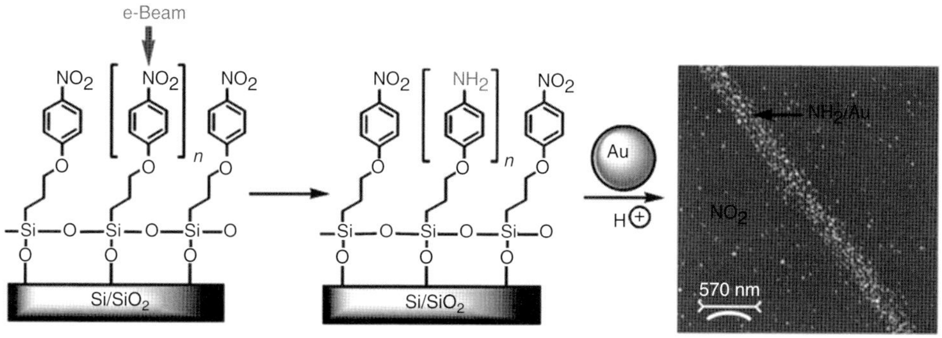

Figure 10.6 The schematic and atomic force microscopy images of fabricating designed Au nanoparticle structures on silica surface. (Mendes *et al.* [11]. Reproduced with permission from American Chemical Society.)

10.5 SPM Lithography

Scanning probe microscopy (SPM) family includes scanning tunneling microscopy (STM), atomic force microscopy (AFM), and magnetic force microscopy (MFM). At the beginning, STM was used to investigate the surface atomic structure. With the development of this technology, it was found that SPM can be used to modify materials, deposit materials, remove and etch materials, and manipulate a single atom. Because SPM-based nanofabrication can overcome the fundamental size limits of conventional lithography techniques, it has become increasingly popular in creating nanoscale structures. In the following, we will discuss the two major technologies within the SPM family: STM and AFM.

The composition of STM and AFM is shown in Figure 10.7. The STM was invented in 1981 by Binnig *et al.* [12]. It has a sharp conducting tip with a bias voltage applied between the tip and the sample. The tip and the sample are placed in an ultrahigh vacuum chamber. When the distance between the tip and the sample is within several nanometers, electrons begin to tunnel from the tip to the sample or from the sample to the tip, which depends on the sign

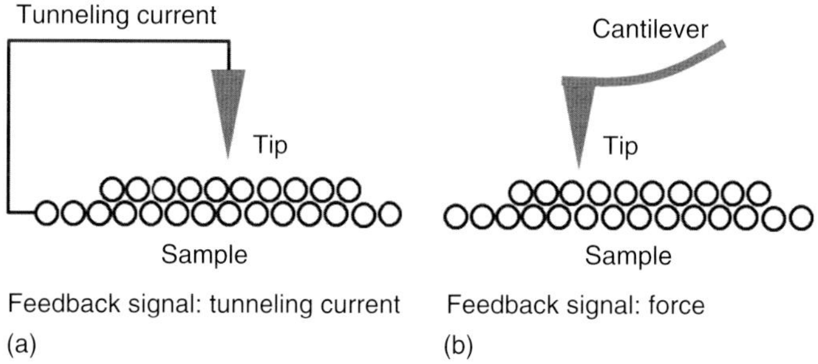

Figure 10.7 The composition of STM (a) and AFM (b).

of the bias voltage. The tunneling current has an exponential dependence on the distance between the tip and the sample, so STM has precision vertically and subnanometer resolution laterally. It has two modes: constant-height and constant-current modes. In STM lithography, the constant-current mode is used more often because the tunneling current is usually used as a lithographic source. AFM was developed from STM, and the attractive or repulsive forces between the tip and the sample are used in the measurement instead of the tunneling current in STM. The attractive or repulsive forces also depend on the space between the tip and the sample. In AFM, the tip is located at the end of a cantilever. The attractive or repulsive forces cause the cantilever to deflect. By measuring the cantilever deflection, the surface topography can be obtained.

Ringger *et al.* [13] and Staufer *et al.* [14] used STM to produce nanoscale patterns of lines and cones, respectively, on an atomically flat surface of a metallic glass in vacuum, which demonstrated the application of STM as an effective lithographic tool. And then Dagata *et al.* [15] used an STM to induce local oxidation on a hydrogen-passivated silicon substrate in air. Their work further broadens the applications of STM lithography. Furthermore, arrays of tips have been developed for parallel lithography [16].

The AFM-based nanolithography techniques are less restrictive than that of STM because AFM can be conducted in a normal room environment and can be used to image any kind of materials. AFM can be equipped with conductive tips to expose resists, induced oxidation, assisted etching, and induced deposition as an STM. The difference is that AFM lithography must operate in a constant voltage mode because the underlying principle and configuration are different from the STM lithography. Moreover, nanostructures can be fabricated with a higher resolution but with a lower scan speed by using STM lithography than by using AFM lithography [16].

The extraordinary spatial sensitivities of the tunneling current mean that STM has its own basic advantages in nanofabrication. The tunneling current can be considered as a miniature electron beam as that in EBL. So it can write patterns on resists or substrates as the process in EBL. And the STM lithography has several advantages over the regular EBL. The tunneling current density of STM is high enough to expose the high-resolution resists, since high-resolution resists are generally less sensitive. Meanwhile, the applied voltages in STM lithography are about several volts, which are four orders of magnitude lower than that in EBL. The low voltage produces less backscattering and minimizes the proximity effect [2], which determines the resolution of EBL.

STMs can also be used to move single atoms to form nanostructures, known as *atom manipulation* [17]. The most famous experiment is controlling single metal atoms to construct a quantum corral nanostructure on a metal surface. Figure 10.8 shows the STM images of the process for making a quantum corral built with 39 Ag atoms on Ag [18].

STM can deposit materials as other lithography techniques. When a voltage bias between the tip and the sample is applied, the atoms or nanoparticles in the STM tip can be transferred to the surface. It has been found that Au, Ni, and Pt tips

(a)

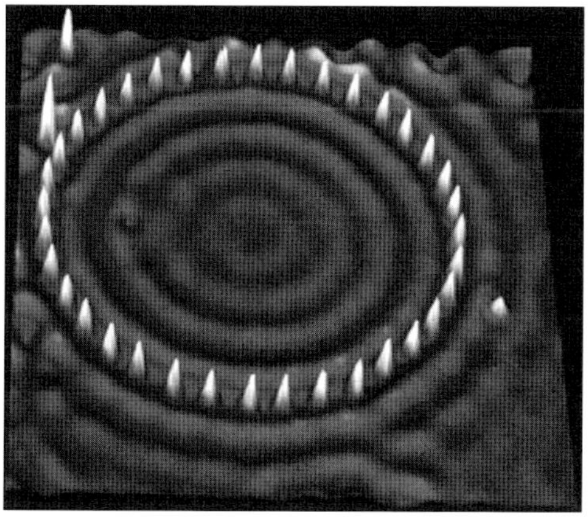

(b)

Figure 10.8 STM images of the process for making a quantum corral built with 39 Ag atoms on Ag. (Hla *et al.* [18]. Reproduced with permission from American Institute of Physics.)

all can create dots on surface [19, 20]. But this process has low reproducibility. The AFM-based deposition is similar to STM-induced deposition, but it has less reliability for material removal or modification and the nanostructure fabrication.

In addition, the electric field between the tip and the sample in STM or AFM can induce local chemical reaction. For example, SPM-induced oxidation has become a major technique for nanofabrication. SPM tip-induced local oxidation is an electrochemical process. Both the tip and the substrate should be somewhat conductive, and substrates should be able to form stable oxides. Avouris *et al.* [21] used a heavily doped Si tip with −10 V bias to induce local oxidation of an n-type Si substrate. Another example is that STM can realize the reversible conductance transitions of organic compound films on the nanometer scale by the application of local electric field pulses [22], which is shown in Figure 10.9.

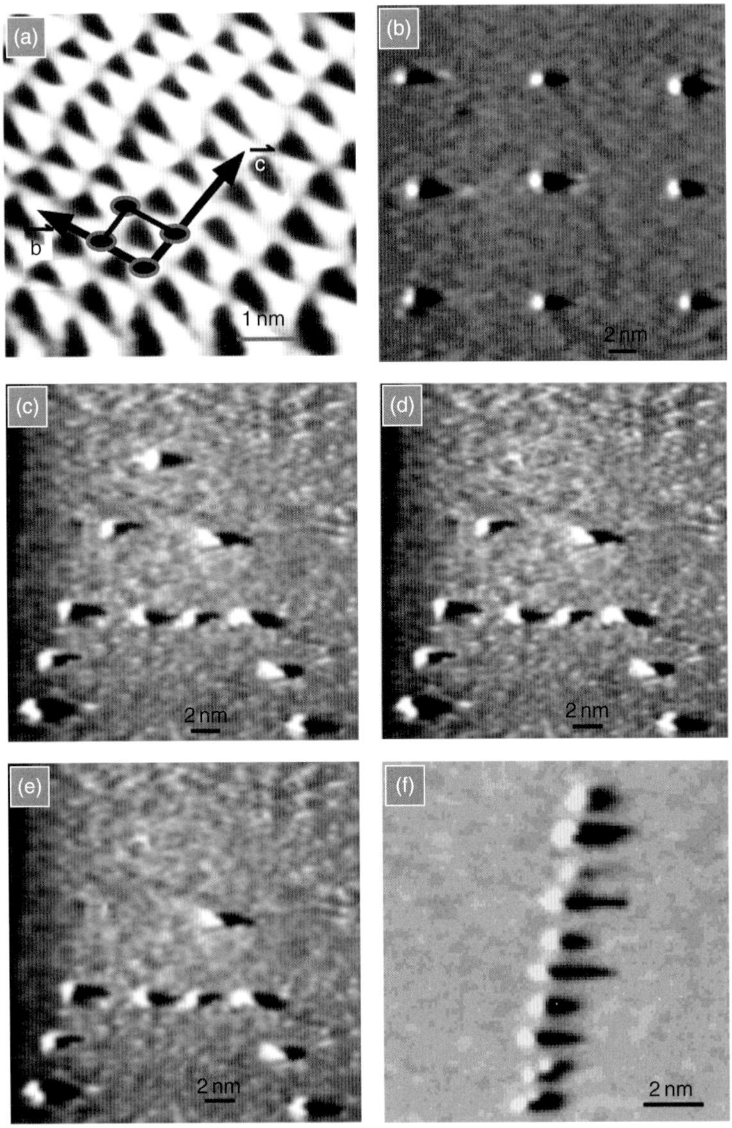

Figure 10.9 (a) The STM images for 3-nitrobenzal malononitrile (NBMN) and 1,4-phenylenediamine (pDA) composite film on HOPG. Panels (b–f) show the reversible conductance transitions of these organic compound films by the application of local electric field pulses by scanning tunneling microscope. (Gao *et al.* [22]. Reproduced with permission from American Institute of Physics.)

10.6
Dip-Pen Lithography

Dip-pen lithography (DPL) was first reported by Mirkin's group [23, 24]. In this process, an AFM tip is used to deposit "ink" molecules on the substrate surface. The AFM tip is coated with ink, and when the tip reaches a surface, a meniscus is formed between the tip and the surface of the substrate. As the tip is moved, ink is pulled off and remains on the surface due to surface tension. The schematic of DPL is shown in Figure 10.10 [25]. Using this method, sub-50-nm lines of material can be deposited either sequentially or in parallel by using multiple tips [26, 27]. Small organic molecules, polymers, DNA, proteins, peptides, colloidal nanoparticles, metal ions, and sols [23, 24, 28–31] can all be patterned by DPL. Meanwhile, substrates are not restricted in this technology. Hard or soft substrate materials, insulating, semiconducting, and metallic substrates can all be patterned. However, the process of writing is dependent on temperature, humidity, writing speed, contact force between tip and substrate, and the interaction between ink and substrates [32, 33]. The size of the water meniscus that bridges the tip and substrate depends on relative humidity.

Figure 10.11 shows an example of a molecular nanopattern of 1-octadecanethiol (ODT) deposited on a gold substrate by DPL [25]. Figure 10.11a,b demonstrates the diffusion properties of the "ink." The opposite contrast images provide additional evidence that the molecules are being transported from the tip to the surface. And DPL could generate arrays and grids in addition to individual lines and dots. An array of 25 ODT dots of 0.46 mm in diameter, spaced 0.54 mm

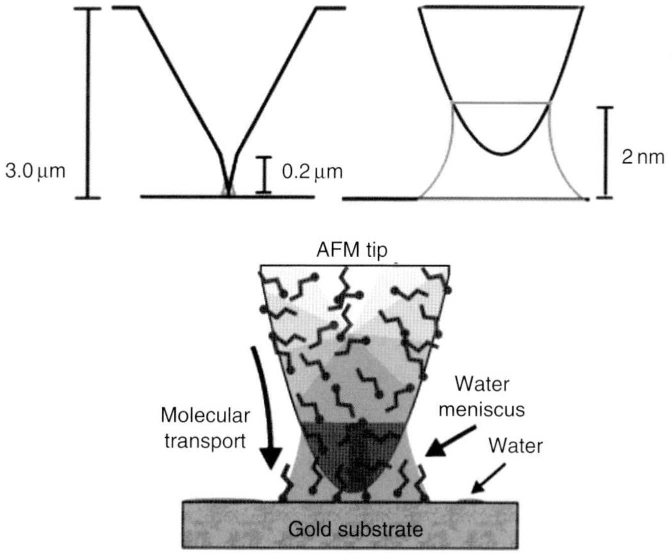

Figure 10.10 Schematic diagram of the DPL process. (Rozhok *et al.* [25]. Reproduced with permission from American Chemical Society.)

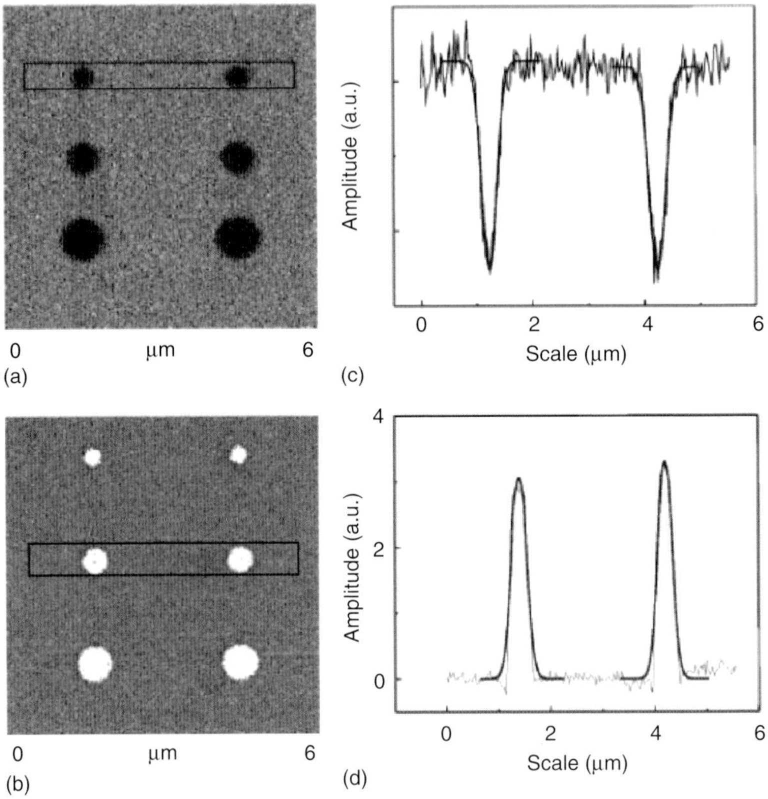

Figure 10.11 LFM images of (a) ODT and (b) MHA dots patterned onto a gold substrate under tip–substrate contact force of 0.5 nN and holding times of 2, 4, and 8 s; the frictional force profiles of (c) ODT and (d) MHA dot patterns. Thick lines represent Gaussian fits. (Rozhok *et al.* [25]. Reproduced with permission from American Chemical Society.)

apart (Figure 10.11c), was generated by holding an ODT-coated tip in contact with the surface without lateral movement to form each dot. A grid consisting of eight intersecting lines of 2 mm in length and 100 nm wide (Figure 10.11d) was generated by sweeping the ODT-coated tip.

10.7 Summary

In summary, nanofabrication is an essential tool for the evolution of electronic, photonic, biomedical, and nanosystem technologies. The direct-writing nanolithography includes several categories. The main differences can be concluded as the writing beams and the materials. Each nanolithography technology has its own advantages and weakness. But all direct-writing nanolithography technologies have a much lower efficiency and higher cost than photolithography, which limited their applications in industries. But with the development of technology and materials, we believe the direct-writing nanolithography will have a great rapid progress and the applications will be broadened.

Acknowledgments

The authors acknowledge the National Basic Research Program of China (973 Program) under Grant No. 2015CB351905 and the National Natural Science Foundation of China under Grant Nos. 61306015 and 51372034.

References

1. Hohenau, A., Ditlbacher, H., Lamprecht, B., Krenn, J.R., Leitner, A., and Aussenegg, F.R. (2006) *Microelectron. Eng.*, **83**, 1464.
2. Tseng, A.A., Chen, K., Chen, C.D., and Ma, K.J. (2003) *IEEE Trans. Electron. Packag. Manuf.*, **26**, 141.
3. Khoury, M. and Ferry, D.K. (1996) *J. Vac. Sci. Technol., B*, **14**, 75.
4. Namatsu, H., Nagase, M., Kurihara, K., Iwadate, K., Furuta, T., and Murase, K. (1995) *J. Vac. Sci. Technol., B*, **13**, 1473.
5. Reyntjens, S. and Puers, R. (2001) *J. Micromech. Microeng.*, **11**, 287.
6. Orloff, J.H., Utlaut, M., and Swanson, L.W. (eds) (2003) *High Resolution Focused Ion Beams*, Kluwer, New York.
7. Tseng, A.A. (2005) *Small*, **1**, 924.
8. Liu, Y., Longo, D.M., and Hull, R. (2003) *Appl. Phys. Lett.*, **82**, 346.
9. Gierak, J., Septier, A., and Vieu, C. (1999) *Nucl. Instrum. Methods Phys. Res., Sect. A*, **427**, 91.
10. van Dorp, W.F. and Hagen, C.W. (2008) *J. Appl. Phys.*, **104**, 081301.
11. Mendes, P.M., Jacke, S., Critchley, K., Plaza, J., Chen, Y., Nikitin, K., Palmer, R.E., Preece, J.A., Evans, S.D., and Fitzmaurice, D. (2004) *Langmuir*, **20**, 3766.
12. Binnig, G., Rohrer, H., Gerber, C., and Weibel, E. (1982) *Phys. Rev. Lett.*, **49**, 57.
13. Ringger, M., Hidber, H.R., Schlögl, R., Oelhafen, P., and Güntherodt, H.J. (1985) *Appl. Phys. Lett.*, **46**, 832.
14. Staufer, U., Wiesendanger, R., Eng, L., Rosenthaler, L., Hidber, H.R., Güntherodt, H.J., and Garcia, N. (1987) *Appl. Phys. Lett.*, **51**, 244.
15. Dagata, J.A., Schneir, J., Harary, H.H., Evans, C.J., Postek, M.T., and Bennett, J. (1990) *Appl. Phys. Lett.*, **56**, 2001.
16. Tseng, A.A., Notargiacomo, A., and Chen, T.P. (2005) *J. Vac. Sci. Technol., B*, **23**, 877.
17. Binnig, G. and Rohrer, H. (1999) *Rev. Mod. Phys.*, **71**, S324.
18. Hla, S.W., Braun, K.F., and Rieder, K.H. (2003) *Phys. Rev. B*, **67**, 201402.
19. Houel, A., Tonneau, D., Bonnail, N., Dallaporta, H., and Safarov, V.I. (2002) *J. Vac. Sci. Technol., B*, **20**, 2337.
20. Fujita, D., Onishi, K., and Kumakura, T. (2003) *Jpn. J. Appl. Phys.*, **42**, 4773.
21. Avouris, P., Hertel, T., and Martel, R. (1997) *Appl. Phys. Lett.*, **71**, 285.
22. Gao, H.J., Sohlberg, K., Xue, Z.Q., Chen, H.Y., Hou, S.M., Ma, L.P., Fang, X.W., Pang, S.J., and Pennycook, S.J. (2000) *Phys. Rev. Lett.*, **84**, 1780.
23. Lee, K.B., Park, S.J., Mirkin, C.A., Smith, J.C., and Mrksich, M. (2002) *Science*, **295**, 1702.
24. Piner, R.D., Zhu, J., Xu, F., Hong, S.H., and Mirkin, C.A. (1999) *Science*, **283**, 661.
25. Rozhok, S., Piner, R., and Mirkin, C.A. (2003) *J. Phys. Chem. B*, **107**, 751.
26. Bullen, D., Chung, S.-W., Wang, X., Zou, J., Mirkin, C.A., and Liu, C. (2004) *Appl. Phys. Lett.*, **84**, 789.
27. Hong, S.H., Zhu, J., and Mirkin, C.A. (1999) *Science*, **286**, 523.

28. Noy, A., Miller, A.E., Klare, J.E., Weeks, B.L., Woods, B.W., and DeYoreo, J.J. (2002) *Nano Lett.*, **2**, 109.
29. Wilson, D.L., Martin, R., Hong, S., Cronin-Golomb, M., Mirkin, C.A., and Kaplan, D.L. (2001) *Proc. Natl. Acad. Sci. U.S.A.*, **98**, 13660.
30. Demers, L.M., Ginger, D.S., Park, S.J., Li, Z., Chung, S.W., and Mirkin, C.A. (2002) *Science*, **296**, 1836.
31. Su, M., Liu, X.G., Li, S.Y., Dravid, V.P., and Mirkin, C.A. (2002) *J. Am. Chem. Soc.*, **124**, 1560.
32. Piner, R.D. and Mirkin, C.A. (1997) *Langmuir*, **13**, 6864.
33. Sheehan, P.E. and Whitman, L.J. (2002) *Phys. Rev. Lett.*, **88**, 156104.

11
3D Printing of Nanostructures

Min Gao

11.1
Introduction

3D printing technology is a kind of rapid prototyping (RP), and it is also described as additive manufacturing (AM). The basic principle of this technology is directly fabricating a model layer by layer in an additive process by using a 3D modeling program. So it can create something quickly. In 1984, Chuck Hull invented stereolithography (SL), in which photopolymers were added layer by layer and cured with ultraviolet (UV) lasers. SL was considered as the birth of 3D printing. After that, fused deposition modeling (FDM) and selective laser sintering (SLS) were reported, which were similar to SL. In 1994, the 3D inkjet printing technology, which is similar to the inkjet technology used in 2D printers, was developed at Massachusetts Institute of Technology (MIT) [1]. They named it "three-dimensional printing techniques." But now, 3D printing technology mainly refers to the layer-by-layer technology, not any of the 3D inkjet technology. It developed very quickly and the depth of one layer was reduced to 0.1 mm. The first parts of the human such as ears and fingers were printed in 2000 by 3D printing technology. Now, many commercial 3D printers have been launched. In 2005, Z Corp. launched Spectrum Z510, which was the first high-definition color 3D printer in the market. For 3D printing technology, the resolution and the printing speed are the two important factors that limit its applications. But these two factors are paradoxical: improving the process of printing and searching new printing materials are considered as effective ways to promote the development of 3D printing technology.

11.2
3D Printing Processes

To printing a 3D object, it involves several steps to transfer the CAD files to the real objects. In the following texts, we will discuss the processes of 3D printing:

Advanced Nano Deposition Methods, First Edition. Edited by Yuan Lin and Xin Chen.

Step 1: *CAD file* 3D printing technology begins with a 3D model, which can be created by a 3D scanner or 3D modeling software. The 3D model describes the geometry of the objects. Then the software of the 3D printer slices the design into thousands of or even more layers. These layers will be printed one by one from bottom to top. Many programs, such as SolidWorks, AutoCAD, and Google SketchUp, can be used to create the 3D model.

Step 2: *Transfer CAD to 3D printer* A 3D printer needs a machine language. We should tell it how to move the step motors. Now, most 3D printers accept the STL file format. This file describes the moving functions of the stepper motors in the *X*, *Y*, and *Z* directions.

Step 3: *Printing* Printing the objects is mainly an automated process. The 3D printer creates it layer by layer by different methods and materials.

Step 4: *Postprocessing* Once the parts are printed, it may need additional steps to be cleaned and polished.

11.3 Types of 3D Printing

There are a variety of 3D printing types. The main differences are in the way how layers are deposited or are in the printing materials. Some 3D printers process powdered materials, such as nylon, plastic, ceramic, and metal, which utilize a light or heat source to melt the materials together in the defined shape layer by layer. Other 3D printers process liquid materials, and a UV laser is used to cure the resin. Jetting droplets is another 3D printing process, which is similar to 2D inkjet printing, but with special materials to ink and a binder to fix layer by layer. The majority of entry-level 3D printers employ deposition, which is the most common process. In the following texts, we will introduce the advantages and disadvantages of different 3D printing methods.

11.3.1 Stereolithography

SL is widely considered as the first 3D printing process. And this printing process was used in the first commercial 3D printer. SL is a laser-based process in which the photopolymer resins react with the laser and then is cured to form solid in a precise way to produce objects. SL machine consists of a perforated platform which is mounted below the surface of a vat of liquid photoresin and a UV laser beam. The UV laser beam traces the first layer of the object using information obtained from the 3D model and causes a thin layer of photoresin to harden. Once the layer has been fabricated, the perforated platform is then lowered to create another layer of the object until a whole object is printed [2, 3]. The principle of SL is shown in Figure 11.1.

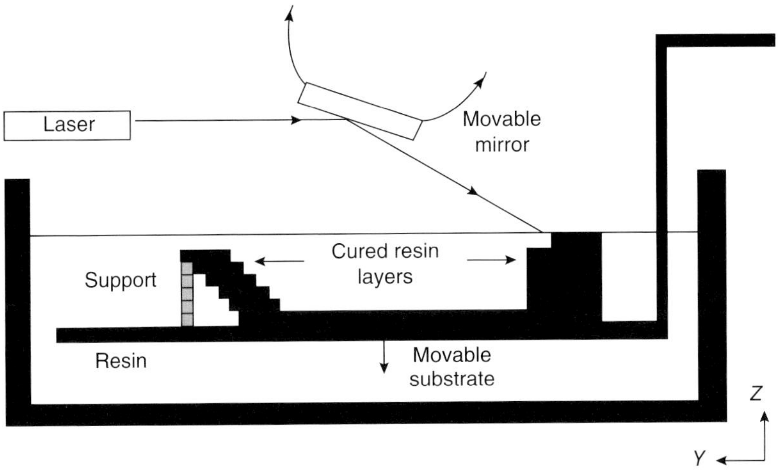

Figure 11.1 The principle of stereolithography. (Reprinted with permission from [3].)

The advantages of SL process include the following. It produces an excellent surface finish [3]. It is one of the most fastest and accurate 3D printing processes [4]. And the SL process can built several parts at once and fully utilize the resin vat.

On the other hand, there are also disadvantages of the SL process. It requires some supporting structures because the parts may be brittle and translucent. After the process is finished, the supporting structures need to be manually removed. Another problem is that the material is expensive, smelly, and toxic [3]. Moreover, many objects 3D printed using SL need to be cleaned.

11.3.2 Fused Deposition Modeling

The FDM technology appeared in the early 1990s. Now, the FDM technology is also an industrial-grade 3D printing process. The FDM machine consists of a movable heater head which deposits molten materials on a substrate. The principle of FDM is shown in Figure 11.2. The build material is heated to a temperature that is a little higher than its melting point so that it can solidify quickly after extrusion from the head. When printing parts with overhanging geometries, the FDM processes require support structures which should be made from a water-soluble material and can be relatively easily washed away once the print is completed.

The advantage of this method is that the build materials can be cheap, nontoxic, and environmentally safe. Now, ABS plastic, medical-grade ABS (MABS), and elastomers can be used in this method. The objects fabricated by this method have a high stability since they are not hygroscopic [3, 5].

The disadvantages of the process include the following: the surface finish of the parts is not as good as that produced using SL because the material extrusion may not be stopped quickly enough [3, 5], the speed of the process may be slow for

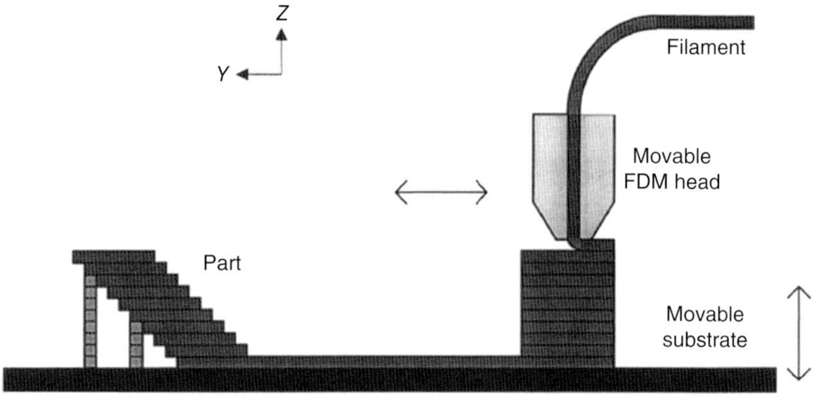

Figure 11.2 The principle of fused deposition modeling. (Reprinted with permission from [3].)

some part, and the layer-to-layer adhesion is also a problem which results in parts that are not watertight. However, postprocessing by using acetone can polish the surface.

11.3.3
Selective Deposition Lamination

Selective deposition lamination (SDL) is a 3D printing process developed and manufactured by Mcor Technologies. Due to printing process in layering and shaping paper, this technology is similar to the laminated object manufacturing (LOM) process developed by Helisys in the 1990s. The SDL 3D printing process builds parts layer by layer using standard copier paper. Each new layer is fixed to the previous layer using an adhesive. The printing process is shown in Figure 11.3. After a new sheet of paper is fed into the 3D printer from the paper feeder machine and adhesive is selectively applied on top of the previous layer, the build plate is moved to a heat plate while applying pressure. The pressure ensures a positive bond between the two sheets of paper. Then an adjustable tungsten carbide blade cuts one sheet of paper at a time to create the edges of the part

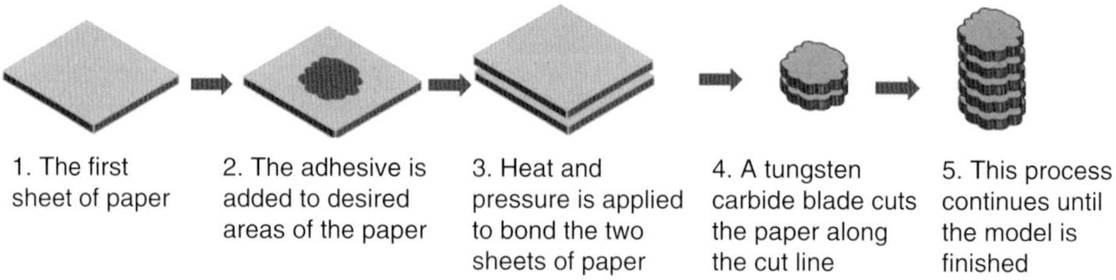

Figure 11.3 The printing process of selective deposition lamination.

by tracing the object outline. When this cutting sequence is completed, the 3D printer deposits the next layer of adhesive and so on until the part is completed.

SDL belongs to a very few 3D printing processes that can produce color 3D printed parts. And because the parts are standard paper, they are wholly safe and eco-friendly and do not require postprocessing. However, the process cannot fabricate objects with complex geometries, and the building size is limited to the size of the feedstock.

11.3.4 Selective Laser Sintering

SLS uses high-energy laser to fuse powder together. The powder can be wax, ceramic, metal, nylon, sand, and polycarbonates. The powder is put in a bed and then is heated to a temperature below the melting point of the powder material in order to avoid the thermal distortion and fusing of the previous layer [6]. The principle of SLS is shown in Figure 11.4. The laser beam traces the cross section of the powder surface to heat up the powder to the sintering temperature, which induces the powder scanned by the laser. The powder that is not scanned by the laser will remain in place as the support to the next layer. When a layer of the cross section is completed, the roller levels another layer of powder for the next pass [2].

The materials for SLS are cheaper than the resin for SL, and they are nontoxic and safe and may be sintered with relatively low-powered lasers [3]. However, the printing produced part needs a long cool time to be removed from the machine. Furthermore, the materials for SLS are sensitive to the heating power, so the laser needs to be set to different parameters for different materials. This process is a time-consuming process, which is not proper for industry. Another disadvantage

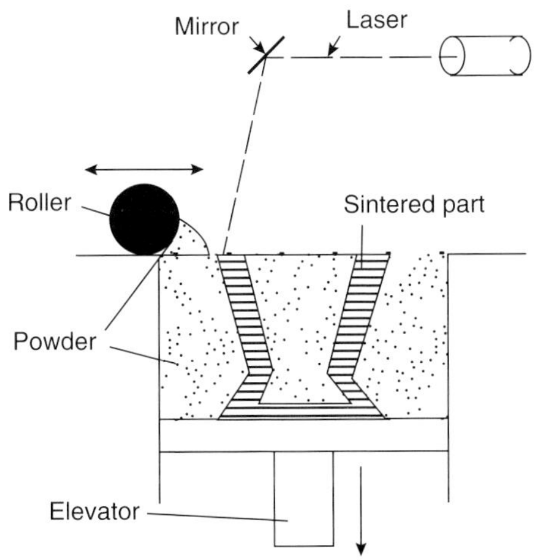

Figure 11.4 The principle of selective laser sintering. (Reprinted with permission from [2].)

is the thermal distortion of the prototype because of the heating of the power. So the printing part needs to be cooled when it becomes too hot. But adding cooling system will further increase the price and volume of the machine.

11.3.5
3D Inkjet Printing

The 3D inkjet printing technology was developed at MIT [1]. The difference between 2D and 3D inkjet printing is that in 3D printing the print head is controlled to move both vertically and horizontally, whereas in 2D it can only move horizontally [7]. The 3D inkjet printer dispenses waxes and polymers to create an object layer by layer. 3D inkjet printing enables low-cost, large-area device construction with high throughput [8]. Since the 3D inkjet printing cannot process at a very high temperature, it is suitable to fabricate devices on plastic substrates. Yamaguchi *et al.* [9] first demonstrated that molten solder can be used in the 3D metal jetting. However, the molten metal jet process is limited because only low-melting-temperature solder can be used in this system. And the jetting process needs to run at a high temperature to keep the metal molten [10].

11.3.6
Multijet Modeling

Multijet modeling (MJM) builds up objects from successive layers of powder using a wide-area head with multiple spray nozzles to spray tiny droplets of melted liquid material. This process allows for the simultaneous deposition of a range of materials, which means that a single part can be produced from multiple materials with different characteristics and properties. MJM is a very precise 3D printing method, producing accurate parts with a very smooth finish.

11.4
3D Direct Laser Writing by Multiphoton Polymerization

Most 3D printing method cannot fabricate nanostructure. In recent years, direct laser writing by multiphoton polymerization (MPP) has been studied to fabricate complex 3D micro-/nanodevices [11–14]. Direct laser writing by MPP has been considered as a promising technique for real 3D nanofabrication. In MPP, when an ultrashort-pulsed laser beam is closely focused on the liquid-state resin, a photosensitizer chromophore is excited by the simultaneous absorption of multiphotons, and then the chain-reaction polymerization occurs [15]. MPP is a nonlinear optical process whose resolution can be less than 100 or 200 nm by using a high numerical aperture (NA) objective lens. Pioneering work on single-photon direct laser writing was performed about 20 years ago [16, 17]. Then direct laser writing by MPP was first demonstrated in 1997 [18].

Two-photon absorption was theoretically predicted in 1931 [19], and it was verified by an experiment in 1961 [20]. Two-photon absorption is defined as the simultaneous absorption of two photons of identical or different frequencies in order to excite a molecule from the ground state to a higher state [21]. The diagram of energy transitions in two-photon absorption is shown in Figure 11.5. Two-photon absorption is a second-order nonlinear optical process, which is most efficient at very high intensities. That is why pulsed lasers are usually used to investigate two-photon absorption [21].

A typical direct laser writing experimental setup is shown in Figure 11.6 [22]. The main components of a direct laser writing system include an ultrafast laser,

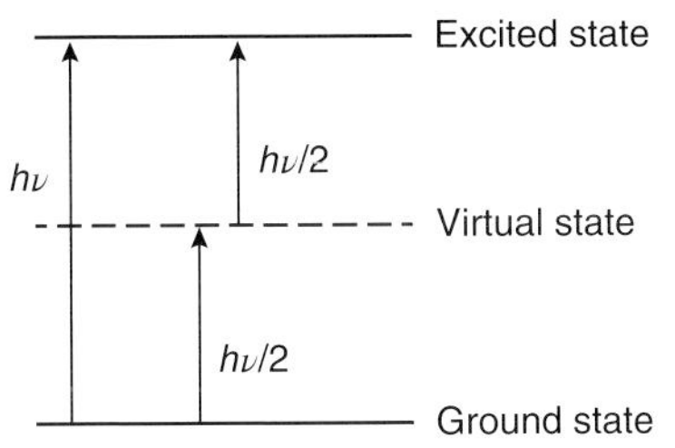

Figure 11.5 The diagram of energy transitions in two-photon absorption.

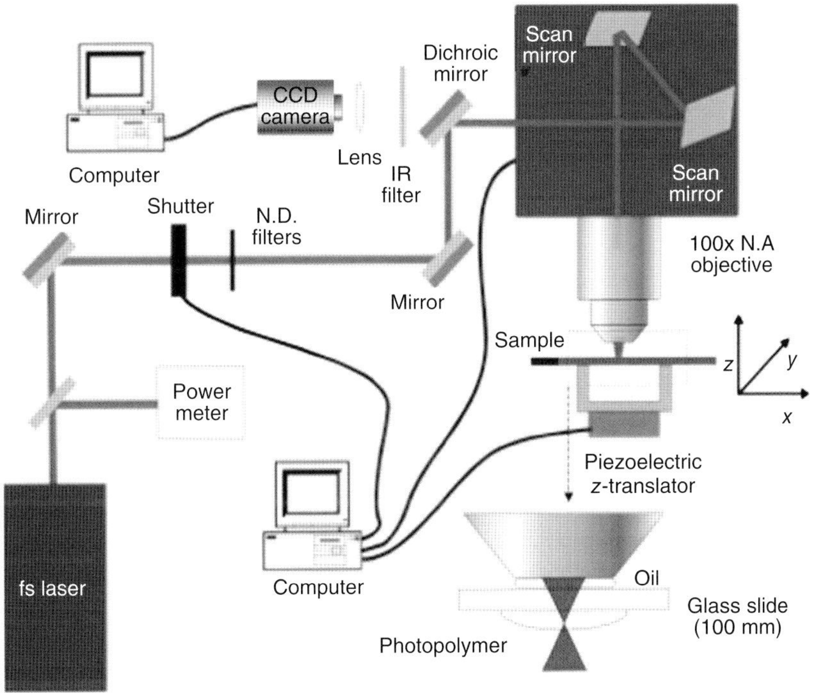

Figure 11.6 A typical experimental setup consisting of a light source, beam and sample movement components, beam control and focusing optics, and a vision system. (Selimis *et al.* [22]. Reproduced with permission from Elsevier.)

a beam/sample motion system, beam focusing optics, beam intensity control and beam shutter, and control software [22]. The laser source is usually a Ti:sapphire femtosecond pulsed laser with a pulse length of less than 200 fs and a repetition rate of 50–80 MHz. The energy of laser is usually in the order of a few nanojoules per pulse which depends on the material, the photoinitiator, and the focusing. To write a 3D nanostructure into the photopolymer using the focused laser beam, the motion system should have high resolution. The laser can be moved with galvanometric scanners, while the sample should be moved with *xyz* stages. Now, piezoelectric stages are usually used as *xyz* stages, which have nanometer-scale accuracy. But it can only move small distances, about a few hundred micrometers, and its moving speed is slow, about at tens of microns per second. So, linear or step-motor stages are also used to finish a large-scale movement. Beam focusing optics is used to induce high intensities of laser which are required for multiphoton absorption and high resolution. In the system, a fast mechanical shutter or an acousto-optic modulator, neutral density filters, a variable attenuator or a combination of a polarizer and a waveplate is used to control the laser beam.

The experimental procedure for fabricating a 3D structure by direct laser writing is shown in Figure 11.7 [22]. First, the laser beam is tightly focused on the material. Second, the focused beam or the sample is moved following a designed pattern. After the laser writing of the structure, the sample is immersed into an appropriate developer. At last, the 3D polymeric freestanding structure is fabricated.

The materials suitable for 3D direct laser wiring consist of at least two components: a monomer or a mixture of monomers/oligomers, which can provide the final polymer, and a photoinitiator, which can absorb the laser light and provide the active species that will promote the polymerization [22]. Hydrogels, acrylate materials [23, 24], the epoxy-based photoresist SU-8 [25],

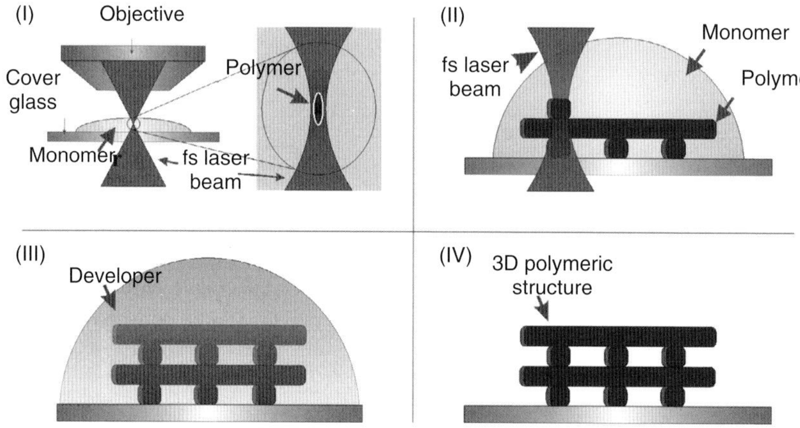

Figure 11.7 The experimental procedure for fabricating a 3D structure by direct laser writing. (Selimis *et al.* [22]. Reproduced with permission from Elsevier.)

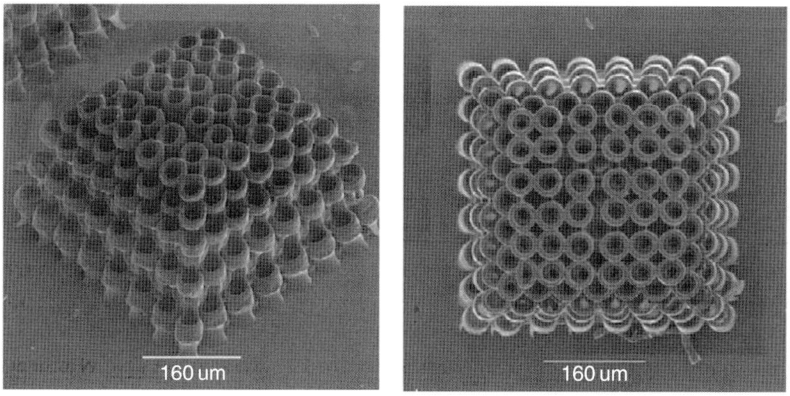

(a) (b)

Figure 11.8 3D scaffolds made by direct laser writing using PEG as a photopolymer. (Ovsianikov *et al.* [30]. Reproduced with permission from Elsevier.)

and hybrid materials [26, 27] are mostly used as negative photoresists. Recently, redox and Diels–Alder photopolymerization have also been reported to be used in 3D printing. Among these materials, hydrogels are very popular for fabricating scaffolds for tissue engineering, as they provide a microenvironment that resembles the extracellular matrix [28, 29]. Hydrogels can be broadly divided into two categories: synthetic hydrogels and natural polymers. The most popular synthetic hydrogel is polyethylene glycol, commonly known as *PEG*. It is cheap and can exist in a lot of different formulations. Figure 11.8 shows the 3D scaffolds made by two-photon absorption of direct laser writing using PEG as a photopolymer [30]. It is composed of 50-μm-tall hollow cylinders with radius of 25 μm in a square arrangement within each layer. Every second layer is shifted by the length of the cylinder diameter along the diagonal.

11.5 3D Printing Applications

11.5.1 Medical Applications

It has been demonstrated that 3D printing technology can be used in medical applications [31]. The use of 3D printing technology for diagnostic purpose was probably the first medical application. Many of surgeons' tasks involve working inside human bodies. And much of their operating procedure is carried out using the sense of touch. So, models that they can both see from any angle and feel with their hands are very useful to them. They use models to understand the complex surgical procedures and assist themselves in this process. Now, models fabricated by 3D printing can help surgeons to reduce time in surgery for complex cases by allowing the surgeons to better plan ahead of time and to better understand the situation during the procedure.

Tissue engineering is an interesting emerging area of medical science. It can be defined as the science and engineering of functional tissues and organs for the repair of diseased body parts [32]. Now, 3D printing techniques have recently been introduced in the tissue engineering field. Data obtained from computed tomography (CT) or magnetic resonance imaging (MRI) medical scans can be used to create customized CAD models. The desired implant area of the patient is scanned by CT or MRI, and the data is imported into CAD software. The software enables a surgeon to design an implant according to individual needs. After the information is transferred to a 3D printing system, a biocompatible and biodegradable scaffold is fabricated layer by layer [32, 33].

The first applications of 3D printing techniques in the field of tissue engineering were primarily implants used in craniomaxillofacial surgery. Levy *et al.* fabricated hydroxyapatite (HA) scaffolds for orbital floor implants. The development of biomimetic bone substitutes is also a growing field of research. A great variety of materials have been used for replacement and repair of damaged bones. These materials include metals, ceramics, polymers, and their combinations. Bioceramics are the preferred materials as bone grafts due to their low density, chemical inertness, high wear resistance, excellent tissue adherence, and similarity with natural bones. Polymeric and ceramic materials can be resorbable in biological environment [34, 35]. A more complex polymeric scaffold fabricated is presented in Figure 11.8.

Although several kinds of polymers have been investigated for bone tissue engineering, no single polymer can satisfy all the requirements of a bone substitute. Composite materials may offer a future solution to the problems. In fact, natural bone matrix is an organic/inorganic composite of collagen [35]. Polymer–ceramic composite materials have the desirable properties of each of the constituent materials, such as high wear resistance of ceramics and toughness of polymers. Nevertheless, 3D printing techniques already offer an efficient way to control the design, fabrication, and modeling of scaffold-based tissue engineering.

11.5.2 Industrial Manufacturing

3D printing can fabricate aerospace sector, the automotive sector, and architectural models. Now, some noncritical parts fabricated by 3D printing are already implemented on aircraft. The users include GE/Morris Technologies, Airbus/EADS, Rolls-Royce, BAE Systems, and Boeing. Academic and research institutes are developing 3D printing process and materials for more aerospace sectors. Many automotive companies, particularly at the cutting edge of motor sports and F1, have utilized the 3D printing techniques like the aerospace companies. They use the technologies for prototyping applications. And many automotive companies are now also looking at the potential of 3D printing for the production of replacement parts rather than holding huge inventories. Architectural models have a long and staple application of 3D printing processes for producing accurate demonstration models of an architect's vision. 3D printing offers a relatively fast, easy, and economically viable method of producing detailed models directly from 3D CAD. Many successful architectural firms now commonly use 3D printing (in-house or as a service) as a critical part of their workflow for increased innovation and improved communication.

11.5.3 Daily Consumption

3D printing has entered into our life and improved it, such as fabricating furniture, jewelry, food, and so on. In terms of resolution and more flexible materials, 3D printed accessories include shoes, headpieces, hats, and bags. Initial forays into 3D printing food were with chocolate and sugar, and these developments have continued apace with specific 3D printers hitting the market. Some other early experiments with food include the 3D printing of "meat" at the cellular protein level. More recently, pasta is another food group that is being researched for 3D printing food.

11.5.4 Limitation of 3D Printing Applications

Although there is no doubt that 3D printing technologies are useful in many fields, there are numerous deficiencies in existing 3D printing technologies. Key issues that may change these deficiencies in favor of using 3D printing technologies include speed, cost, accuracy, materials, and operation. 3D printing is of relatively slow speed. The models can often take a day or even longer to fabricate. Although many applications currently do not require higher accuracy, most 3D printing methods can only manufacture features at scales larger than 100 μm. And another important factor limits its applications for medical models is that only a few 3D printing polymer materials are safe for being placed inside the body.

11.6 Summary

In summary, 3D printing can create complete models in a single process, which has been proven to save time, manpower, and cost. This technology has broad applications in our future life, which will bring significant changes in many fields. But microscale and nanoscale 3D printing technologies still need more efforts on the materials and the process of printing to improve the printing resolution and speed.

Acknowledgments

The authors acknowledge the National Basic Research Program of China (973 Program) under Grant No. 2015CB351905 and the National Natural Science Foundation of China under Grant Nos. 61306015 and 51372034.

References

1. Cima, M., Sachs, E., Fan, T.L., Bredt, J.F., Michaels, S.P., Khanuja, S., Lauder, A., Lee, S.J., Brancazio, D., Curodeau, A., and Tuerck, H. (1995) Three-dimensional printing techniques. US Patent 5387380.
2. Yan, X. and Gu, P. (1996) *Comput.-Aided Des.*, **28**, 307.
3. Pham, D.T. and Gault, R.S. (1998) *Int. J. Mach. Tools Manuf.*, **38**, 1257.
4. Dickens, P.M. (1995) *Proc. Inst. Mech. Eng. Part B-J. Eng. Manuf.*, **209**, 261.
5. Jacobs, P.F. (ed.) (1996) *Stereolithography and Other RP and M Techniques*, ASME Press, New York.
6. Klocke, F., Celiker, T., and Song, Y.A. (1995) *Rapid Prototyping J.*, **1**, 32.
7. Ru, C., Luo, J., Xie, S., and Sun, Y. (2014) *J. Micromech. Microeng.*, **24**, 053001.
8. Kullmann, C., Schirmer, N.C., Lee, M.-T., Ko, S.H., Hotz, N., Grigoropoulos, C.P., and Poulikakos, D. (2012) *J. Micromech. Microeng.*, **22**, 055022.
9. Yamaguchi, K., Sakai, K., Yamanaka, T., and Hirayama, T. (2000) *Precis. Eng.*, **24**, 2.
10. Ko, S.H., Chung, J., Hotz, N., Nam, K.H., and Grigoropoulos, C.P. (2010) *J. Micromech. Microeng.*, **20**, 125010.
11. Zhou, W.H., Kuebler, S.M., Braun, K.L., Yu, T.Y., Cammack, J.K., Ober, C.K., Perry, J.W., and Marder, S.R. (2002) *Science*, **296**, 1106.
12. Boiko, Y., Costa, J.M., Wang, M., and Esener, S. (2001) *Opt. Express*, **8**, 571.
13. Belfield, K.D., Schafer, K.J., Liu, Y.U., Liu, J., Ren, X.B., and Van Stryland, E.W. (2000) *J. Phys. Org. Chem.*, **13**, 837.
14. Albota, M., Beljonne, D., Bredas, J.L., Ehrlich, J.E., Fu, J.Y., Heikal, A.A., Hess, S.E., Kogej, T., Levin, M.D., Marder, S.R., McCord-Maughon, D., Perry, J.W., Rockel, H., Rumi, M., Subramaniam, C., Webb, W.W., Wu, X.L., and Xu, C. (1998) *Science*, **281**, 1653.
15. Lee, K.-S., Kim, R.H., Yang, D.-Y., and Park, S.H. (2008) *Prog. Polym. Sci.*, **33**, 631.
16. Gale, M.T., Rossi, M., Pedersen, J., and Schuetz, H. (1994) *Opt. Eng.*, **33**, 3556.
17. Pelli, S., Righini, G.C., Scaglione, A., Guglielmi, M., and Martucci, A. (1996) *Opt. Mater.*, **5**, 119.
18. Maruo, S., Nakamura, O., and Kawata, S. (1997) *Opt. Lett.*, **22**, 132.
19. Göppert-Mayer, M. (1931) *Ann. Phys.*, **401**, 273.
20. Kaiser, W. and Garrett, C.G.B. (1961) *Phys. Rev. Lett.*, **7**, 229.
21. Boyd, R.W. (ed.) (2008) *Nonlinear Optics*, Academic Press.
22. Selimis, A., Mironov, V., and Farsari, M. (2015) *Microelectron. Eng.*, **132**, 83.
23. LaFratta, C.N., Fourkas, J.T., Baldacchini, T., and Farrer, R.A. (2007) *Angew. Chem. Int. Ed.*, **46**, 6238.
24. Farsari, M., Filippidis, G., Sambani, K., Drakakis, T.S., and Fotakis, C. (2006) *J. Photochem. Photobiol., A*, **181**, 132.
25. Juodkazis, S., Mizeikis, V., and Misawa, H. (2009) *J. Appl. Phys.*, **106**, 051101.
26. Farsari, M. and Chichkov, B.N. (2009) *Nat. Photonics*, **3**, 450.
27. Farsari, M., Vamvakaki, M., and Chichkov, B.N. (2010) *J. Opt.*, **12**, 124001.
28. Torgersen, J., Qin, X.-H., Li, Z., Ovsianikov, A., Liska, R., and Stampfl, J. (2013) *Adv. Funct. Mater.*, **23**, 4542.
29. Drury, J.L. and Mooney, D.J. (2003) *Biomaterials*, **24**, 4337.
30. Ovsianikov, A., Malinauskas, M., Schlie, S., Chichkov, B., Gittard, S., Narayan, R., Löbler, M., Sternberg, K., Schmitz, K.P., and Haverich, A. (2011) *Acta Biomater.*, **7**, 967.
31. Webb, P.A. (2000) *J. Med. Eng. Technol.*, **24**, 149.

32. Peltola, S.M., Melchels, F.P.W., Grijpma, D.W., and Kellomaki, M. (2008) *Ann. Med.*, **40**, 268.
33. Landers, R., Pfister, A., Hubner, U., John, H., Schmelzeisen, R., and Mulhaupt, R. (2002) *J. Mater. Sci.*, **37**, 3107.
34. Kalita, S.J., Bose, S., Hosick, H.L., and Bandyopadhyay, A. (2003) *Mater. Sci. Eng., C-Biomimetic Supramol. Syst.*, **23**, 611.
35. Liu, X.H. and Ma, P.X. (2004) *Ann. Biomed. Eng.*, **32**, 477.

12
Nanostructured Thin Film Solid Oxide Fuel Cells

Alex Ignatiev, Rabi Ebrahim, Mukhtar Yeleuov, Daniel Fisher, Xin Chen, Naijuan Wu, and Serekbol Tokmoldin

12.1
Introduction

The increasing demand on energy has stimulated extensive research on renewable and alternative energy utilization. Within this area, research on fuel cells has accelerated due to their limiting emission of only water vapor. A number of different types of fuel cells have been developed; however, recent attention has focused on solid oxide fuel cells (SOFCs) due to their high efficiency (60–70%) [1–5]. Fuel cells directly convert the chemical energy of fuel into electricity. As a result, their energy conversion efficiency is high as they are electrochemical systems and not thermal systems and hence are not limited by Carnot. Thus, the usual losses involved in the conversion of fuel heat to mechanical energy and then to electrical energy are avoided [1]. Fuel cells are capable of using practical fuels with minimal environmental pollution. Fuel cells also have the advantages of modular construction and low operating noise. However, utilization of one of the most efficient fuel cells, the SOFC, has not fully reached the commercial stage due to the very high temperature of operation of SOFCs and the resultant relatively expensive large-scale manufacturing processes [3, 6, 7]. Reduction of operating temperature is critical to the full commercial utilization of SOFCs and the realization of their fuel efficiency and environmental benefits. This can be achieved by the development and application of SOFCs with ultrathin electrolyte layers that have resultant operating temperatures in the 450–550 °C range.

12.2
Solid Oxide Fuel Cells

The principle of SOFC operation is shown in Figure 12.1. An SOFC consists of an electrolyte and two electrodes: cathode and anode. The typical electrolyte is an oxygen ion conductor such as yttria-stabilized zirconia (YSZ), one of the most commonly used electrolyte materials. Oxygen ion conductors have oxygen

Advanced Nano Deposition Methods, First Edition. Edited by Yuan Lin and Xin Chen.

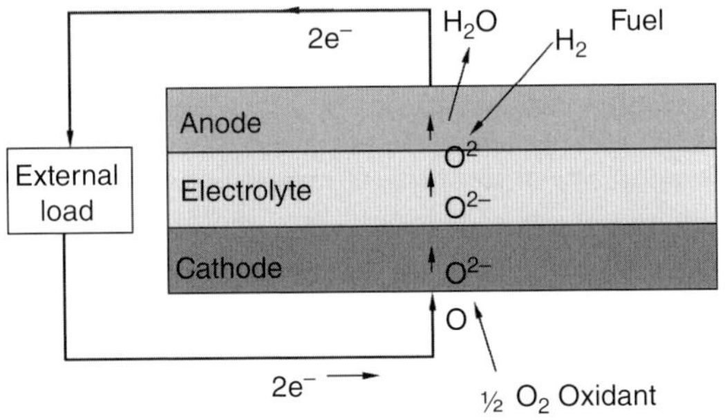

Figure 12.1 A schematic diagram of the principle of solid oxide fuel cell operation.

vacancies in the crystal structure, so that oxygen ions can readily transport through them at elevated temperatures. The cathode side of the SOFC is exposed to an oxidant (air or O_2). The oxygen molecules fed to the cathode undergo reduction reactions, capture electrons, and become oxygen ions. The cathode must not only have high ionic conductivity but must also have high electronic conductivity in that it is used as an electrode for the cell. Hence, oxide materials used as cathodes in SOFCs are identified as mixed ionic/electronic conductors (MIECs) of which $LaSrCoO_3$ (LSCO) is a typical cathode material. The oxygen ions move from the cathode to the anode through the electrolyte by means of diffusion and combine with the fuel (H_2) at the anode side. The anode is typically a metal (nickel) and must be porous for the transport of fuel and for the release of water. The combination of hydrogen and oxygen ions at the anode/electrolyte interface releases electrons into an external circuit, which then can do work, and results in water as the reaction product that is released to the environment.

As noted, the high operating temperatures for SOFCs (800–950 °C) [5, 6] lead to complications in SOFC manufacture and application including large size and high cost per kilowatt. Significant benefits would accrue to SOFC design and utilization if the temperature of operation could be lowered. This can be addressed in two principal ways: (i) development of a new oxide electrolyte material with higher oxygen ion conductivity and hence higher ion currents and (ii) thinning the electrolyte layer to yield reduced transit time for oxygen ions and hence higher ion currents.

A number of oxide electrolytes have been developed for increased ionic conductivity with resultant lowering of operating temperature [8]; however, they generally have drawbacks that limit their application including interdiffusion, reduction to an electronic conducting state, and high expense [9]. Reduction of electrolyte thickness, however, could be effective in reducing operating temperature.

It is projected that in order to reduce operating temperature to $\leq$500 °C, the electrolyte thickness should approach 1 μm. Such thicknesses for the electrolyte layer would generally be obtained by physical or chemical vapor deposition. Such deposition would then typically take place on the anode as the substrate. It should be noted that for the metallic anode to be an effective fuel cell component, it must be porous. This presents a challenge for vapor deposition as the electrolyte layer must have complete coverage of the porous anode and thus must be impervious to gas flow. Porous media are difficult to completely cover with thin films since the vapor deposition process is somewhat conformal and will generally require a thickness level equal to the largest pore size for complete coverage. Hence, in order to utilize a 1-μm-thick electrolyte layer that must completely cover a porous anode, the anode pore structure must have surface pore sizes of less than 1 μm. Such nanoporosity is typically not available in standard porous metallic materials, and thus another approach is required to fabricate an SOFC with a thin electrolyte layer.

Vapor deposition of oxide thin films is quite direct on metallic substrates, and nickel substrates have been used in a number of oxide thin film applications including high-temperature superconducting wire fabrication [10]. Further, nickel foil substrates have been processed to yield atomically ordered substrates for thin film oxide deposition [11]. Thus the growth of oxide thin films on atomically ordered nickel foil could result in atomically ordered oxide thin films. This point is of importance for SOFCs as atomic order for the grown oxide electrolyte on the ordered nickel anode would yield an anode/electrolyte interface with minimal defects, thus reducing interface resistance and increasing cell output.

For the thin film SOFC with projected lowered operating temperature, a thin (~1 μm) electrolyte layer of YSZ would need to be grown on an atomically ordered nickel foil substrate. In this thin film structured fuel cell, the LSCO oxide cathode would then be grown on the YSZ electrolyte, and an anode-supported thin film fuel cell would be obtained. The challenge with this design is the need for the nickel foil anode to be porous. To realize porosity in the foil anode, a lithographically based patterning process was developed for the nickel foil substrate to reveal closely spaced micron-sized pores for fuel feed through the nickel.

12.2.1
Thin Film Solid Oxide Fuel Cell Fabrication

Ten-micron-thick atomically ordered roll-textured nickel foil [11] (1.6 cm × 1.6 cm) was prepared for use as the substrate/anode upon which the cell would be fabricated. The nickel foil was treated in a mixture of acetic, nitric, sulfuric, and phosphoric acids to remove the rolling marks, after which the foil was annealed at 650 °C for 2 h in an argon atmosphere to reinforce the (100) lattice orientation normal to the foil surface. The YSZ electrolyte layer of about 1.5 μm thickness was deposited onto the nickel foil using pulsed laser deposition (248 nm KrF excimer laser) at a temperature of 650 °C. This yielded a smooth and highly atomically ordered YSZ film on the roll-textured nickel foil as shown by the SEM micrograph and the XRD 2θ scan in Figure 12.2b.

Figure 12.2b shows YSZ highly ordered in the normal direction to the foil. Figure 12.3 shows that the YSZ film grown on the atomically ordered nickel foil is also ordered in the "in-plane" direction as revealed by the sharp YSZ peaks in the GAADS ϕ-scan of Figure 12.3b. It can be seen that the YSZ film lattice is rotated at 45° with respect to the nickel lattice due to the lattice constant mismatch of

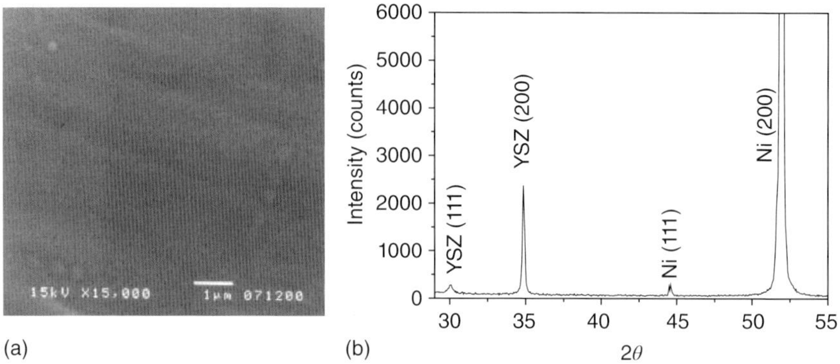

Figure 12.2 (a) SEM micrograph of the YSZ film deposited on the nickel foil substrate indicating a smooth YSZ surface. (b) XRD 2θ scan indicating a YSZ film (100) ordered in the film normal direction.

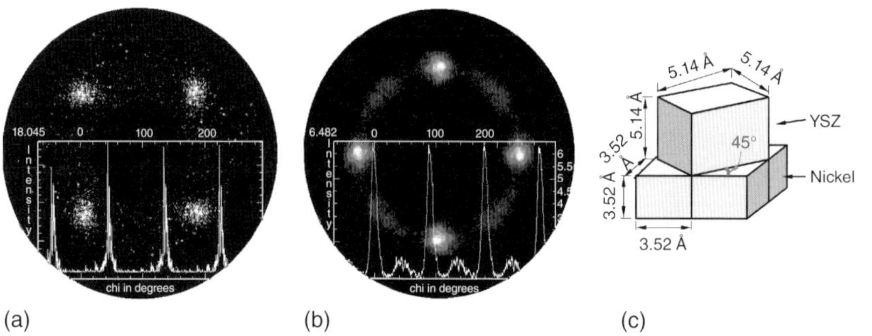

Figure 12.3 (a) GAADS ϕ-scan of the roll-textured nickel foil showing strong in-plane atomic ordering of the nickel; (b) GAADS ϕ-scan of YSZ on the nickel foil showing the 45° rotation of the YSZ lattice with respect to the nickel lattice; and (c) lattice cell diagram of the lattice structure of the nickel and YSZ materials with indication of the 45° rotation of the YSZ lattice to the nickel lattice.

the two materials ($a_{Ni} = 3.52$ Å vs. $a_{YSZ} = 5.14$ Å with $a_{YSZ} \approx 2^{1/2} a_{Ni}$). This high degree of atomic order of the YSZ on nickel ensures a well-ordered interface and thus reduced interface resistance due to the lack of grain boundaries. The growth of the YSZ layer was then followed by pulsed laser deposition growth of the LSCO cathode layer of about 2 μm thickness onto the YSZ/nickel system.

As the nickel foil to be used as an anode was not porous, porosity was generated by patterning and etching the foil. A hexagonal array of pores of about 50 μm diameter with 50 μm spacing was formed in the nickel foil by photolithographic patterning followed by electrochemical etching (using 6 M H_2SO_4) at room temperature [8] so as to form a porous nickel anode for the fuel cell elements. The YSZ layer in the material stack acted as an effective etch stop in the etching process with excellent etching of the pores in the nickel as can be seen in Figure 12.4. Further, the YSZ/LSCO layer (~3 μm thick) originally on top of the nickel foil was well

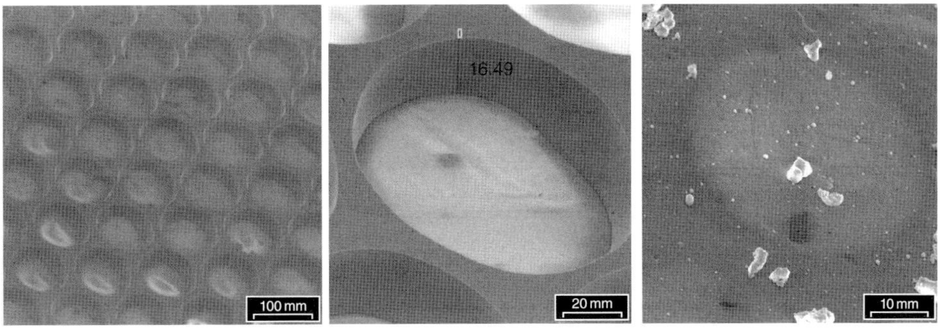

(a) (b) (c)

Figure 12.4 SEM micrographs of photolithographically patterned and etched nickel foil substrate; (a) hexagonal array of etched pores; (b) single pore showing a continuous membrane at the bottom of the pore; and (c) the LSCO/YSZ side of the etched sample showing a continuous and intact LSCO/YSZ coating with the shadow of the pore in the nickel layer below.

preserved after the etching process with a continuous and robust film membrane seen at the bottom of the pore as shown in Figure 12.4b,c.

Vapor deposition is very effective in the fabrication of high density films, which bodes well for the YSZ electrolyte layer as it needs to be impervious to gas flow and electron flow. However, with respect to the cathode layer, high density is not overly beneficial. The cathode is an MIEC, and to maximize ion current, it may be of benefit to increase surface area for enhanced oxygen dissociation through an increase in porosity of the cathode. This was accomplished by growing the LSCO films at room temperature to yield columnar growth and resultant increased surface area, and then annealing it to increase conductivity. Figure 12.5 shows the area-specific resistance data for LSCO films grown at room temperature and annealed in air at 450, 550, 650, and 750 °C and compared to a high-density film grown at 550 °C. It can be seen that the LSCO film grown at room temperature and annealed at 550 °C yields the best resistance value for the cathode.

The SEM micrographs of the surface of the LSCO films processed under different anneal temperatures are shown in Figure 12.6 where it is clear that surface porosity is obtained under all of the process conditions. The 550 °C annealed films have adequate surface porosity along with the good value of area-specific resistance seen in Figure 12.5 and was hence chosen as the cathode for the thin film fuel cell. There was however concern that the porosity of the LSCO could lead to gas transport through the cathode to the electrolyte layer if the cathode was totally porous. To mitigate such a possibility, a dense LSCO layer of about 200 nm was first grown at 550 °C on the YSZ followed by the room temperature growth of an additional 500-nm-thick LSCO layer and annealed at 550 °C [7].

This resulted in a two-layered structure (dense bottom layer and porous top layer) for the cathode, thus preventing gas transport through the LSCO. Figure 12.7 shows the SEM micrographs of the initial LSCO dense layer and the final annealed LSCO porous layer after processing.

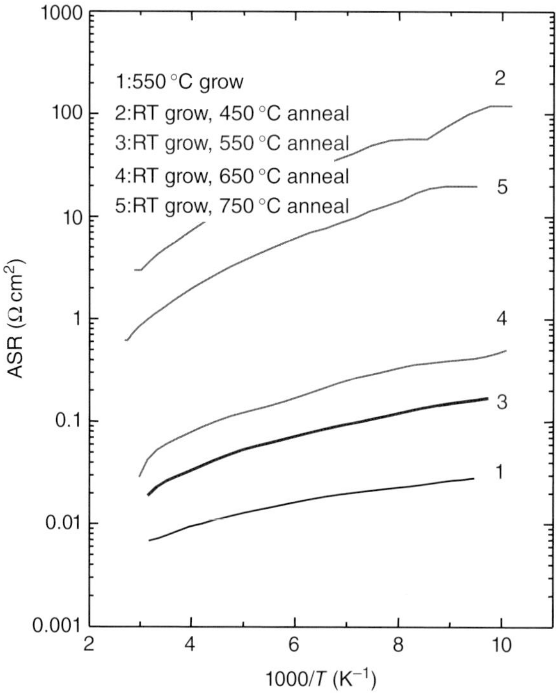

Figure 12.5 The area-specific resistance (ASR) for LSCO films grown at room temperature and annealed at various temperatures. Note that the 550 °C annealed film has the lowest resistance of the annealed group of samples.

Figure 12.6 SEM micrographs of LSCO films grown at room temperature and annealed at various temperatures: (a) 450 °C, (b) 550 °C, (c) 650 °C, and (d) 750 °C. The film annealed at 550 °C was chosen for application as a cathode for the thin film fuel cell.

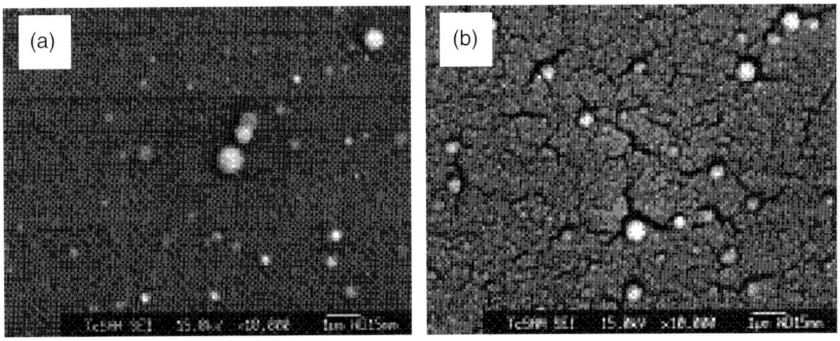

Figure 12.7 SEM of (a) the dense bottom LSCO layer and (b) the top of the annealed porous LSCO layer. The particles seen are due to the laser deposition process.

The annealing process for the LSCO cathode layer also enhanced the surface exchange coefficient for the cathode. The oxygen transport through a mixed conductor is determined by both the diffusion coefficient D_o and the oxygen surface exchange coefficient k [12]. The value of D_o/k is L_c. For samples with thickness L much greater than L_c, the transport is limited by the diffusion inside the film; for thin films with L much smaller than L_c, the transport is limited by the surface exchange rate [13]. L is much smaller than L_c for the thin film LSCO cathode; hence, it is critical to optimize surface exchange. To this end, it was identified that postannealed LSCO films had a higher surface exchange coefficient than preannealed LSCO films as can be seen in Figure 12.8. The significant increase in surface exchange was linked to an increase in surface roughness of the cathode film after annealing, which leads to an increase in overall surface area and resultant increase in oxygen dissociation rate at the cathode. Thus, not only does the annealing process improve the area-specific resistance of the LSCO cathode, but it also enhances the surface exchange of oxygen.

As previously noted, it is projected that the atomically ordered nature of the nickel substrate and the YSZ and LSCO films should lead to reduced interface resistance in the thin film fuel cell due to the reduction of defects at the atomically ordered dense bottom layer of LSCO on the YSZ electrolyte. To confirm this premise, impedance spectroscopy analysis was performed on an LSCO/YSZ/LSCO sample in the temperature range from 400 to 650 °C in air. A representative impedance plot obtained on the sample and the analysis on it are shown in Figure 12.9. The distinctive features of the ionic conduction of the YSZ electrolyte and the oxygen exchange on the LSCO electrode surface were observed on the impedance spectra. As shown in Figure 12.9a, the dotted curve is the measured data of the sample taken at 450 °C. In the plot, the left semicircle corresponds to the YSZ bulk ion conductivity, and the right semicircle is due to the LSCO composite electrode. The interfacial impedance between the bulk and the electrode is small and hardly discerned in the diagram. Figure 12.9b is the equivalent circuit of the sample, where Rs is the resistance of the sample leads, R1 is the resistance of the YSZ bulk, and CPE is a constant-phase element

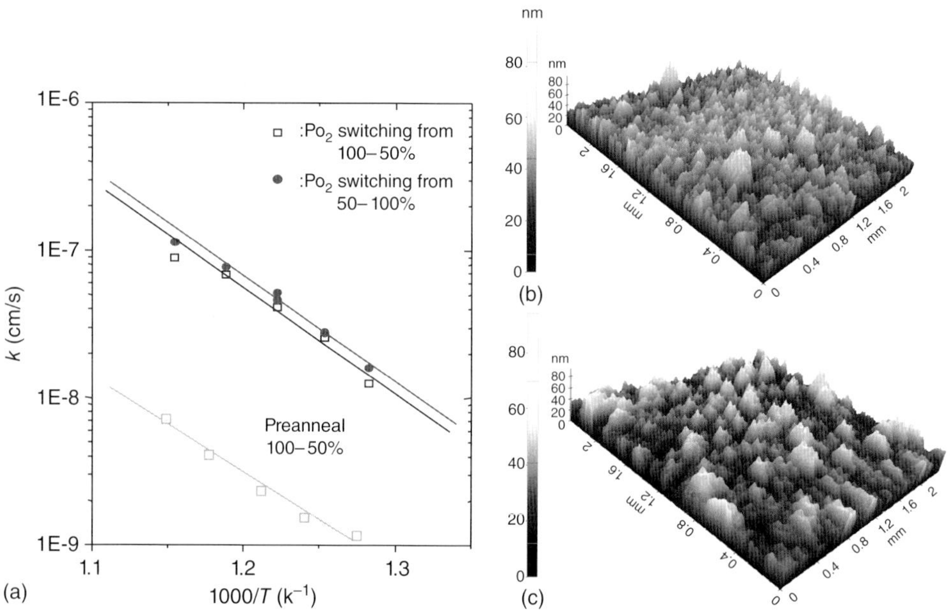

Figure 12.8 (a) Surface exchange coefficient (k) for LSCO films pre- and postannealed. Note the more than 1 order of magnitude increased in surface exchange coefficient upon annealing; (b) AFM scan of LSCO film preannealed; and (c) AFM scan of the LSCO film postannealed. Note the increase in surface roughness of the postannealed film.

representing the capacitive contribution of the YSZ bulk. Due to the low interface impedance and the low LSCO dense layer impedance, the total electrode impedance can be principally represented by the impedance of the LSCO porous electrode alone, which has the form [14]

$$Z_{\text{chem}} = R_{\text{chem}}(1 + jwt_{\text{chem}})^{-1/2} \tag{12.1}$$

where R_{chem} and t_{chem} can be defined by

$$t_{\text{chem}} = 1/\text{GE}_P \tag{12.2}$$

$$R_{\text{chem}} = \frac{1}{\text{GE}_T\text{GE}_P^{1/2}} \tag{12.3}$$

and GE_P and GE_T are the Gerischer fitting elements. Choosing appropriate values for GE_P and GE_T yields the fit shown as the solid line in Figure 12.9a. The excellent fit of the model to the data indicates confirmation of the model with negligible interface impedance for the YSZ/LSCO interface.

The impedance measurements also yielded resistance of the composite LSCO layer as a function of temperature (Figure 12.10). The composite electrode showed reduced resistance with an increase in temperature and a 1.5 orders of magnitude lower resistance than a dense LSCO film sample, showing a good property of the composite LSCO for the use as an SOFC cathode.

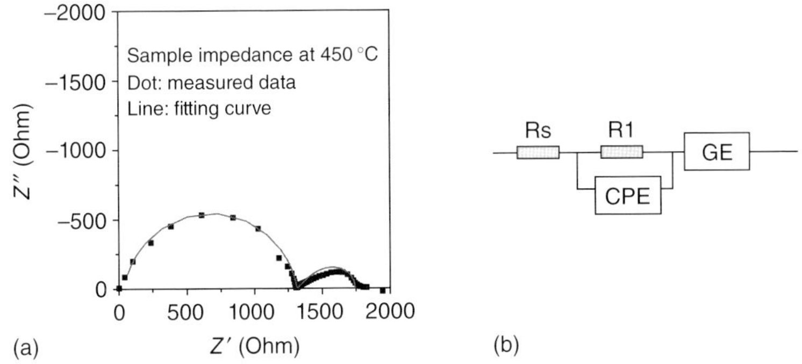

Figure 12.9 (a) LSCO/YSZ/LSCO sample impedance analysis at 450 °C and (b) its equivalent circuit-fitting model.

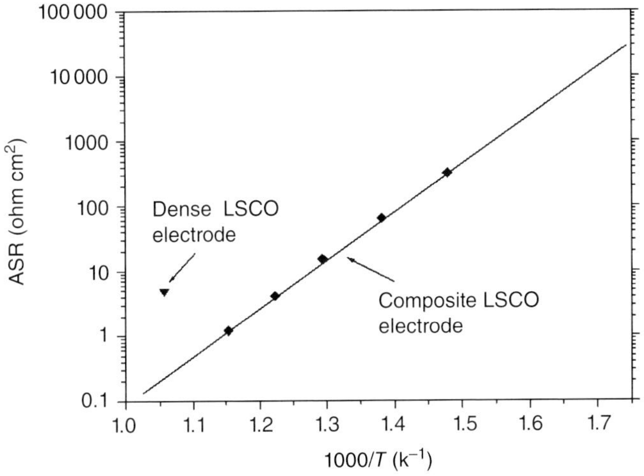

Figure 12.10 Electrode area-specific resistance obtained from impedance measurements. Note the lower ASR for the composite LSCO cathode.

12.2.2
Thin Film Solid Oxide Fuel Cell Testing

The fuel cells were tested over a temperature range of 475–575 °C with the data displayed in Figure 12.11. This figure clearly shows strong power output of the thin film SOFC under much reduced operating temperatures. Significant output is measured at temperatures as low as 480 °C with output as high as 120 mW/cm^2 at 570 °C. Although the areal power density is lower than that reported for bulk SOFCs, the operating temperate is reduced by more than 300 °C due to the thin film structure of the cell. The volume power density, however, is projected to be greater than 1 W/cm^3 due to the thin film structure of the cell.

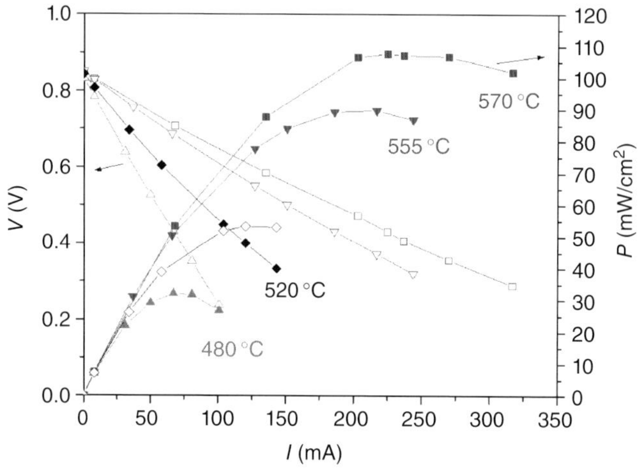

Figure 12.11 Thin film fuel cell performance test at different temperatures.

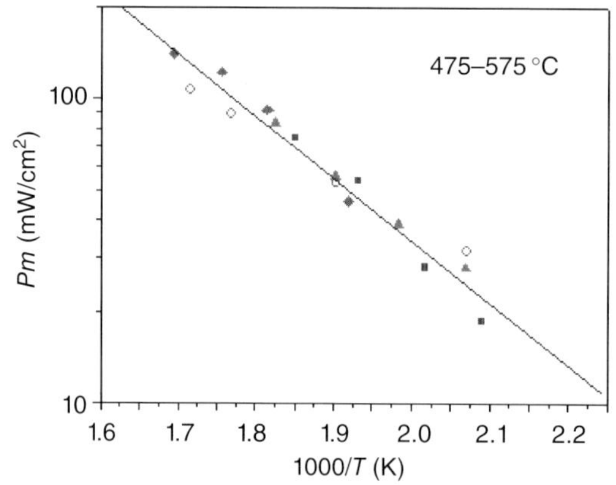

Figure 12.12 SOFC maximum power output at different temperatures from several fuel cells indicated by different shaped data points.

The fuel cell maximum output with temperature shows an Arrhenius relation, and the output of a number of the fuel cells tested at various temperatures can be represented with a single straight line in the maximum power versus reciprocal of temperature Arrhenius plot (Figure 12.12).

As noted, the reduced temperature of operation is principally driven by the low thickness of the YSZ electrolyte. It should therefore be expected that at any one temperature, the cell output should be dependent on electrolyte thickness. Figure 12.13 shows the dependence of cell current on YSZ electrolyte thickness where it can be seen that as YSZ thickness decreases, the cell current increases. This data indicates that increased performance of the thin film SOFC can be realized by further decreasing electrolyte thickness from the nominal 1.5 μm thickness used for the majority of the fabricated cell element.

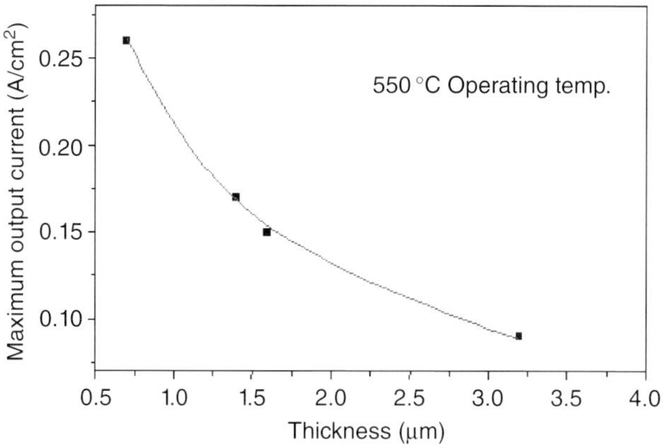

Figure 12.13 Cell output current as a function of YSZ electrolyte thickness at an operating temperature of 550 °C.

Preliminary measure of the efficiency of the thin film SOFC has yielded a value of approximately 50%. Given the nonoptimized nature of the cell, it is expected that this figure will rise with cell optimization.

Since one of the major impacts on fuel cell performance is the anode triple-phase boundary (nickel anode–electrolyte–fuel), and since the photolithographic patterning and etching process can in principle modify the nickel anode–electrolyte contact regions, pore shape in the nickel anode was modified to maximize cell output. An approach to modifying the anode–electrolyte contact region was to specifically etch the nickel so as to not completely etch the pores. This led to the pores in the nickel anode having central islands within the hexagonal frames of the pores [15] (Figure 12.14). The islands are connected to the hexagonal frame at the middle of each side and hence have electrical contact with the nickel anode, therefore yielding good electron transport from the region. They also have contact with the YSZ electrolyte and hence yield a longer nickel–electrolyte contact region that can result in increase of the anode triple-phase boundary. It should be noted that the extent of the anode

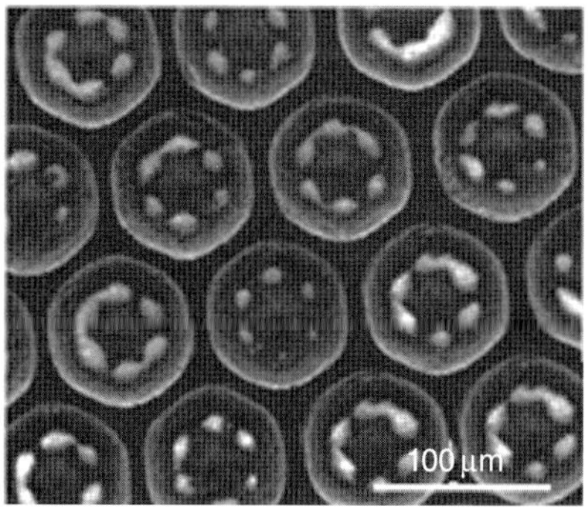

Figure 12.14 SEM micrograph of the nickel side of the LSCO/YSZ/Ni cells after controlled electrochemical etching on the nickel anode. Note the nickel islands in the center of the pores (dark region) connected to the pore sides.

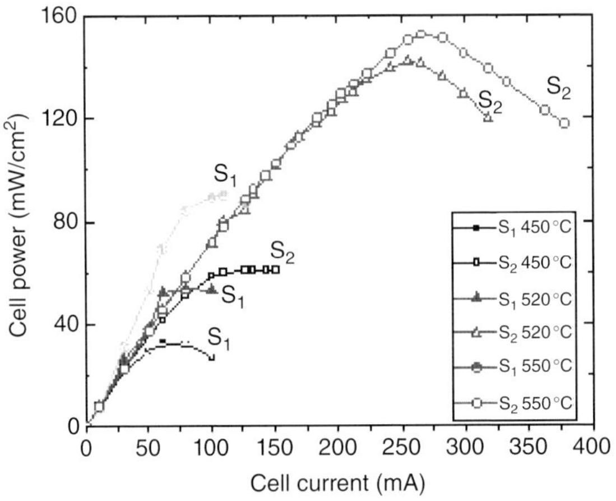

Figure 12.15 Thin film cell output for cells having pores with clean bottoms (S_1) and cells having pores with nickel islands at the bottom (S_2).

triple-phase boundary is not well known; however, recent work [16] has indicated a 5–7 μm reach of the boundary from the physical nickel–YSZ contact point. The presence of the islands does in fact increase output of the cells (Figure 12.15), indicating that increased nickel–YSZ contact perimeter enhances cell performance. Figure 12.15 shows that cell performance is increased by more than a factor of 2 for cells with nickel islands at the bottom of the pores as compared to cells without islands. This process of designing the etch process can therefore be important in enhancing cell performance.

12.2.3
Thin Film Fuel Cell Stack Development and Testing

A number of cell elements are "stacked" together to increase cell output (voltage and current). For this purpose, microformed stainless steel interconnects were placed between the cell elements that were connected in parallel. Although the parallel connection stacking design is unusual as most stacks have cells connected in series, it was utilized to simplify the interconnect design and implementation. Ten individual fuel cell elements were stacked on each other separated by interconnects. A frame for the stack was machined from machinable glass ceramic (Macor) and fitted with inlets for H_2 and air (Figure 12.16a). The interconnects of size 1.6 cm × 1.6 cm were fabricated in a zigzag accordion-like shape from 0.12-mm-thick stamped stainless steel foil and were stacked orthogonally every two cell elements. The stainless foil was coated on both sides with a 200-nm-thick gold layer deposited by e-beam evaporation so that the stainless steel would not oxidize and form high-resistance interface with the cathode (Figure 12.16b). Each square interconnect had a 0.3 cm × 0.3 cm extended tab, which was used for the parallel electrical connection.

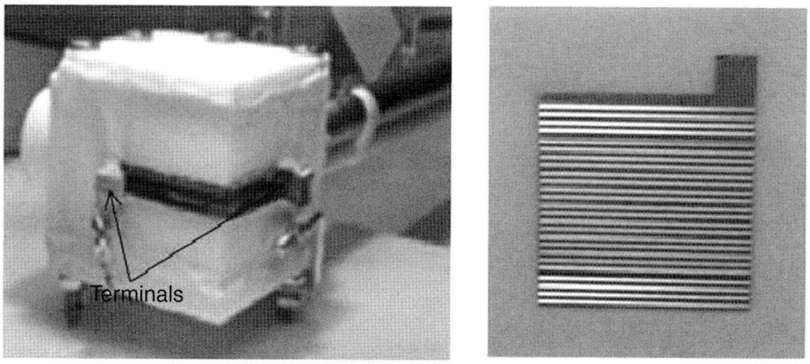

(a) (b)

Figure 12.16 (a) The ceramic frame of the fuel cell stack with cells stacked in parallel and (b) fuel cell stainless steel interconnect coated with gold.

The zigzag channels of the interconnect were set to be orthogonal to subsequent interconnects so as to allow for orthogonal H_2 and airflow. The interconnect channel sides were sealed using liquid ceramic to avoid gas leaks. After stacking of 10 SOFC/interconnect pairs, 2 Macor square caps were fixed tightly on the stack frame by stainless steel bolts to contain the stack. Two stainless steel 1/4 in. tubes were attached orthogonal to the stack so as to deliver air and hydrogen to the stack. The total size of the stack was 2.5 cm × 2.5 cm × 2.5 cm. Finally, the orthogonally oriented interconnect tabs were electrically connected using gold paste to form the "+" and "−" electrical poles of the cell. The assembled stack was electrically connected to the testing setup using nickel wires, and the gases were delivered to the stainless tubes at the anode and the cathode sides of the stack.

The hydrogen air performance of the 10 SOFC element stack and of a single thin film SOFC element used in this stack was measured at fixed temperatures of 450, 520, and 550 °C and at a fixed H_2/airflow rate. The single SOFC cell showed an open-circuit voltage up to 0.9 V, a closed-circuit current of up to 390 mA, and a power density of up to 155 mW/cm^2 (Figure 12.17). For the case of the SOFC stack, an open-circuit voltage of 0.71 V was measured along with a closed-circuit current of 3390 mA and a stack power density of approximately 1270 mW/cm^2 – about 8.6 times higher than that of a single cell. The drop of open-circuit voltage is believed to be due to some leakage of gas in the stack, and the lower power density is related to the lower voltage.

The stack was tested contentiously at 550 °C for 24 h, with no noticeable change in the output power. However, cycling the stack resulted in successive decrease in output which by the 6th or 7th cycle resulted in loss of performance of the stack. Subsequent evaluation of the stack showed fracture of the YSZ/LSCO membrane at the bottom of the nickel pores (Figure 12.18a). This problem occurred due to thermal expansion differences in the nickel and YSZ/LSCO layer of the cell. Calculations of oxide membrane failure as a function of nickel pore size yielded the data of Figure 12.18b, which shows that to minimize membrane failure for 1-μm-thick

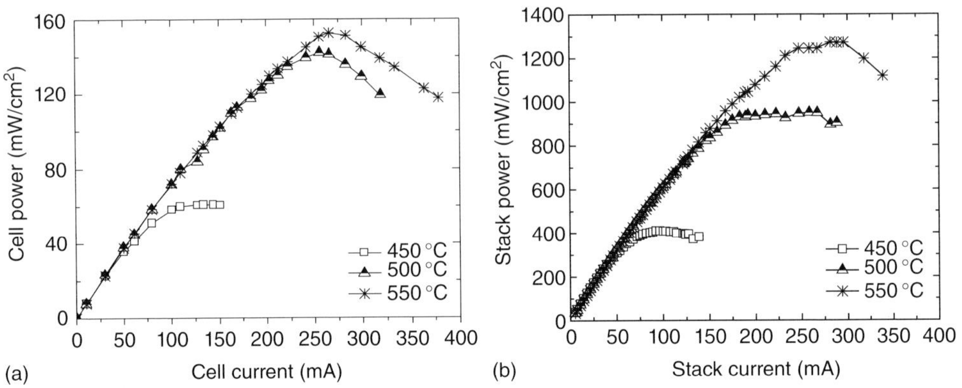

Figure 12.17 Thin film SOFC performance as a function of operating temperatures. (a) Single-cell current versus produced power and (b) 10-cell stack current versus produced power.

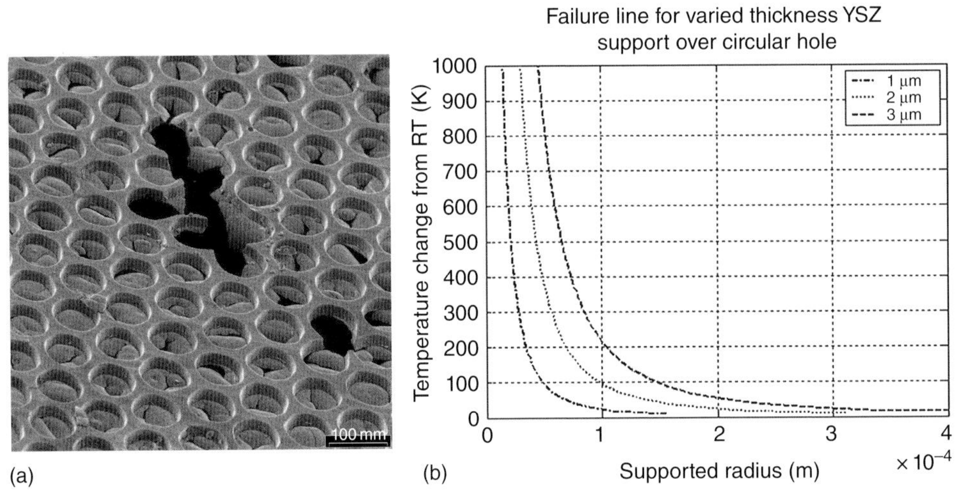

Figure 12.18 (a) SEM micrograph of cracking of the self-support electrolyte membrane after cell testing and (b) calculations of oxide membrane failure as a function of nickel pore diameter.

YSZ/LSCO membrane, the nickel pore size needs to be less than approximately 5 μm [16]. This will be instituted in future cell designs.

It is well to note that the reduced operating temperature of the thin film SOFC can be a key to the use of a broader variety of fuels from the hydrocarbon family. Currently hydrogen gas is obtained from the processing of hydrocarbon fuels – usually as a by-product, but it is hoped and expected that in the future hydrogen will be obtained from the efficient dissociation of water. In the meantime, hydrocarbon fuels will be used to extract hydrogen. However, fuel cells have been shown to operate on hydrocarbon fuels with the need for expensive catalysts (reformers) to dissociate the hydrocarbon to yield hydrogen. Nickel (the anode of our thin film SOFC) is a catalyst for hydrocarbon dissociation but only at low temperatures. At the temperatures of operation of bulk SOFCs, the nickel anode gets "coked" – covered with carbon as a function of time [17] – and then becomes compromised both as a catalyst and as a porous anode. For the thin film SOFC, the lower temperature of operation reduces or eliminates "coking," and thus the thin film SOFC could be self-reforming not requiring expensive catalysts. This could allow thin film SOFCs to operate on plentiful natural gas, but this approach needs to be further developed for thin film SOFCs.

12.3
Summary

The development of thin film SOFCs has resulted in the reduction the SOFC operating temperature due to the much thinner electrolyte layer. The use of thin film deposition coupled with photolithography and electrochemical etching has allowed for the fabrication of cells with ultrathin electrolyte layers and patterned nickel foil substrates yielding porosity for the foil so that it can act as a cell anode. This has resulted in a thin film SOFC that operates at 470–550 °C – a major reduction of operating temperature that not only reduces long-term chemical interactions at the interfaces of the cell layers and therefore increases reliability but also can result in the use of inexpensive stainless steels for the support structure of the cell thus lowering cell costs. The thin film nature of the cells further enhances their applicability as these result in high-volume power densities and thus low cell footprints. The initial performance data for thin film SOFC stacks is encouraging with further nickel anode pore nanostructuring, promising to increase reliability under temperature cycling. Nanostructured thin film SOFCs promise to be an efficient and effective energy producer for the future.

Acknowledgments

The authors acknowledge partial support for this work from the State of Texas through the Center for Advanced Materials at the University of Houston.

References

1. Lynd, L.R., Cushman, J.H., Nichols, R.J., and Wyman, C.E. (1991) Fuel ethanol from cellulosic biomass. *Science*, **251**, 1318.
2. Minh, N. (2004) Solid oxide fuel cell technology – Features and applications. *Solid State Ionics*, **174**, 271.
3. Kordesch, K.V. and Simader, G.R. (1995) Environmental impact of fuel cell technology. *Chem. Rev.*, **95**, 191.
4. Casanova, A. (1998) A consortium approach to commercialized Westinghouse solid oxide fuel cell technology. *J. Power Sources*, **71** (1–2), 65–70.

5. Chen, X., Wu, N.J., Smith, L., and Ignatiev, A. (2004) Thin-film heterostructure solid oxide fuel cells. *Appl. Phys. Lett.*, **84**, 2700.
6. de Souza, S., Visco, S.J., and De Jonghe, L.C. (1997) Thin-film solid oxide fuel cell with high performance at low-temperature. *Solid State Ionics*, **98**, 57.
7. Ignatiev, A., Chen, X., Wu, N., Lu, Z., and Smith, L. (2008) Nanostructured thin solid oxide fuel cells with high power density. *Dalton Trans.*, **40**, 5501.
8. Badwal, S.P.S. and Ciacchi, F.T. (2000) Oxygen-ion conducting materials for solid oxide fuel cells. *Ionics*, **6**, 1.
9. Badwal, S.P.S., Giddey, S., Munnings, C., and Kulkarni, A. (2014) Review of progress in high temperature solid oxide fuel cells. *J. Aust. Ceram. Soc.*, **50**, 23.
10. Zeng, J., Ignatiev, A., Zhou, Y., and Salama, K. (2006) A single oxide buffer layer on cube textured Ni substrate for the development of YBCO coated conductors by photo-assisted MOCVD. *Supercond. Sci. Technol.*, **19**, 1.
11. Goyal, A., Norton, D.P., Budai, J.D., Paranthaman, M., Specht, E.D., Kroeger, D.M., Christen, D.K., He, Q., Saffian, B., List, F.A., Lee, D.F., Martin, P.M., Klabunde, C.E., Hartfield, E., and Sikka, V.K. (1996) High critical current density superconducting tapes by epitaxial deposition of $YBa_2Cu_3O_x$ thick films on biaxially textured metals. *Appl. Phys. Lett.*, **69**, 1795.
12. Chen, X., Wang, S., Yang, Y.L., Smith, L., Wu, N.J., Jacobson, A., and Ignatiev, A. (2000) A study of surface exchange coefficient on La0.5Sr0.5Co3-δ thin films. *MRS Proc.*, **606**, 275.
13. Steele, B.C.H. (1995) Interfacial reactions associated with ceramic ion transport membranes. *Solid State Ionics*, **75**, 157.
14. Adler, S.B. (1998) Mechanism and kinetics of oxygen reduction on porous La1-XSrxCoO3-δ electrodes. *Solid State Ionics*, **111**, 125.
15. Park, J., Cho, S., and Hawthorne, J. (2013) Electrochemical induced pitting defects at gate oxide patterning. *IEEE Trans. Semicond. Manuf.*, **26**, 315.
16. Fisher, D. (2012) PhD thesis, Fundamental Physical Processes Active in the Development of Thin Film Solid Oxide Fuel Cells. University of Houston (Unpublished).
17. Monnerat, B., Kiwi-Minsker, L., and Renken, A. (2001) Hydrogen production by catalytic cracking of methane over nickel gauze under periodic reactor operation. *Chem. Eng. Sci.*, **56**, 633.

13
Nanostructured Magnetic Thin Films and Coatings

Goran Rasic

13.1
Introduction

Magnetic materials have been known since ancient times starting with magnetite (Fe_3O_4) over 2500 years ago. Since then, they have been a crucial part of human history, beginning with the compass, which allowed explorers to travel the globe and discover new worlds. Later, the discovery of electromagnets allowed for the creation of large magnetic fields. Today, magnetic materials can be found virtually everywhere from magnetic storage technology, credit cards, magnetic brakes in hybrid vehicles, power transformers to even biomedical applications. With such versatility and omnipresence, the field of magnetism continues to provide scientific appeal, as new magnetic phenomena emerge and many older ones are yet to be fully understood.

Owing to such broad applications, magnetic materials also have a large business appeal. Unfortunately, if we look at the periodic table, we see that only a small number of elements are ferromagnetic. Those are, not counting some rare earth elements, Ni, Co, and Fe. Thus, it comes as no surprise that a lot of research has been done in an effort to modify the intrinsic magnetic properties of those materials in order to make them better suited for a desired application. Traditionally this involved doping and alloying, where a combination of different elements would give rise to an altered magnetic property of interest [1–3]. This method, however, is limited to the number of material combinations that are possible and, more importantly, useful.

Luckily, magnetic properties of a material change significantly with reduced dimensions. While magnetic thin films still represent the foundation of most device applications, creating nanoscale magnetic structures has enabled new breakthroughs. By reducing dimensions, materials can be tailored to improve desired properties for a specific application. This has given us a new method of controllably changing magnetic properties. With reducing dimensions, a wide range of nanostructures such as thin films, nanowires, particles, junctions, tubes, and others can be created. When traditional methods of magnetic

Advanced Nano Deposition Methods, First Edition. Edited by Yuan Lin and Xin Chen.

property control (doping and alloying) are combined with a broad range of nanoscale geometries, control over intrinsic magnetic properties such as total magnetization, coercivity, Curie point, and anisotropy is possible. In addition, by combining materials with different magnetic properties, new functionalities such as giant magnetoresistance [4, 5], exchange bias [6, 7], multiferroism [8, 9], and magnetoelectricity [10, 11] can be achieved.

The field of magnetic materials is vast and certainly beyond the scope of this chapter. The goal of this chapter is to give an overview of some interesting ways of how the magnetic properties of a material can be tailored for a specific application via size reduction. The chapter will focus on nanostructuring thin films as a way to enhance the desired magnetic properties of the material.

This chapter is composed of two sections. The first section discusses losses in soft magnetic materials for high-frequency application, an increasingly important field in today's technological push toward faster, smaller, and more efficient devices. Magnetic losses in magnetic materials arise due to the magnetic hysteresis and are thus intrinsic to the material [1–3]. This means that to reduce the hysteresis losses, the material needs to be altered. This, however, is not always desirable as other properties tend to be negatively affected as well. Fortunately, nanotechnology has offered a solution to one of the long-standing problems in magnetism – changing the properties without changing the material. Patterning the surface of a thin magnetic film in a specific way showed a remarkable reduction in coercivity, allowing for reduction in the hysteresis losses [12–14].

The second section discusses magnetic information storage technology, a prominent example of the benefits of nanostructured materials. With continuous progress in technology and over 3 billion estimated Internet users [15], the amount of data created daily is enormous and growing at a rapid pace. As information production increases so does the need for information storage. It comes as no surprise then that magnetic data storage is a several billion dollar industry. Unfortunately, the current data storage technology is approaching its limits and a solution is desperately needed. Once again, nanotechnology offers a solution. By creating nanoscale-patterned media, it is possible to increase the amount of information that can be stored per unit area and thus increase capacity.

13.2 High-Frequency Devices

Today's technological trends are continuously pushing toward smaller, faster, and more energy-efficient devices. Continuous progress in science and engineering has allowed us to build nanoscale devices. Many of these devices, such as sensors and antennas, operate at frequencies above 1 GHz. However, at those high frequencies, electrical and magnetic losses become significant [1, 2]. Electrical losses, which arise from eddy current, can be reduced by selecting materials with high electrical resistivity and forming laminate structures [1–3, 16]. Magnetic losses, however, are intrinsic to the material and grow proportional to the frequency [16, 17]. At high frequencies, these losses can be immense and pose a serious constraint for high-frequency device development. In order to reduce the hysteresis losses, magnetic materials with high electrical resistivity and small coercivity values are desired. Ferrites, ferrimagnetic metal oxides, are one example of such materials. Ferrites combine high electrical resistivity with ferromagnetic-like material properties, making them one of the technologically most important materials. However, even ferrites usually have considerable coercivity values. The solutions are sought after by doping or alloying, but those methods usually alter other materials properties as well. Clearly, a method to reduce magnetic losses without losing other desirable properties is needed. Nanoscience has offered a way to selectively change the individual material properties. Through proper selection of the surface pattern, reduction in coercivity is possible.

13.2.1 Ferromagnets

When we think and talk about magnets we are usually talking about ferromagnets, the most common and the technologically most important magnetic materials. In them, the magnetic dipole moments are strongly coupled causing them to align parallel to each other and to the direction of an external magnetic field. This preferential alignment of magnetic dipoles makes ferromagnetic materials easy to magnetize with relatively low applied magnetic field. Furthermore, unlike diamagnets and paramagnets, ferromagnets have a maximum value of magnetization called *saturation magnetization* (M_S). The maximum value of magnetization is reached when all the magnetic dipole moments in the material are oriented parallel to the externally applied magnetic field. Increasing the external field further will not increase the magnetization of the material.

Another distinguishing feature of ferromagnets is that they retain some magnetization even when the external field is removed. The amount of retained magnetization is called *remnant magnetization* (M_R) and is a consequence of the strong interaction of magnetic dipoles. Due to the fact that ferromagnetic materials have remnant magnetization, a magnetic field applied in the opposite direction of the magnetization is required to reduce the total magnetization of the material to zero. The amount of magnetic field required to achieve zero magnetization is called the *coercivity* (H_C) of a material. Coercivity is a crucial property of a magnetic material, and one of the most basic divisions of magnets is based on the size of their coercivity. Materials with a large coercivity require large magnetic field to demagnetize and are called *hard magnets*, while materials with a small coercivity are called *soft magnets*. Hard magnets are of interest in applications where the retention of magnetization is required, such as magnetic information storage. Soft magnets are more useful for applications where we want to be able to quickly reverse the magnetization of a material, such as high-frequency devices [1, 2].

As a result of the aforementioned phenomena, ferromagnetic materials have a nonlinear and history-dependent magnetization with respect to the applied field.

This behavior is characterized by a hysteresis curve. A hysteresis curve is obtained by plotting the values of magnetization with respect to the applied magnetic field as we saturate the material in opposite directions. Shown in Figure 13.1 is a schematic of a typical hysteresis curve for a ferromagnetic material.

To explain why a hysteresis curve is obtained, we look back at the magnetic dipole moments of ferromagnets. Again, the exchange energy between the magnetic moments wants them all to point in the same direction, and this is achieved when a material has reached saturation magnetization. When the external magnetic field is removed, the magnetostatic energy of the material is large, and in order to reduce it, some of the magnetic dipole moments reorient in the opposite direction. This competition between exchange energy wanting to point neighboring dipoles in the same direction and the magnetostatic energy attempting to reduce the amount of dipoles pointing in same direction results in the creation of magnetic domains [1–3]. Magnetic domains represent regions in a material in which all magnetic dipoles are pointing in the same direction. The direction in which the dipoles are pointing and the size of the domain depend on the lowest energy configuration of the material itself. Magnetic domains are separated by domain walls, in which the magnetic dipoles are oriented such that they bridge the different dipole orientations between the adjacent domains. However, there is an energy cost associated with creating a domain wall, and so the final magnetic configuration of a material is a balance between the energy gain from reducing the size of a magnetic domain and the energy cost of creating a domain wall. The final domain configuration results in a nonzero net magnetization and, to reduce it to zero, additional field needs to be applied. As a result, ferromagnetic materials will trace out a hysteresis curve when alternating magnetic field is applied.

13.2.2 Coercivity

In ferromagnetic materials, magnetic domain structure is determined by the competition between the magnetostatic energy trying to reduce the size of magnetic domains and the exchange interaction trying to point all the magnetic dipoles in the same direction. The theoretically expected domain structure would consist of alternating parallel domains pointing in the opposite direction. As a result, these domains would cancel each other out, resulting in a remnant magnetization and

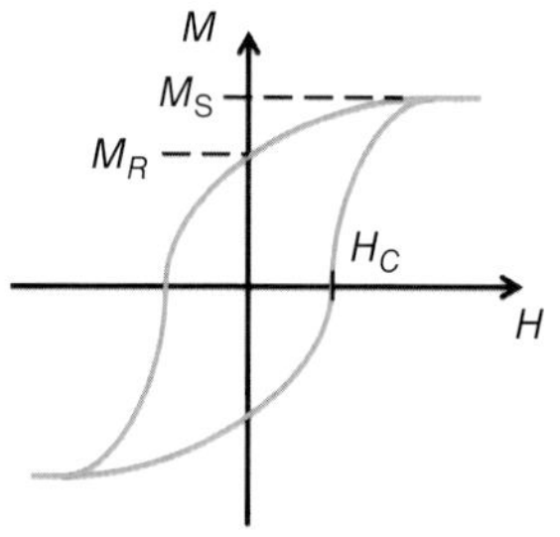

Figure 13.1 Hysteresis curve for a ferromagnetic material.

coercivity approaching zero. This configuration is called a *parallel stripe domain*, and it is the lowest energy configuration for an ideal material [1–3]. In real materials, however, there are vacancies, particles, inclusions, and many other material defects that need to be taken into account. A magnetic material changes its total magnetization by expanding domains pointing in the direction of the external field and shrinking ones pointing in the opposite direction. The primary method through which this is achieved is domain wall movement. However, this domain wall movement can be hampered by any material defects, inclusions, or even surface roughness. Coercivity is then a measure of how easily domain walls move through a material. For this reason, in real materials, we do not observe a parallel stripe domain. Instead, as a result of domain walls sticking to inclusions, we observe a sort of "deformed" parallel stripe domain pattern resembling a maze or a fingerprint.

13.2.3 Magnetic Losses

As mentioned earlier, soft ferromagnets are especially interesting for high-frequency devices. However, magnetic losses are dependent on the material and directly proportional to the frequency:

$$P_h = W_h \times f \tag{13.1}$$

where f is the frequency of cycling through the hysteresis loop and W_h is the hysteresis loss equal to the area under the hysteresis curve [17].

From the equation, it is easy to see that going to megahertz and larger frequencies drastically increases losses. The solution is clear – the area of the hysteresis loop needs to be decreased by reducing either the magnetization or the coercivity.

However, in high-frequency device applications, large saturation magnetization is usually desired. In addition, we would like to reach saturation magnetization easily via a small applied magnetic field. This is achieved by growing the magnetic material in such a way as to align the magnetic easy axis with a certain crystallographic plane in the material. In the case of polycrystalline materials, this means growing the material so that the majority of grains are oriented in the desired crystallographic direction (called *texturing the material*).

The second option is to reduce the coercivity. This appears simple in theory, but coercivity is an intrinsic property of a magnetic material. The traditional approach was, and to a large extent still is, to change the properties of a magnetic material by simply changing its composition. This however is merely a half-measure as only a handful of elements in the periodic table are ferromagnetic, which places a significant constraint on material selection [1, 2]. Thus, the whole field of magnetism and magnetic materials is limited to different combinations of alloys and dopants of these elements [1, 2, 18–20]. This problem of materials constraint is further amplified by the fact that no new soft magnetic materials (aside from amorphous alloys) have been developed since the 1950s [1].

13.2.4 Nanoscale Methods of Loss Reduction

Recently, however, a method of reducing coercivity without reduction in other desirable properties was reported [12–14]. To address the issue of modifying the properties of materials without changing their composition, they need to be examined at the nanoscale. Using surface nanostructures we can alter the system's lowest energy configuration, effectively changing domain size and distribution. If the pattern is chosen correctly, coercivity can be reduced. By understanding how domain walls propagate through a material and considering material energetics, an elegant solution can be envisioned. As discussed previously, parallel stripe domain pattern represents the lowest energy state in an ideal material, but in real materials domain walls "stick" to defects in the material and alter the domain structure. The solution is to engineer the material in such a way as to induce a desired domain configuration. By nanostructuring the surface of the material with periodic parallel trenches across the length of the sample, an artificial parallel stripe domain pattern can be created.

13.2.5 Manufacturing Considerations

Due to the fact that magnetic properties are dictated by the material, when discussing applications of magnets, the first consideration is certainly what material to use. For high-frequency application, a group of magnetic materials called *ferrites* is preferred. Ferrites are ferrimagnetic transition metal oxides. They behave magnetically very similar to ferromagnets and are also characterized by a hysteresis loop. The key difference comes in their electrical properties. While a typical ferromagnet is conductive, a ferrite is electrically resistive. This helps minimalize the electrical losses and makes ferrites useful in a number of applications such as high-frequency [1, 2], microwave [21], and spin filter devices [22, 23]; antenna and sensors [24]; memory [16, 25]; power electronics [22, 26, 27]; and even magnetoelectric composites [11, 28].

One of the technologically most important ferrites is nickel ferrite (NFO) with a chemical formula $NiFe_2O_4$. NFO is an inverse spinel cubic ferrite that is magnetically soft and electrically insulating. This combination of favorable properties makes NFO interesting for high-frequency applications. Of course, to maximize the usefulness of the material, large saturation magnetization and low coercivity are desired. As ferrites are typically produced from powders using traditional ceramic techniques, they tend to be polycrystalline materials. This means that the degree of grain orientation (texture) becomes crucial. Increasing texture along the magnetic easy axis direction of a material, <111> in case of the NFO, ensures maximal saturation and minimal coercivity.

While many methods to fabricate NFO such as pulsed laser deposition [29, 30], sputtering [31], and chemical vapor deposition [32, 33] have been reported, chemical solution deposition (CSD) offers a fast, low-cost method for large-scale thin film fabrication. It has also been demonstrated to be a viable method for fabrication of highly textured NFO thin films [27]. In addition, CSD can easily be combined with nanoimprint lithography (NIL) to create nanoscale structures.

In a conventional NIL procedure, the hard ceramic film is coated with a soft material to which the pattern of interest is transferred using a stamp. This imprinted pattern is then transferred to the underlying hard material layer by etching, and the remaining soft coating is removed [34–36]. This process is lengthy, complex, and, depending on the materials used for etching, can be expensive. A simpler method of directly imprinting the film has been shown to give excellent results [12–14]. A schematic of the surface-patterned NFO thin film fabrication procedure using direct NIL is shown in Figure 13.2. First, the NIL stamps were made from pattern masters using polydimethylsiloxane (PDMS).Then, unlike a conventional NIL, the stamp with the features was imprinted onto the material immediately following spin coating while simultaneously heating the film. After the solvent evaporated, the PDMS stamp was removed, and the surface-patterned amorphous NFO thin film was further annealed until fully crystallized.

13.2.6
Coercivity Reduction in Surface-Patterned Magnetic Thin Films

The first step is to evaluate if the direct imprint technique that was described previously produced a viable pattern transfer. The patterned NFO thin film was fabricated using the procedure described in Figure 13.2. To induce a parallel stripe domain pattern formation in the material, a corduroy pattern was chosen. For pattern masters, a compact disk (CD) was initially selected. The large scale and high quality of the groove pattern of the CD made it an excellent low-cost alternative to lithographically fabricated masters.

Figure 13.3 shows the atomic force microscopy (AFM) images of the NIL master (a), stamp (b), and final patterned NFO thin films (c, d) as well as the roughness images (bottom of a–d). AFM images show periodic corduroy pattern of a CD master (Figure 13.3a) being very precisely and cleanly transferred to the PDMS stamp (Figure 13.3b) and the NFO thin film (Figure 13.3c,d). Figure 13.3c shows that the pattern was transferred without any defects commonly associated with

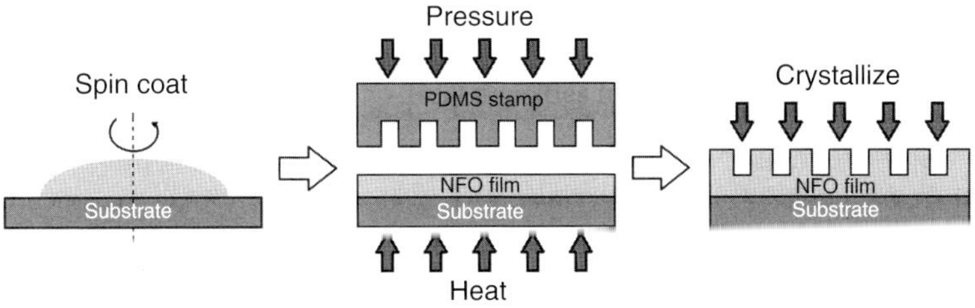

Figure 13.2 Schematic of $NiFe_2O_4$ thin film fabrication procedure.

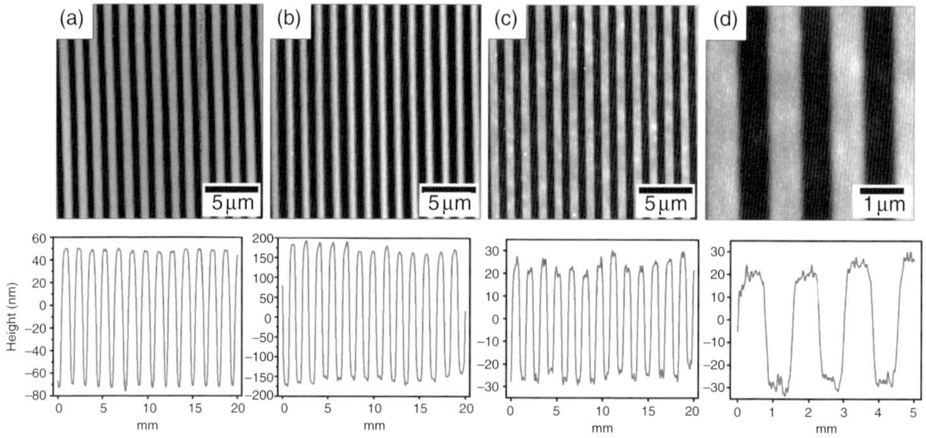

Figure 13.3 Atomic force microscopy images of (a) CD master, (b) PDMS stamp, and (c,d) patterned $NiFe_2O_4$ thin films. (Rasic and Schwartz [12]. Reproduced with permission of Cambridge University Press.)

NIL techniques such as air gaps, cracking, bending, misalignment, or feature collapse. A higher magnification image of the NFO thin film is shown in Figure 13.3d. Roughness analyses of AFM images are shown at the bottom parts of the corresponding images. The noticeable reduction in feature height of NFO thin films as compared to the stamp is due to the low thickness of the NFO thin film after the spin-coating process.

Results of the AFM scans revealed the direct imprinting to be a very elegant and simple way that NIL can be combined with CSD to create precise surface nanostructures over a large area. However, magnetic properties are closely related to the crystallographic properties of a material, and it is imperative that the texture of the thin film is not affected by the process of imprinting.

To evaluate the effects of imprinting on the crystallographic properties, X-ray diffraction (XRD) was used. For better evaluation of the effect of patterning, an additional NFO thin film without any surface pattern was made. XRD measurements, shown in Figure 13.4, of both planar and surface-patterned NFO thin films were taken for comparison. XRD $\theta - 2\theta$ measurements show both planar and patterned films to be inverse spinel NFO with no impurities present confirming the quality of the films. In addition, both show only peaks corresponding to [111] family of planes confirming that the films are textured along the NFO easy axis direction. Comparing the peaks present and their intensity in Figure 13.4, we notice that the texture is similar in both cases, with no grain misorientation or new phase formation in the patterned films. This, combined with transmission electron microscopy (TEM) images [12] and AFM images (Figure 13.3), confirms there was no grain growth disruption by the PDMS stamp during the NIL process. Similar results were obtained regardless of film thickness or pattern period demonstrating that it is possible to directly imprint the material before crystallization without any

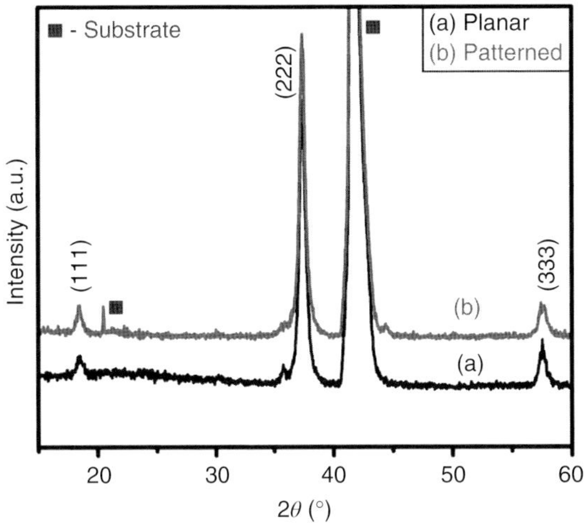

Figure 13.4 X-ray diffraction scans of (a) plain and (b) patterned $NiFe_2O_4$ thin films. (Rasic and Schwartz [12]. Reproduced with permission of Cambridge University Press.)

loss in texture. Direct imprinting greatly simplifies the process of NIL with added benefit of reduced time and cost.

Evaluation of the magnetic properties was conducted at room temperature using a laboratory superconducting quantum interference device (SQUID). The patterned film revealed an 89% reduction in coercivity as compared to the planar one [12]. This extraordinary result was attributed to the patterning. However, the results were not conclusive.

For better evaluation of the effect of nanopatterning the surface of the films on the magnetic properties, an additional step was taken. First, a layer of NFO was spin coated and fully crystallized without any patterning. On top of that layer, another layer of NFO was spin coated and fully crystallized without any patterning. These two planar layers of NFO were designated "base." This base film was then cut in half and then a third, and final, layer of NFO was deposited on each half. On one half, the final layer is again deposited without any patterning ("planar"), and on the other half, the final layer is patterned ("patterned"). This ensures maximal similarity between the samples and allows for easy evaluation of the effects of the patterning. AFM and XRD measurements on these films were similar to the measurements for single-layered NFO thin films shown in Figures 13.3 and 13.4. The details can be found in [13]. The magnetic measurements of these films were taken both parallel to the plane of the film (in-plane) and perpendicular to the film plane (out-of-plane). The results of the measurement are listed in Table 13.1. The normalized SQUID magnetometry measurements of NFO thin films measured in plane of the film are shown in Figure 13.5.

The similar shape of the magnetic hysteresis curve for the base, planar, and patterned NFO thin films indicates that they have similar texture. This confirms that there is no loss in texture with patterning. More importantly, similar curve shapes

Table 13.1 Results of magnetic measurements taken on multilayered $Nife_2O_4$ thin films.

Sample	Saturation magnetization (emu/cm³)	In-plane coercivity (Oe)	Out-of-plane coercivity (Oe)
Base	95	143	332
Planar	106	168	448
Patterned	101	85	119

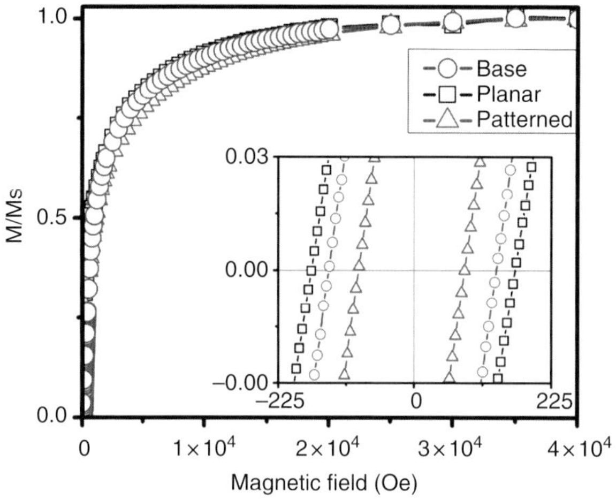

Figure 13.5 Magnetic hysteresis of $NiFe_2O_4$ thin films measured parallel to the plane of the film. Higher magnification image of the origin is shown in inset.

demonstrate similar magnetic response of the NFO films. Taking into account similar saturation magnetization values between the planar and the pattern film, this validates that magnetic thin films can be patterned without losing desirable magnetic properties such as saturation magnetization or curve shape. The out-of-plane measurements showed similar results [13].

For easier evaluation of the in-plane coercivity, the inset in Figure 13.5 shows a higher magnification image centered on the origin. Comparing the base and the planar film, we notice that the addition of another layer of NFO increases the coercivity. This is to be expected since with increasing thin film thickness, the material starts behaving more like a bulk material. However, if the final layer is patterned, the coercivity drops substantially, even below the value for the starting base film. This coercivity reduction with patterning is even more pronounced in the out-of-plane measurements, clearly demonstrating that the reduced coercivity is a direct consequence of the surface patterning of the final NFO thin film layer.

Large coercivity reduction with surface patterning was observed in both single [12] and multilayered [13] NFO thin films measured out-of-plane as well as in-plane. Furthermore, the effect of changing the pattern period was also explored [14]. A series of pattern masters with periods ranging 500–1500 nm were fabricated using interference lithography. In all cases, similar results were obtained. In addition, there was no clear dependence of coercivity reduction on the period size, indicating a universal nature of the effect.

To understand the origin of the coercivity reduction, the magnetic domain structure of the NFO thin films was directly observed using magnetic force microscopy (MFM) [14]. The results of the MFM imaging on planar and patterned NFO thin films are shown in Figure 13.6. The MFM measurement is shown as a color overlay over the AFM topography scan. The two different magnetic domain orientations are colored light and dark gray for ease of analysis. The MFM scan of the planar NFO film in Figure 13.6a shows a typical domain structure consisting of random domains extending over multiple grains. Conversely, in the MFM image of the patterned NFO film in Figure 13.6b, the domains are ordered and follow the surface morphology of the NFO thin films. The protruding features have single domains pointing in the same direction. In the trench between the protruding features, there are also single domains but they are pointed in the opposite direction to that of the protruding features. This alternating domain arrangement creates the parallel stripe domain pattern as predicted. Theoretical examination of the energetics of the system agrees with these findings [14].

The surface topography creates natural places for the domain walls to rest, namely, at the edge of the protruding features. Following the surface topography, the domain walls create a parallel periodic pattern stretching the length of the sample and dividing up the magnetic domains. The parallel stripe pattern is the lowest energy state even in the planar films, but many pinning centers in the sample cause the magnetic domains to form a more random structure as shown in Figure 13.6a. By nanostructuring the surface of the material in a parallel stripe pattern, a new lowest energy state consistent with the theoretical predictions is created.

Another noticeable difference between the planar and pattern NFO films is the size of the magnetic domain. While the magnetic domains in the planar NFO film have an average size of approximately 150 nm, the domains in the patterned film are much larger. This is a direct consequence of the patterning. After the domain walls are resting in their preferred position, between the trench and the protruding feature, new domain walls would have to be created in order to reduce the size of the domain. This however requires additional energy, and the energy requirement for the additional domain walls grows exponentially. At the same time, the energy required to increase the size of a magnetic domain has a slow linear increase. As a result, it is energetically more favorable to increase the size of a domain than to create additional domain walls [14].

Coercivity reduction then is a direct consequence of the surface patterning or, to be more precise, the parallel stripe domain pattern. As the oppositely oriented domains are similar in size, the net magnetization at the film surface is zero.

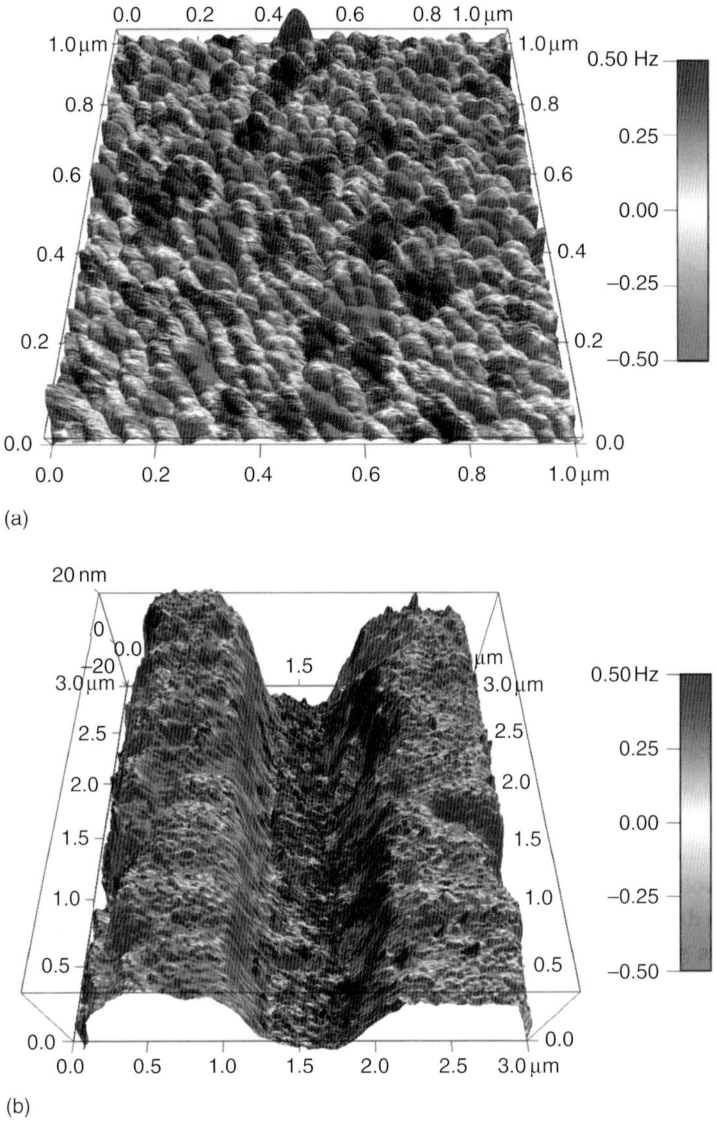

Figure 13.6 Magnetic force microscopy images of (a) a planar $NiFe_2O_4$ thin film and (b) a patterned $NiFe_2O_4$ thin film magnetized out-of-plane. The light and dark gray overlay indicates magnetic field orientation. (Rasic and Schwartz [14]. Reproduced with permission of Wiley.).

This alternating domain arrangement at the surface penetrates into the material causing the overall coercivity to be reduced below that of the base layer. The extent to which the coercivity is reduced depends on the extent to which the surface domain arrangement is transferred into the material. By reducing only the coercivity and retaining other desirable properties, the material selection constraint is eliminated. This allows for larger material selection and ultimately improved performance with reduced losses.

13.3 Magnetic Information Storage Devices

Since its invention over 100 years ago [37], magnetic recording has played a key role in information storage. Initially, magnetic information storage was used for storing analog audio and video content on magnetic tapes. With the invention of computers, it did not take long for people to realize that magnetic materials were good candidates for binary data storage. And so, in 1956, IBM produced the first commercial disk drive. It contained fifty 24″ disks and, at areal density of 2 kb/in.^{2}, was able to store only 5 MB of memory [2]. From this beginning, magnetic information storage became a key component in the technological revolution we have witnessed over the last several decades. Without the ability to store, retrieve, and manipulate a large amount of data, most (if not all) technological progress would not be possible. It comes as no surprise then that the business of data storage is highly profitable and highly competitive.

Advances in technology, scientific understanding, and scale reduction have allowed us to follow Moore's law [38] for an unbelievable 60 years and counting. The measure by which technological advances in magnetic storage systems are evaluated is the areal density. This refers to the amount of "bits" of information that can be stored per unit area of the material. The larger the areal density, the more the information we can store on a hard disk drive (HDD).

An HDD consists of one or more magnetic disks that rotate at a high speed. A magnetic write head is used to write the information onto the magnetic disk. One bit of information is stored as a magnetic bit and consists of multiple grains of the material with the same magnetic field orientation. The read operation is done by reading the stored information from the disk using a read head.

At first, magnetic bits were oriented in the plane of the magnetic disk. This is called *longitudinal magnetic recording* (LMR) and is shown in Figure 13.7a. The inductive magnet of the write head was used to orient the magnetic domains in the plane of the magnetic disk. Increasing capacity then simply became a matter of reducing the size of individual packet of information (the magnetic bit) to increase areal density. Over the years, the size of the individual bits was reduced, but the basic idea remained the same. By reducing the size of the magnetic bit more of them can be placed on an HDD, thus increasing the total capacity. This approach was followed for nearly 50 years with research focusing on reducing the size of the grains and read/write head, moving the read/write head closer to the surface, increasing the speed of rotation of the magnetic disks, and other scaling considerations. By the turn of millennia, following incredible progress, LMR was approaching its superparamagnetic limit (a physical phenomenon where the magnetic bit loses its long-term stability with a reduction in size) initially

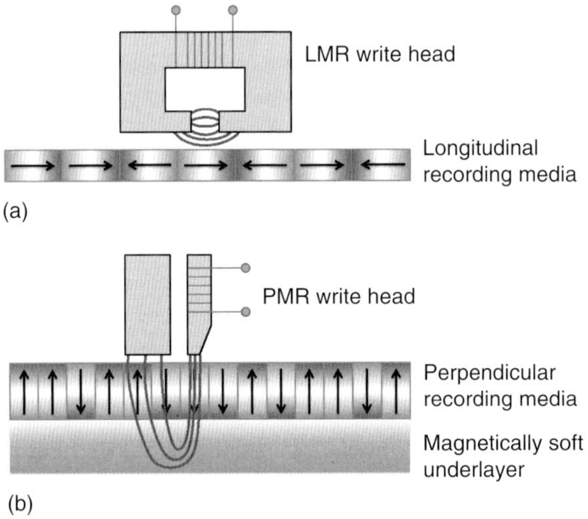

Figure 13.7 Schematic drawing of (a) longitudinal magnetic recording and (b) perpendicular magnetic recording.

estimated at 40 Gbit/in.^{2} [39]. By utilizing materials with higher coercivity, this limit was increased to the 100 Gbit/in.^{2} range [40], but a new approach was needed.

In 2005, the first HDD utilizing perpendicular magnetic recording (PMR) was introduced. In it, as the name suggests, the orientation of magnetic domains was changed from in-plane to perpendicular to the magnetic disk as shown in Figure 13.7b [41]. This changed the interaction of adjacent bits from repulsive to attractive, increasing thermal stabilization. In addition, a magnetically soft underlayer was added, which helped increase the effective write field, allowing for the use of larger coercivity materials. As a result, magnetic domains with significantly smaller size in the plane of the disk could be created, increasing areal density and storage capacity. To commemorate this achievement, Hitachi even created a musical flash animation [42]. By changing to PMR, the superparamagnetic limit was increased to areal density of approximately 1 Tbit/in.^{2} [43, 44].

13.3.1
Superparamagnetic Limit

The biggest constraint and thus most important consideration for future magnetic storage is the superparamagnetic limit. Nanoparticles with single magnetic domain generally have two stable orientations of the magnetic moment that are separated by an energy barrier. To increase the areal density, the size of the individual bits must be reduced. However, as we reduce the bit size, the magnetic anisotropy lowers and the energy barrier becomes comparable to the thermal energy causing instability in the recorded data. This thermally activated switching is described by the Néel–Arrhenius equation [45]:

$$\tau = \frac{1}{f_0} \exp \frac{\Delta E}{k_B T} \quad (13.2)$$

where k_B is the Boltzmann constant and τ is the average time required for the magnetization to change its direction due to thermal fluctuation called the *Néel relaxation time.* Attempt frequency f_0 is on the order of $10^9 - 10^{12}$ Hz [46]. Assuming coherent rotation mechanism, the energy barrier ΔE for a magnetic particle without the presence of external magnetic field is

$$\Delta E = KV \quad (13.3)$$

where V is the volume of the particle and K is the magnetic anisotropy energy density. The value ΔE required will vary depending on parameters such as the material, desired time of magnetization retention, and the working temperature. The general consensus is that $KV > 50 k_B T$ is required for at least 5 years of information retention [39, 40, 47].

From Equations 13.2 and 13.3, it is clear that further reduction of V is possible by increasing the magnetic anisotropy of the material K [48]. Unfortunately this leads to a problem of writing the information as there is a limit to the maximum magnetic field a write head can produce [49, 50].

13.3.2
Signal-to-Noise Ratio

Another option is to reduce the volume V by reducing the number of grains in a magnetic bit. This leads to the issue of the signal-to-noise ratio (SNR) [51]. In a continuous granular media (CGM), the SNR is the result of averaging over many grains, the more grains we average over the higher SNR. By using a grain-counting method, the SNR can be approximated as [52]

$$\text{SNR} \sim 10 \log \frac{V_{\text{bit}}}{V_{\text{grain}}} = 10 \log N \quad (13.4)$$

where N is the number of grains in a magnetic bit, V_{bit} is the total volume of all the grains within the magnetic bit, and V_{grain} is the volume of the individual grain in the magnetic bit. Thus, in order to continue increasing areal density, while maintaining a sufficient number of grains to average over, the size of the grain needs to be reduced. This causes a problem as the size of individual grains reaches their superparamagnetic limit. This contrast in the requirements for large SNR, high thermal stability, and sufficient magnetic field produced by the write head is known as the *magnetic recording trilemma* [53, 54].

13.3.3
Present-Day Solutions

In 2012, IBM estimated that 2.5×10^{18} bytes of data are created everyday and expected to grow exponentially [55]. The continuation of technological progress depends on continuous increase in the areal density. As of 2015, the areal densities of consumer HDDs are exceeding 800 Gbit/in.^{2} and rapidly approaching the 1 Tbit/in.^{2} limit of PMR. To continue increasing storage capacities, novel designs such as helium-filled disks [56] and shingled magnetic recording (SMR) [57, 58] are used. However, this is a temporary solution to a long-standing problem, and the scientific community is still searching for new ways to increase capacity and improve performance.

Looking at the Néel–Arrhenius equation (3), we see that to delay the superparamagnetic limit we have two options. We can either use materials with higher magnetic anisotropy constants or increase the single domain volume. Both of these approaches are, in fact, being heavily investigated.

The first approach, using materials with a higher K constant, leaves a major engineering problem to be solved. As the size of the magnetic bit gets reduced, the size of magnetic write head must be reduced as well. This creates a problem since the maximum effective field generated by such a small read head is limited, making the write operation difficult, if not impossible. To solve this issue, a method called heat-assisted magnetic recording (HAMR) has been proposed [59–63]. HAMR is based on the temperature dependence of coercivity. By temporarily heating the material close to its Curie temperature, coercivity is reduced and a lower field is required for writing. To heat the material locally for HAMR, a laser beam is generally proposed [64]. After heating, the material cools quickly in the presence of the magnetic field generated by the write head and is permanently stored in the material. This allows for reduction in the size of the magnetic bit by using materials with larger magnetic anisotropy constants. Using HAMR on FePt, areal densities exceeding 1 Tbit/in.^{2} have been reported [65].

While expected to be the next step in increased storage capacity, with consumer HDDs promised by 2016, the ultimate limit to HAMR has been estimated to be between 3 and 5 Tbit/in.^{2} [66–70]. This makes it a temporary measure with a more permanent solution needed.

13.3.4
Bit Patterned Media

According to the Néel–Arrhenius equation (3), there is only one option left, to increase the volume of the magnetic bit. This seems contraintuitive as smaller bits allow for higher areal densities. However, nanoscience has offered a solution.

In traditional CGM, the magnetic film consists of grains with weak exchange coupling. Because the grains are virtually noninteracting, they have to be considered individually, which lowers the KV value. In addition, one magnetic bit has to consist of many grains to ensure stability.

To solve this issue, patterned media has been suggested. Originally, two main forms were proposed – discrete track media (DTM) [71] and bit patterned media (BPM) [72–74]. DTM is simply a modified form of the current technology where the individual magnetic domains have been lithographically patterned into tracks, separated by a nonmagnetic material. Over the years, however, this idea has fallen out of interest in favor of BPM that offers higher densities.

In BPM, the CGM is replaced with lithographically patterned magnetic islands. While the magnetic island can consist of several grains, these grains have strong exchange coupling and act as a single magnetic entity effectively increasing the volume of the magnetic bit. Combined with the separation between the magnetic bits, magnetic islands have a much larger thermal stability, permitting a smaller size of individual magnetic bit [75]. This allows for a significant increase in areal density, beyond the superparamagnetic limit of a material, without the need for drastically changed HDD operation (SMR) or design (HAMR).

The predictions of limits to BPM are approximately 5 Tbit/in^2 [76]. While this limit puts BPM on par with HAMR, BPM can also serve as a base for future technologies that would allow further increases in areal density. By combining BPM with shingled recording, it is estimated that areal density of approximately 10 Tbit/in^2 can be achieved [77–80]. Furthermore, combining BPM with HAMR, it could be possible to achieve areal densities as high as approximately 100 Tbit/in^2 [61, 81].

13.3.5 Manufacturing Considerations

In BPM, the critical challenge is fabrication. Designing and manufacturing actual BPM that can be used in HDD involve considering many aspects of the process. While this chapter focuses on the manufacturing process, it is enlightening to look at the design process to help us understand why the manufacturing demands are so high.

The first consideration is the one that initiated the idea of moving to BPM in the first place – thermal instability of magnetic bits at a reduced scale. The second consideration is the SNR. For BPM, requirements of high thermal stability and large SNR translate to fabrication tolerances for the magnetic islands. For the goal of 1 Tbit/in.^{2} and beyond, the fabrication tolerances are on the order of just 1 nm [82].

Another consideration is the magnetostatic interaction between the islands, which depends on the magnetic configuration. The magnetic configuration is given by the information written. The more bits are written in the same direction, the stronger the magnetostatic interaction [43, 72]. Some other considerations include write/read synchronization, fabrication cost, throughput, and materials used [72, 83, 84]. Lastly, we have to consider the shape of the magnetic islands. There are many factors to consider when deciding on the shape of the final bits. Aside from the areal density, lithographic requirements, suitability for integration into an HDD, and thermal stability switching field distribution are some of the additional parameters that have to be considered [82]. One might expect then that hexagonal close-packed arrays of nanoscale spheres or cylinders would be the natural choice as they offer the highest packing density [85]. However, using them would require a major redesign of the read/write heads [86]. For this reason, shapes with higher bit aspect ratio (BAR) are preferred [87]. BAR is the ratio between the cross-track pitch and the down-track pitch of BPM. While better for integration, higher BAR shapes require smaller features that increase lithographic requirements. To balance between the lithographic and design requirements, shapes with BAR between 1 and 2 are usually selected.

13.3.6 Patterned Media for Magnetic Data Storage

The most common method of manufacturing large-scale nanostructures is lithography. However, usual lithographic methods such as mask or interference lithography are not able to reproducibly and reliably produce features down to 10 nm. In addition, they are costly and so not a good candidate for large-scale manufacturing. E-beam lithography (EBL), however, has been successfully used to create sub-10-nm features. Unfortunately, EBL is a slow, serial, and expensive technique. As such, it is not practical for large-scale manufacturing but could foreseeably be used to manufacture master templates. These master templates could then be copied many times using NIL [88, 89].

The second approach is bottom-up assembly. This would allow for smaller feature size and cheaper manufacturing. The most promising candidates are block copolymers as they are known to produce periodic patterns after annealing [90, 91]. Simply put, block copolymer (BCP), is a polymer made up of at least two chemically different monomers (blocks). These monomers segregate into a variety of ordered structures upon heating. Due to the differences in physical and/or chemical properties of separate polymeric chains, these developed BCPs can then be used for a variety of applications such as nanofabrication, fuel cells, and optoelectronic devices [91].

However, bottom-up assembly is plagued with issues related to maintaining large-scale ordering. In addition, to conform to the current requirements of HDD, the tracks would have to be circular, adding an additional degree of difficulty to the process.

Clearly to increase areal densities of BPM to their theoretical limits, a combination of several different methods has to be used. By using top-down approaches to template the bottom-up assembly, the pattern can be extended over large areas with improved placement precision and feature size distribution [92, 93].

The benefit to combining top-down and bottom-up approaches is twofold. The first is called *pattern rectification*, and as the name suggests, it refers to the fact that the BCP film will smooth out the small imperfections in the EBL pattern. The second advantage is called *density multiplication*, which allows us to write a looser EBL, considerably reducing write time of the EBL. EBL written pattern then serves as a guide for the bottom-up assembly of BCP, which creates the pattern at the final resolution [89].

While many patterning methods have been explored, the two most promising are graphoepitaxy [94] and chemoepitaxy [95]. In graphoepitaxy, the patterns used to direct self-assembly are written physically into the substrate, usually in the form of trenches or posts. The physical pattern then spatially directs the assembly of BCP. While graphoepitaxy is not limited in the shape of the pattern that we can create, it does create height variation in the magnetic islands that is problematic for the HDD operation [94]. As such, the focus of the industry is, at least currently, on chemoepitaxy. In chemoepitaxy, the pattern is placed on the substrate chemically by using two different chemical environments, which then attract different components of the BCP directing the assembly. In this way, the BCP is guided in 2D without change in the height of the magnetic islands. In addition, it is capable of producing rectangular bits with BAR >1.

The current industry standard BCP is polystyrene-*block*-poly(methyl methacrylate) (PS-*b*-PMMA). PS-*b*-PMMA is a polymer that forms perpendicular lamellae, can be developed in such a way to remove only one block of the copolymer, self-assembles with thermal annealing, and is chemically contrast guided. Unfortunately, the ultimate resolution of PS-*b*-PMMA for industry requirement is limited to approximately 25-nm full pitch [96].

One option is to use a line-doubling technique with the PS-*b*-PMMA copolymer to reduce the pitch size. Two common types are double lithographic patterning [97] and spacer-defined double patterning [98]. Promising results have been demonstrated by using line-doubling technique called *sidewall spacer doubling* after producing PS-*b*-PMMA stripes [99, 100]. A schematic of the procedure is shown in Figure 13.8.

The first step is to create stripes of PS-*b*-PMMA on top of a silicon dioxide capping layer at twice the final target linewidth (Figure 13.8a). The SiO_2 capping layer is used as it is compatible with BCP deposition process and to transfer the pattern to the carbon underlayer. Hydroxyl terminated polystyrene (PS-OH) is grafted on the SiO_2 layer and annealed. Then a layer of PMMA e-beam resist is coated on top, and a parallel stripe pattern is written using an e-beam. After developing the PMMA, oxygen plasma is used to chemically modify the exposed PS-OH. The rest

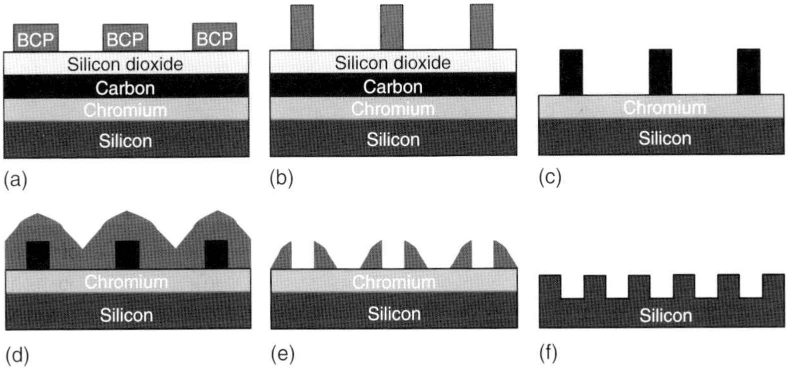

Figure 13.8 Schematic of line-doubling technique: (a) PS-*b*-PMMA BCP pattern at twice the final period, (b) trimmed BCP stripes, (c) pattern transfer to the carbon layer, (d) spacer deposition, (e) removal of the spacer material and the carbon walls, and (f) pattern transfer into the final silicon layer.

of the PMMA is then removed and a layer of PS-*b*-PMMA is coated and annealed. This results in the PMMA, part of the BCP, wetting the oxygen plasma exposed parts, aligning the BCP with the e-beam written pattern.

After development, the pattern stripes are trimmed by roughly 25% of the starting width as illustrated in Figure 13.8b. The pattern is then etched through the SiO_2 layer and into the sacrificial carbon layer using reactive-ion etching (RIE) as shown in Figure 13.8c. The carbon layer is used due it its highly vertical wall profile. The next step is depositing spacers on carbon sidewalls (Figure 13.8d). Anisotropic fluorine-based RIE is then used to remove spacer material from the top of carbon walls and the central region between them as well as any remaining SiO_2. Next, the carbon walls are removed leaving spacers at twice the starting period as shown in Figure 13.8e. The spacers then serve as a etch mask for the underlying Cr layer, which in turn is used to etch the features into the final layer (usually silicon) used for the template (Figure 13.8f).

Using this method, two separate patterned submasters were made – one with circumferential and one with radial lines. The final master with rectangular islands is formed by imprinting with both submasters as shown in Figure 13.9 [86]. This approach reduces fabrication demands of nanoscale features as the two submasters are easier to manufacture.

After the master is created, the pattern is transferred using NIL [34, 101]. NIL is chosen due to its ability to transfer a nanoscale pattern over large areas at high volumes and low cost. Reverse tone NIL is preferred as it allows faster and better spreading of the resist on BPM patterns, superior etch contrast for pattern transfer, and is less sensitive to residual layer thickness [84]. In addition, the reverse tone image is mechanically more robust eliminating pattern collapse. The details of the NIL process and the entire BPM fabrication process can be found in [84]. Following the NIL and a series of RIE steps, the final magnetic BPM was created. The scanning electron microscopy (SEM) images of fabricated BPM for several different dimensions of the magnetic dots are shown in Figure 13.10 [84]. Figure 13.10a shows magnetic dots in a hexagonal array at an areal density of 1 Tbit/in.^{2}. Figure 13.10b,c shows rectangular arrays at areal densities of 1.2 and 1.6 Tbit/in.^{2}, respectively. The arrow in Figure 13.10a–c indicates down-track direction and the higher magnification images are given in Figure 13.10d–f. The details of the fabricated BPM are given in Table 13.2.

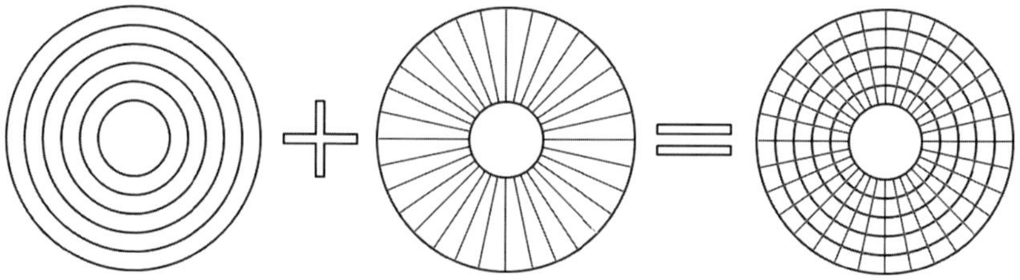

Figure 13.9 Schematic of combining two pattern submasters to create a final master.

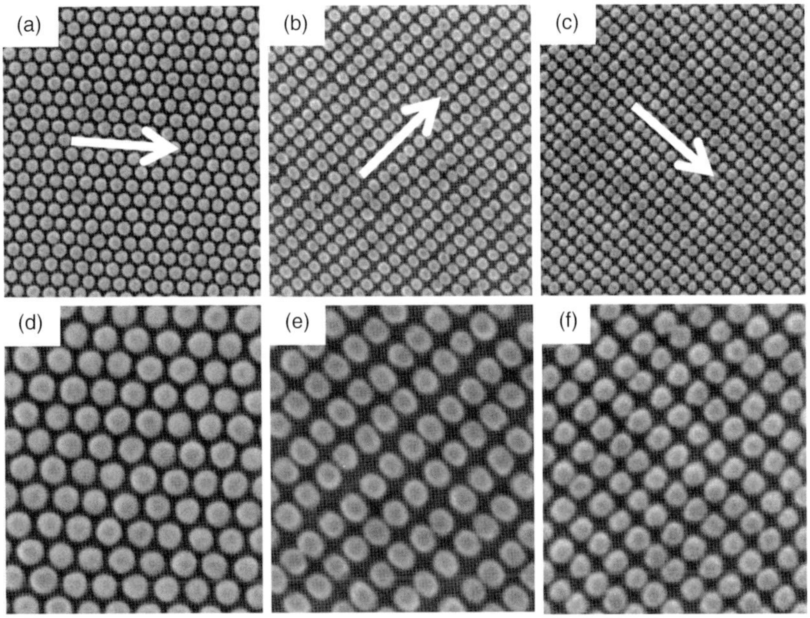

Figure 13.10 Scanning electron microscopy images of BPM magnetic dots in (a) 1 Tdot/in.^{2} hexagonal array, (b) 1.2 Tdot/in.^{2} rectangular array, and (c) 1.6 Tdot/in.^{2} rectangular array. (d–f) Higher magnification images of (a–c). (Albrecht *et al.* [84]. Reproduced with permission of IEEE.)

Table 13.2 Comparison of BPM at different areal densities.

Areal density (Tdot/in.^{2})	**1**	**1.2**	**1.6**
Array type	**Hexagonal**	**Rectangular**	**Rectangular**
Circumferential (nm)	27.3	27	22
Radial (nm)	—	20.5	18.5
BAR	1	1.3	1.2
Placement 1σ	1.68	1.86	1.31

In Figure 13.10 and Table 13.2, the dot size and placement attest to the quality of the fabricated BPM. It is interesting to compare the hexagonal array (fabricated in 2012) and the 1.6 Tdot/in.^{2} rectangular array (fabricated in 2014). The improvements of the fabrication process of BPM resulted in the rectangular array having smaller size, higher areal density, and better placement accuracy of the magnetic dots as compared to the hexagonal array. The results show that BPM, at least from the fabrication perspective, is a viable technology.

The recording performance of these BPM films was thoroughly examined both experimentally and computationally. To test the recording performance of the BPM, a drag tester is used. This involves performing write and read operations using a conventional PMR head and then extracting and examining the results [102]. The error rate of the signal was found to be less than 10^{-2} [84]. This is sufficient for modern signal processing and error correction methods in HDD [82, 83].

In addition to increasing areal density by switching from PMR to BPM, it is predicted that by shingle writing on BPM further increases in areal densities are possible [77–80]. Thus, the recording performance of shingled magnetic recording on 1.6 Tdot/in.^{2} BPM was also evaluated. The error rate of less than 10^{-2} was obtained, demonstrating that 1.6 Tdot/in.^{2} BPM can indeed be combined with SMR. An example of SMR on 1.6 Tdot/in.^{2} BPM is shown in Figure 13.11. Figure 13.11a shows a 2D readback image of an AC field erased 1.6 Tdot/in.^{2} BPM. We observe that the magnetic dots have a random magnetization state. The same area was then written using SMR. After SMR, the information was

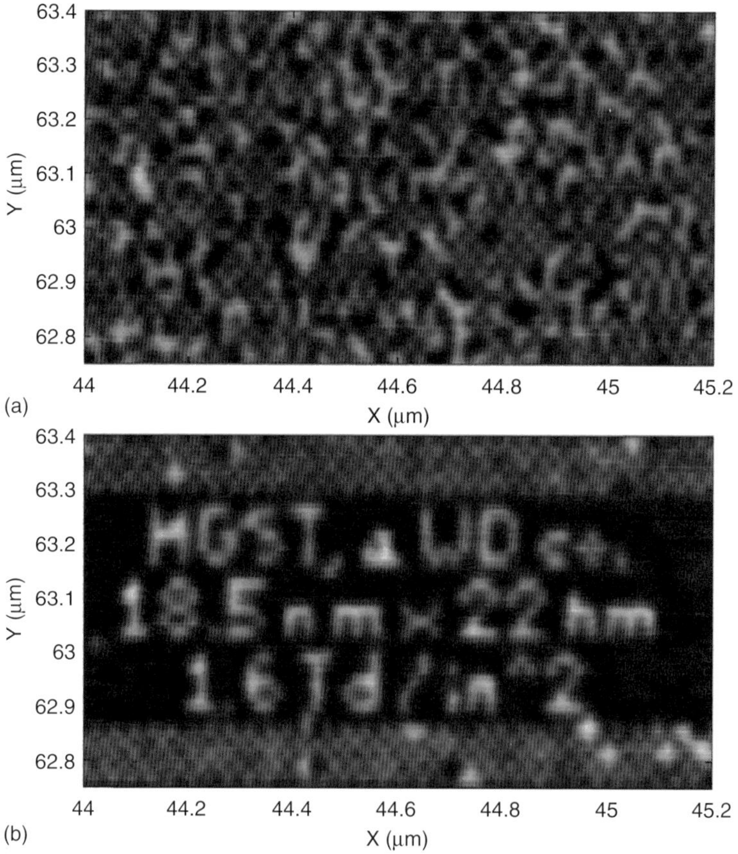

Figure 13.11 2D readback image of shingled magnetic recording on bit patterned media with areal density of 1.6 Tdot/in.^{2}: (a) before and (b) after. (Albrecht *et al.* [84]. Reproduced with permission of IEEE.)

successfully written onto the magnetic dots as shown in the 2D readback image in Figure 13.11b.

The major hurdles in the fabrication of high-density BPM have mostly been overcome, and recording on areal densities exceeding the limits of PMR has already been demonstrated. While the research is still ongoing and there are still some problems to solve, BPM certainly seems to be the future of magnetic information storage.

13.4 Summary

Nanoscience and nanotechnology have had a tremendous impact on the scientific community. The fact that the fundamental material properties change in the nanoscale regime has enabled new and innovative applications of materials. Magnetic materials are a good example of how nanoscience has resulted in greatly improved material properties and given a new boost to the field. In this chapter, an overview of some interesting ways of how the magnetic properties of a material can be enhanced through reducing dimensions was given.

In the first section, a novel method of reducing magnetic losses in materials through patterning the surface was described. This finding breaks the dependence of magnetic properties on the material used. Instead, the best-suited materials can be selected based on the properties of interest (saturation magnetization, permeability, mechanical properties, and cost) and then nanostructured to reduce the coercivity and thus magnetic losses. This result could have a tremendous impact on a plethora of high-frequency devices such as magnetic sensors, antenna, microwave, and spin filter devices.

In the second section, current progress in magnetic information storage technology is reviewed. The current technology has reached its limits, and new methods that would enable us to continue the technological progress are being explored. While several design changes such as HAMR and SMR are considered as options, BPM holds the long-term promise for the future of magnetic recording. The lithographic challenges have been overcome with innovative combinations of different techniques, and early device demonstrations have been successful. However, much more work will need to be done to push BPM to its predicted theoretical limits, and the field is certainly full of challenges and opportunities for new and innovative ideas. Through combining with HAMR and SMR, BPM promises to keep areal densities increasing for decades to come.

Acknowledgments

This work is supported by the NSF (HRD-0833184), NSF-PREM (1523617) and NASA (NNX09AV07A). The author would like to thank Dr. Branislav Vlahovic and Mrs Mary Rasic for their support in writing this book chapter.

References

1. Cullity, B.D. and Graham, C.D. (2009) *Introduction to Magnetic Materials*, 2nd edn, IEEE/Wiley, Hoboken, NJ.
2. Spaldin, N.A. (2011) *Magnetic Materials: Fundamentals and Applications*, 2nd edn, Cambridge University Press, Cambridge, New York.
3. Kronmüller, H. and Parkin, S.S. (2007) *Handbook of Magnetism and Advanced Magnetic Materials*, vol. **2**, John Wiley & Sons, Ltd.
4. Baibich, M.N., Broto, J.M., Fert, A., Van Dau, F.N., Petroff, F., Etienne, P., Creuzet, G., Friederich, A., and Chazelas, J. (1988) *Phys. Rev. Lett.*, **61**, 2472.
5. Moodera, J.S., Kinder, L.R., Wong, T.M., and Meservey, R. (1995) *Phys. Rev. Lett.*, **74**, 3273.
6. Nogués, J., Sort, J., Langlais, V., Skumryev, V., Suriñach, S., Muñoz, J.S., and Baró, M.D. (2005) *Phys. Rep.*, **422**, 65.
7. Nogués, J. and Schuller, I.K. (1999) *J. Magn. Magn. Mater.*, **192**, 203.
8. Ramesh, R. and Spaldin, N.A. (2007) *Nat. Mater.*, **6**, 21.
9. Cheong, S.-W. and Mostovoy, M. (2007) *Nat. Mater.*, **6**, 13.
10. Martin, L., Crane, S.P., Chu, Y.H., Holcomb, M.B., Gajek, M., Huijben, M., Yang, C.H., Balke, N., and Ramesh, R. (2008) *J. Phys. Condens. Matter*, **20**, 434220.
11. Fiebig, M. (2005) *J. Phys. D: Appl. Phys.*, **38**, R123.
12. Rasic, G. and Schwartz, J. (2013) *MRS Commun.*, **3**, 207.
13. Rasic, G. and Schwartz, J. (2014) *IEEE Magn. Lett.*, **5**, 1.
14. Rasic, G. and Schwartz, J. (2015) *Phys. Status Solidi A*, **212**, 213.
15. ITU (2015) ICT Facts and Figures – The World in 2015.
16. Goodenough, J.B. (2002) *IEEE Trans. Magn.*, **38**, 3398.
17. Steinmetz, C.P. (1984) *Proc. IEEE*, **72**, 197.
18. Nie, J., Li, H., Feng, Z., and He, H. (2003) *J. Magn. Magn. Mater.*, **265**, 172.
19. Otsuki, E., Yamada, S., Otsuka, T., Shoji, K., and Sato, T. (1991) *J. Appl. Phys.*, **69**, 5942.
20. Zaspalis, V.T., Antoniadis, E., Papazoglou, E., Tsakaloudi, V., Nalbandian, L., and Sikalidis, C.A. (2002) *J. Magn. Magn. Mater.*, **250**, 98.
21. Adam, J.D., Krishnaswamy, S.V., Talisa, S.H., and Yoo, K.C. (1990) *J. Magn. Magn. Mater.*, **83**, 419.
22. Luders, U., Barthelemy, A., Bibes, M., Bouzehouane, K., Fusil, S., Jacquet, E., Contour, J.P., Bobo, J.F., Fontcuberta, J., and Fert, A. (2006) *Adv. Mater.*, **18**, 1733.
23. Chapline, M.G. and Wang, S.X. (2006) *J. Appl. Phys.*, **100**, 123909.
24. Zhao, P., Zhao, Z.L., Hunter, D., Suchoski, R., Gao, C., Mathews, S., Wuttig, M., and Takeuchi, I. (2009) *Appl. Phys. Lett.*, **94**, 243507.
25. Ross, C.A., Haratani, S., Castano, F.J., Hao, Y., Hwang, M., Shima, M., Cheng, J.Y., Vogeli, B., Farhoud, M., Walsh, M., and Smith, H.I. (2002) *J. Appl. Phys.*, **91**, 6848.
26. Rai, R.C., Wilser, S., Guminiak, M., Cai, B., and Nakarmi, M.L. (2012) *Appl. Phys. A*, **106**, 207.
27. Seifikar, S., Calandro, B., Rasic, G., Deeb, E., Yang, J., Bassiri-Gharb, N., and Schwartz, J. (2013) *J. Am. Ceram. Soc.*, **96**, 3050.
28. Priya, S., Islam, R., Dong, S.X., and Viehland, D. (2007) *J. Electroceram.*, **19**, 149.
29. Chinnasamy, C.N., Yoon, S.D., Yang, A., Baraskar, A., Vittoria, C., and Harris, V.G. (2007) *J. Appl. Phys.*, **101**, 09M517.
30. Jaffari, G.H., Rumaiz, A.K., Woicik, J.C., and Shah, S.I. (2012) *J. Appl. Phys.*, **111**, 093906.
31. Rigato, F., Estrade, S., Arbiol, J., Peiro, F., Luders, U., Marti, X., Sanchez, F., and Fontcuberta, J. (2007) *Mater. Sci. Eng., B*, **144**, 43.
32. Datta, R., Loukya, B., Li, N., and Gupta, A. (2012) *J. Cryst. Growth*, **345**, 44.
33. Li, N., Wang, Y.-H.A., Iliev, M.N., Klein, T.M., and Gupta, A. (2011) *Chem. Vap. Deposition*, **17**, 261.

34. Xia, Y.N. and Whitesides, G.M. (1998) *Angew. Chem. Int. Ed. Engl.*, **37**, 551.
35. Peroz, C., Chauveau, V., Barthel, E., and Sondergard, E. (2009) *Adv. Mater.*, **21**, 555.
36. Zhou, W. (2013) *Nanoimprint Lithography: An Enabling Process for Nanofabrication*, Springer Science & Business Media.
37. Daniel, E.D., Mee, C.D., and Clark, M.H. (1999) *Magnetic Recording: the First 100 Years*, IEEE Press.
38. Moore, G.E. (1998) *Proc. IEEE*, **86**, 82.
39. Charap, S.H., Lu, P.-L., and He, Y. (1997) *IEEE Trans. Magn.*, **33**, 978.
40. Weller, D. and Moser, A. (1999) *IEEE Trans. Magn.*, **35**, 4423.
41. Wood, R., Hsu, Y., and Schultz, M. (2007) Perpendicular Magnetic Recording Technology.
42. H. G. S. Technologies (2005) Get Perpendicular. https://www1.hgst.com/hdd/research/recording_head/pr/PerpendicularAnimation.html
43. Richter, H.J., Dobin, A.Y., Heinonen, O., Gao, K.Z., Veerdonk, R.J.M.v.d., Lynch, R.T., Xue, J., Weller, D., Asselin, P., Erden, M.F., and Brockie, R.M. (2006) *IEEE Trans. Magn.*, **42**, 2255.
44. Khizroev, S. and Litvinov, D. (2006) *Perpendicular Magnetic Recording*, Springer Science & Business Media.
45. Néel, L. (1949) *Ann. Geophys.*, **5**, 99.
46. Brown, W. Jr., (1979) *IEEE Trans. Magn.*, **15**, 1196.
47. Andreas, M., Kentaro, T., David, T.M., Manfred, A., Yoshiaki, S., Yoshihiro, I., Shouheng, S., and Eric, E.F. (2002) *J. Phys. D: Appl. Phys.*, **35**, R157.
48. Weller, D., Moser, A., Folks, L., Best, M.E., Lee, W., Toney, M.F., Schwickert, M., Thiele, J.-U., and Doerner, M.F. (2000) *IEEE Trans. Magn.*, **36**, 10.
49. Piramanayagam, S.N. and Srinivasan, K. (2009) *J. Magn. Magn. Mater.*, **321**, 485.
50. Dong, Y. (2011) Micromagnetic study of perpendicular magnetic recording media. PhD dissertation. University of Minnesota.
51. Bertram, H.N. and Williams, M. (2000) *IEEE Trans. Magn.*, **36**, 4.
52. Piramanayagam, S. (2007) *J. Appl. Phys.*, **102**, 011301.
53. Richter, H.J. (2007) *J. Phys. D: Appl. Phys.*, **40**, R149.
54. Thompson, D.A. and Best, J.S. (2000) *IBM J. Res. Dev.*, **44**, 311.
55. IBM (2011) Global Chief Marketing Officer Study.
56. Yang, J., Tan, C., and Ong, E. (2010) *Microsyst. Technol.*, **16**, 1699.
57. Greaves, S., Kanai, Y., and Muraoka, H. (2009) *IEEE Trans. Magn.*, **45**, 3823.
58. Salo, M., Olson, T., Galbraith, R., Brockie, R., Lengsfield, B., Katada, H., and Nishida, Y. (2014) *IEEE Trans. Magn.*, **50**, 18.
59. Richter, H.J., Parker, G., Staffaroni, M., Grobis, M., and Stipe, B.C. (2014) *IEEE Trans. Magn.*, **50**, 1.
60. Rottmayer, R.E., Batra, S., Buechel, D., Challener, W.A., Hohlfeld, J., Kubota, Y., Li, L., Bin, L., Mihalcea, C., Mountfield, K., Pelhos, K., Peng, C., Rausch, T., Seigler, M.A., Weller, D., and Yang, X. (2006) *IEEE Trans. Magn.*, **42**, 2417.
61. Kryder, M.H., Gage, E.C., McDaniel, T.W., Challener, W.A., Rottmayer, R.E., Ganping, J., Hsia, Y.-T., and Erden, M.F. (2008) *Proc. IEEE*, **96**, 1810.
62. Rausch, T., Gage, E., and Dykes, J. (2015) in *Ultrafast Magnetism I*, vol. **159** (eds J.-Y. Bigot, W. Hübner, T. Rasing, and R. Chantrell), Springer International Publishing, p. 200.
63. Weller, D., Mosendz, O., Richter, H.J., Parker, G., Pisana, S., Santos, T.S., Reiner, J., Hellwig, O., Stipe, B., and Terris, B. (2015) in *Ultrafast Magnetism I*, vol. **159** (eds J.-Y. Bigot, W. Hübner, T. Rasing, and R. Chantrell), Springer International Publishing, p. 228.
64. Ruigrok, J., Coehoorn, R., Cumpson, S., and Kesteren, H. (2000) *J. Appl. Phys.*, **87**, 5398.
65. Wu, A.Q., Kubota, Y., Klemmer, T., Rausch, T., Chubing, P., Yingguo, P., Karns, D., Xiaobin, Z., Yinfeng, D., Chang, E.K.C., Yongjun, Z., Hua, Z., Kaizhong, G., Thiele, J.U., Seigler, M., Ganping, J., and Gage, E. (2013) *IEEE Trans. Magn.*, **49**, 779.
66. Wang, X., Gao, K., Zhou, H., Itagi, A., Seigler, M., and Gage, E. (2013) *IEEE Trans. Magn.*, **49**, 686.

67. Weller, D., Mosendz, O., Parker, G., Pisana, S., and Santos, T.S. (2013) *Phys. Status Solidi A*, **210**, 1245.
68. Weller, D., Parker, G., Mosendz, O., Champion, E., Stipe, B., Wang, X., Klemmer, T., Ju, G., and Ajan, A. (2014) *IEEE Trans. Magn.*, **50**, 3100108.
69. Xu, B., Wang, H., Cen, Z., and Liu, Z. (2014) *IEEE Trans. Magn.*, **51**, 1.
70. Baoxi, X., Hongtao, W., Zhanhong, C., Zhejie, L., Jianming, L., Toh, Y.T., Kaidong, Y., Jing, Z., and Hongzhi, Y. (2014) *IEEE Trans. Magn.*, **50**, 1.
71. Moon, K., Che, X., Tang, Y., Lee, H., Tang, Y., Moneck, M.T., Zhu, J.-G., and Takahashi, N. (2008) *IEEE Trans. Magn.*, **44**, 3442.
72. Albrecht, T.R., Hellwing, O., Ruiz, R., Schabes, M.E., Terris, B.D., and Wu, X.Z. (2009) *Nanoscale Magnetic Materials and Applications*, (eds J. Liu, E. Fullerton, O. Gutfleisch, and D. J. Sellmyer), Springer, p. 237.
73. Richter, H. and Harkness, S. (2006) *MRS Bull.*, **31**, 384.
74. Terris, B. and Thomson, T. (2005) *J. Phys. D: Appl. Phys.*, **38**, R199.
75. Ross, C.A. (2001) *Annu. Rev. Mater. Res.*, **31**, 203.
76. Shiroishi, Y., Fukuda, K., Tagawa, I., Iwasaki, H., Takenoiri, S., Tanaka, H., Mutoh, H., and Yoshikawa, N. (2009) *IEEE Trans. Magn.*, **45**, 3816.
77. Yao, W. and Victora, R.H. (2013) *IEEE Trans. Magn.*, **49**, 5208.
78. Sumei, W., Yao, W., and Victora, R.H. (2013) *IEEE Trans. Magn.*, **49**, 3644.
79. Shu, X., Gongye, Z., Jincai, C., and Bo, L. (2012) *IEEE Trans. Magn.*, **48**, 3891.
80. Greaves, S.J., Muraoka, H., and Kanai, Y. (2012) *J. Magn. Magn. Mater.*, **324**, 314.
81. McDaniel, T.W. (2005) *J. Phys. Condens. Matter*, **17**, R315.
82. Schabes, M.E. (2008) *J. Magn. Magn. Mater.*, **320**, 2880.
83. Grobis, M.K., Hellwig, O., Hauet, T., Dobisz, E., and Albrecht, T.R. (2011) *IEEE Trans. Magn.*, **47**, 6.
84. Albrecht, T., Arora, H., Ayanoor-Vitikkate, V., Beaujour, J.M., Bedau, D., Berman, D., Bogdanov, A., Chapuis, Y.-A., Cushen, J., Dobisz, E.E., Doerk, G., Gao, H., Grobis, M., Gurney, B., Hanson, W., Hellwig, O., Hirano, T., Jubert, P.O., Kercher, D., Lille, J., Liu, Z., Mate, C.M., Obukhov, Y., Patel, K.C., Rubin, K., Ruiz, R., Schabes, M., Wan, L., Weller, D., Wu, T.-W., and Yang, E. (2015) *IEEE Trans. Magn.*, **51**, 1.
85. Ross, C.A. and Cheng, J.Y. (2008) *MRS Bull.*, **33**, 838.
86. Wan, L., Ruiz, R., Gao, H., Patel, K.C., Lille, J., Zeltzer, G., Dobisz, E.A., Bogdanov, A., Nealey, P.F., and Albrecht, T.R. (2012) *J. Micro/Nanolithogr. MEMS MOEMS*, **11**, 031405.
87. Albrecht, T.R., Bedau, D., Dobisz, E., Gao, H., Grobis, M., Hellwig, O., Kercher, D., Lille, J., Marinero, E., Patel, K., Ruiz, R., Schabes, M.E., Wan, L., Weller, D., and Wu, T.W. (2013) *IEEE Trans. Magn.*, **49**, 773.
88. Chou, S.Y., Krauss, P.R., and Renstrom, P.J. (1996) *J. Vac. Sci. Technol., B*, **14**, 4129.
89. Griffiths, R.A., Williams, A., Oakland, C., Roberts, J., Vijayaraghavan, A., and Thomson, T. (2013) *J. Phys. D: Appl. Phys.*, **46**, 503001.
90. Bates, F.S. and Fredrickson, G.H. (1990) *Annu. Rev. Phys. Chem.*, **41**, 525.
91. Segalman, R.A. (2005) *Mater. Sci. Eng., R*, **48**, 191.
92. Liu, C.-C., Han, E., Onses, M.S., Thode, C.J., Ji, S., Gopalan, P., and Nealey, P.F. (2011) *Macromolecules*, **44**, 1876.
93. Chevalier, X., Nicolet, C., Tiron, R., Gharbi, A., Argoud, M., Pradelles, J., Delalande, M., Cunge, G., Fleury, G., Hadziioannou, G., and Navarro, C. (2013) *Proc. SPIE Int. Soc. Opt. Eng.*, **8680**, 868006.
94. Kim, S., Shin, D.O., Choi, D.-G., Jeong, J.-R., Mun, J.H., Yang, Y.-B., Kim, J.U., Kim, S.O., and Jeong, J.-H. (2012) *Small*, **8**, 1458.
95. Ouk Kim, S., Solak, H.H., Stoykovich, M.P., Ferrier, N.J., de Pablo, J.J., and Nealey, P.F. (2003) *Nature*, **424**, 411.
96. Bates, C.M., Maher, M.J., Janes, D.W., Ellison, C.J., and Willson, C.G. (2014) *Macromolecules*, **47**, 2.
97. Bunday, B.D., Bishop, M., McCormack, J.D.W., Villarrubia, J.S., Vladar, A.E., Dixson, R., Vorburger, T.V., Orji, N.G.,

and Allgair, J.A. (2004) *Proc. SPIE Int. Soc. Opt. Eng.*, **5375**, 515.

98. Guarini, K.W., Black, C.T., Milkove, K.R., and Sandstrom, R.L. (2001) *J. Vac. Sci. Technol., B*, **19**, 2784.

99. Shuaigang, X., XiaoMin, Y., Kim, Y.L., Rene, J.M.v.d.V., David, K., and Thomas, P.R. (2011) *Nanotechnology*, **22**, 305302.

100. Patel, K., Ruiz, R., Lille, J., Wan, L., Dobiz, E., Gao, H., Robertson, N., and Albrecht, T. (2012) *SPIE Adv. Lithography*, **8323**, 83230U.

101. Chou, S.Y. (1997) *Proc. IEEE*, **85**, 652.

102. Grobis, M., Dobisz, E., Hellwig, O., Schabes, M.E., Zeltzer, G., Hauet, T., and Albrecht, T.R. (2010) *Appl. Phys. Lett.*, **96**, 052509.

14
Phase Change Materials for Memory Application

Liangcai Wu and Zhitang Song

14.1
Introduction

Phase change random access memory (PCRAM), owing to its nonvolatility, scalability, and compatibility with CMOS technology, has attracted growing attention as one of the most promising candidates for the next-generation nonvolatile memory (NVM) technology [1–4]. In addition, PCRAM has the potential to meet the demands of storage system as well as dynamic random-access memory (DRAM), given the continued power efficiency and storage capacity improvement through scaling. However, current features of $Ge_2Sb_2Te_5$ (GST)-based PCRAM are insufficient to meet the needs of DRAM, especially on operation speed, programming cycles, and power consumption. Although the data retention ability of GST can meet the requirement of consumer electronics, the data retention of GST should be improved to meet the needs of other fields that require higher temperature. Furthermore, in the practical production process, a layer of dielectric such as SiN_x is required in order to avoid oxidation of phase change material. The dielectric film deposition is always performed at a temperature above 400 °C for a relatively long period of time. Long-time heat treatment at high temperature could lead to interface issues, such as poor contact between phase change materials and electrode. In addition, for high-density memory arrays, thermal cross-talk becomes more severe. From these points of view, the improvement of thermal stability of GST phase change material or developing alternative novel phase change material is significant for device reliability and PCRAM applications.

In this chapter, firstly, the GST phase change material is briefly introduced and then the improvement in the properties of GST such as N- and C-doped GST is discussed. C doping is considered as one of the most feasible schemes to improve the properties of GST, and C-doped GST (GSTC) material becomes one of the most promising materials. The thermal stability of GST is significantly enhanced with C incorporation, and the voids formation at high temperature is also suppressed. Most of the C atoms exist randomly in the amorphous cluster around the crystal grains, confining the growth of crystals, and the rest of the C atoms

Advanced Nano Deposition Methods, First Edition. Edited by Yuan Lin and Xin Chen.

are trapped in the grains, inclining to replace Te atoms or take the interstitial site. The local environments of Ge-Te tetrahedral units have been disturbed by the stable C–C chains with much more tetrahedral Ge sites and shorter bond length of Ge-Te, leading to slower crystallization procedure and larger amorphous stability. Lower reset power is promised for the PCRAM cells with GSTC material. More than 2×10^4 cycles suggest that GSTC film is a potential phase change material for high-density PCRAM application.

In Section 14.3, our recently researched typical novel phase change material Ti-Sb-Te (TST) is talked about. Compared with widely researched Ge-Sb-Te ternary alloy, Sb-Te binary alloys, which have been successfully applied in phase change optical recording, show higher crystallization speed than GST. The high crystallization speed of Sb-Te binary alloys is attributed to the growth-dominated crystallization mechanism. However, due to the low crystallization temperature and poor data retention, Sb-Te binary alloys themselves are not suitable for PCRAM application. In this part, Ti-doped Sb-Te alloys were introduced as new phase change materials. Firstly, Ti-doped Sb_2Te alloys are discussed. Compared with GST, $Ti_{10}Sb_{60}Te_{30}$ possesses higher crystallization temperature, better data retention, and rapid crystallization speed. Then Ti-doped Sb_2Te_3 alloys are discussed. For pure Sb_2Te_3 alloys, the crystallization temperature of Sb_2Te_3 is too low (<100 °C) and the data retention is very poor, which makes pure Sb_2Te_3 not appropriate for PCRAM application. While for Ti-doped Sb_2Te_3 alloy, it has larger resistance contrast and higher crystallization temperature. More importantly, uniform phase is observed in crystalline Ti-doped Sb_2Te_3 film and the grain size is sharply decreased, benefiting for accelerating the crystallization rate. Ti-doped Sb_2Te_3 film-based PCRAM cells have fast operation speed with low power consumption, and more than 10^6 cycles were obtained with stable reset and set resistances.

14.2 $Ge_2Sb_2Te_5$ and Its Properties' Improvement

14.2.1 $Ge_2Sb_2Te_5$ Phase Change Material

As mentioned earlier, GST is one of the most widely studied phase change materials for PCRAM because of its outstanding overall phase change performances. It is well known that the structure of the metastable phase of GST is NaCl type in the space (*Fm*-3*m*), as shown in Figure 14.1, where the 4a sites are wholly occupied by only Te atom and the 4b sites are randomly occupied by Ge atom, Sb atom, and about 20% vacancy [5]. Figure 14.2 shows the sheet resistance of GST film. As shown in Figure 14.1, sheet resistance of the GST film decreases gradually in the initial heating stage until a sudden drop in sheet resistance occurs at around the crystallization temperature (T_1), which is attributed to the phase transformation from amorphous to face-centered cubic (fcc) crystalline state. Then another drop

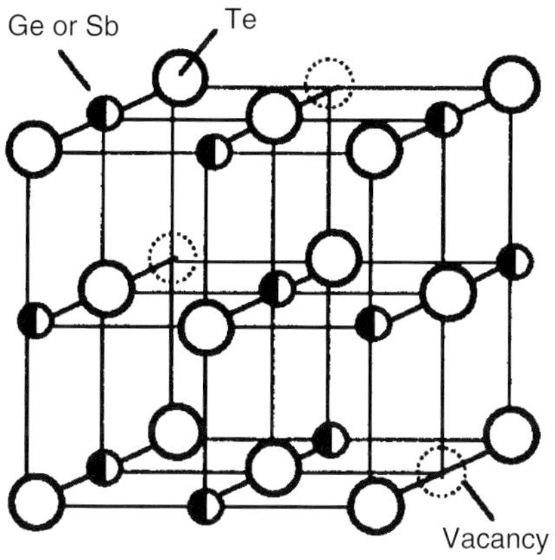

Figure 14.1 Atomic arrangement of the metastable phase of GST. (Liu *et al.* [6]. Reproduced with permission of Elsevier.)

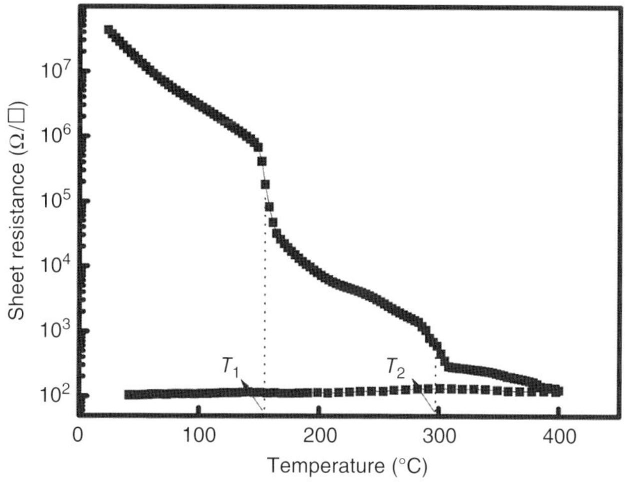

Figure 14.2 Sheet resistances of the GST film.

appears at higher temperature (T_2), corresponding to the second phase transition from fcc to stable hexagonal structure [7].

For commercial PCRAM application, the reset current needs to be lowered to reduce the power consumption of the device. In order to reduce the writing current of memory device, increasing the resistance of phase change material may be a simple approach. Various researches have been carried out to address these issues by doping GST with O, Si, or Ag atoms. O-doped GST films were usually deposited by reactive sputtering method, and O contents were controlled by changing the ratio of O_2 and Ar gas flow rate. Our group has put forward implantation method to doping GST. It was found that O implantation has great effect on GST film [6, 8, 9]. The Rs of O-implanted film is higher than that of GST film when the annealing temperature is higher than 340 °C, which may be caused by phase separation and carrier scattering by grain boundaries. O implantation can also suppress the phase transition from fcc to hexagonal structure and refine the grain size.

14.2.2 N-Doped $Ge_2Sb_2Te_5$ Material

However, the implanted O element will inevitably deteriorate the phase change ability of GST film [10, 11]. N doping was selected to improve the properties of GST. The N-doped GST film (GSTN) was deposited by sputtering of GST targets using a magnetron sputtering deposition system at room temperature under a N_2/Ar atmosphere. The temperature-dependent sheet resistance ($R-T$) for the GSTN film is given in Figure 14.3. The phase transition characteristics of GST were associated with conversion from an amorphous to an fcc structure and from an fcc to a hexagonal structure, respectively. The sheet resistances of GST decreased with an increase in temperature, showing a quick drop at the crystallization temperature (T_c). While for GSTN film, the T_c tends to shift to higher temperature. The doping of nitrogen atoms into GST film has promoted T_c obviously upon the first phase transition. Nevertheless, the temperature for the second structure transition of GSTN film is much lower than that of GST (~300 °C), which suggests that nitrogen addition in GST film suppresses the formation of metastable fcc phase greatly, helping to obtain the stable hexagonal phase. The increase in R_s should be attributed to the generation of nitrides. The increase in crystalline resistance is helpful to reduce the reset current in PCRAM cell.

Figure 14.4 shows the structure of GSTN material revealed by XRD. Figure 14.4a shows the XRD results of GSTN annealed at different temperature, and

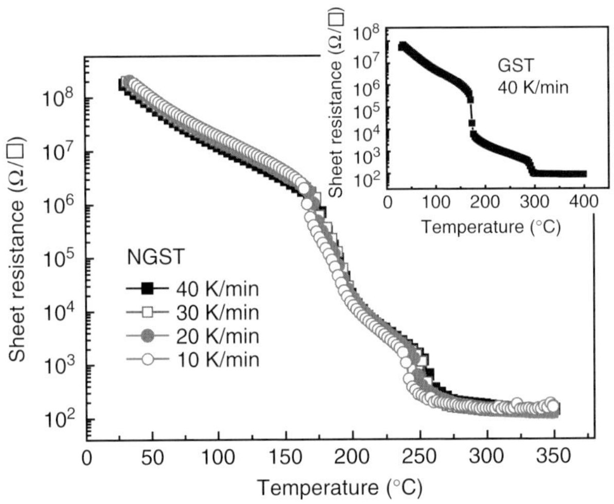

Figure 14.3 Dependence of sheet resistances (R_s) on the annealing temperature for GSTN film.

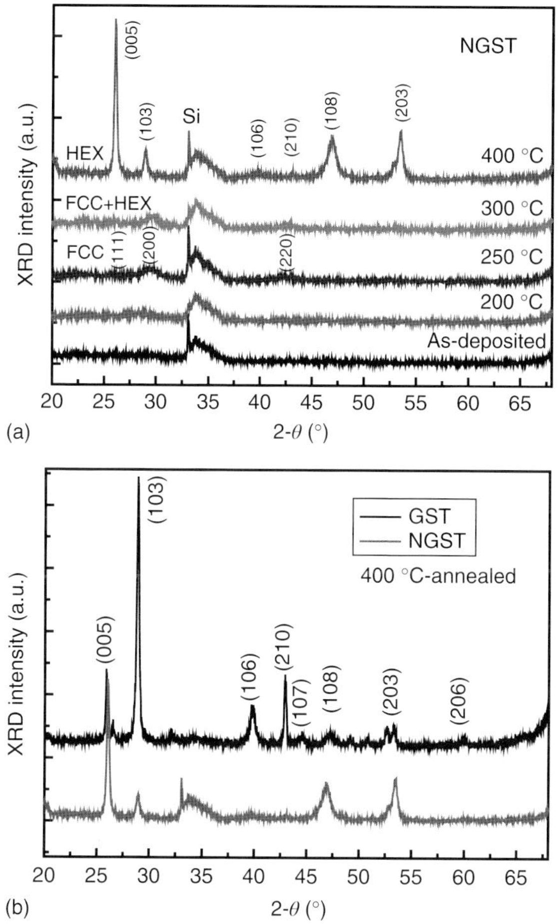

Figure 14.4 XRD results of GSTN and GST films: (a) GSTN at different annealing temperature and (b) comparison of GSTN and GST films at the same annealing temperature (400 °C).

Figure 14.4b gives the comparison of GSTN and GST films at the same annealing temperature of 400 °C. The as-deposited and 200 °C-annealed GSTN films are found to be amorphous since no detectable diffraction peaks of the film appear, as shown in Figure 14.4a. As annealed at 250 °C for 2 min, the diffraction peaks of fcc crystalline phase are observed in the film. A mixed fcc and hexagonal structure tends to shape when heating the film at 300 °C, and the emergence of new peaks at 400 °C indicates a structural transformation from fcc to hexagonal phase.

The suppression of grain growth of GST by nitrogen addition can be confirmed by comparing the XRD pattern of NGST and GST films annealed at the same temperature of 400 °C, as seen in Figure 14.4b, where the most intense peak of NGST film at around 29° ((*103*) peak) is weaker and broader than that of GST film, indicating the reduction of grain size.

14.2.3 C-Doped $Ge_2Sb_2Te_5$ Material

Apart from N doping, C doping is another promising method to improve the properties of GST material. According to the theoretical *ab initio* investigation, the C dopants are found to be more effective on the atomic and electronic structures in amorphous GST material. Actually, C atom had been doped in many phase change materials, such as GeTe and Sb, to gain better properties. Here, we talk about our results of C-doped GST material focused on thermal stability, microstructure characteristics, as well as reversible phase change characteristics [12, 13].

14.2.3.1 Film Properties and Microstructure Characteristics

GSTC films are deposited by cosputtering of GST and C targets using magnetron sputtering system at room temperature. Figure 14.5a shows the temperature dependence of R_s for the GSTC films with various C contents, and the inset is the activation energy analysis for electrical conduction of the amorphous films. A continuous decrease in R_s with annealing temperature is observed for each film until a sharp resistance drop occurs when the temperature reaches the first transition point (T_c) of around 200 °C, which is attributed to the phase transition from amorphous to fcc structure. It is obvious that the crystallization temperature T_c increases with C-doping content, indicating a better thermal stability of the amorphous phase, thus improving data retention of PCRAM. A second sudden drop in R_s appears at around 290 °C for GST and larger than 330 °C for the C-doped films as marked in the figure. The GSTC film has a higher transition temperature in the second step than in the first one. Hence, the fcc structure that is employed in the reversible phase switching process can be preserved at a higher temperature, promising a more robust compatibility with manufacturing technology. Furthermore, the two-step decrease in R_s, abrupt for GST, shows a rather gradual profile for the C-doped materials, suggesting that carbon incorporation slows down the crystallization process of the material.

Figure 14.5b shows the Kissinger plots for E_a calculation for the GSTC films with various C contents. The slope of the linear plot determines the E_a to be 3.03, 3.24, 3.36, and 3.56 eV for GST, GSTC9%, GSTC18%, and GSTC25% films, respectively.

Figure 14.6 shows the XRD results for GST and GSTC18% films. Figure 14.6a,b gives the programmed and measured temperature profiles of GST and GSTC18% films during the XRD measurement, in which the actual temperature value fits well with the programmed one. The XRD patterns reveal a crystallization behavior of the films as a function of temperature as presented in Figure 14.6c,d. It confirms that both of the as-deposited films are amorphous. The diffraction peaks appear as the temperature reaches 160 and 180 °C for GST and GSTC18% films, respectively, and can both be indexed to the fcc phase. At around 250 °C, the GST film undergoes a cubic-hexagonal (HEX) phase transition in Figure 14.6c, which takes place at approximately 290 °C for the GSTC18% film in Figure 14.6d.

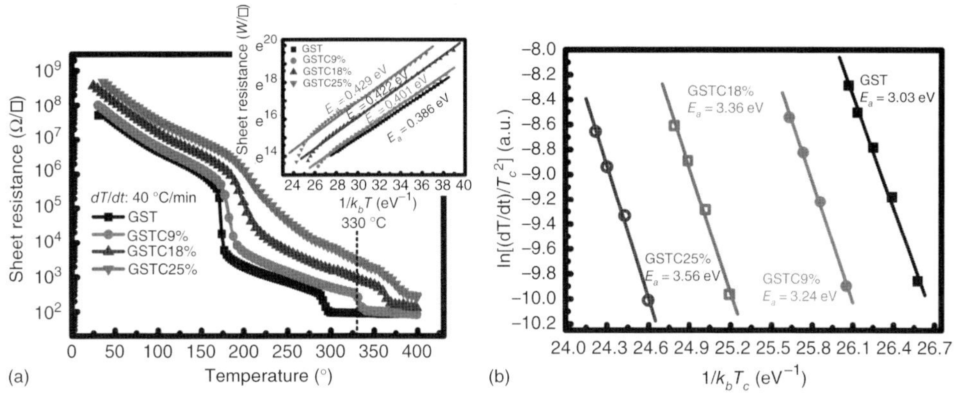

Figure 14.5 (a) Temperature dependence of R_s for the GSTC films (inset is the activation energy analysis for electrical conduction of the films) and (b) Kissinger plots for E_a calculation of the GSTC films with various C contents. (Zhou *et al.* [12]. Reproduced with permission of American Institute of Physics.)

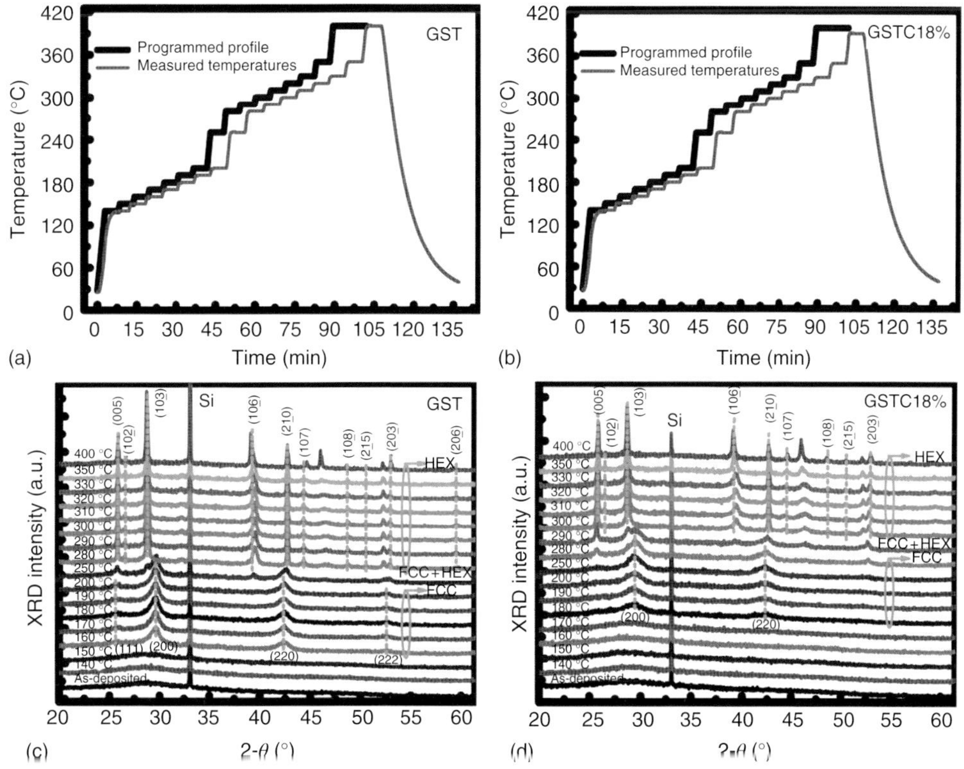

Figure 14.6 Programmed and measured temperature as a function of time (a) GST and (b) GSTC18% films; XRD results for (c) GST and (d) GSTC18% films. (Zhou *et al.* [12]. Reproduced with permission of American Institute of Physics.)

The temperature of phase transition (amorphous–fcc, fcc–HEX) has been enhanced obviously by carbon addition. Furthermore, the relatively weaker diffraction intensity of the XRD patterns of GSTC18% suggests that the crystal grains have been confined significantly by the carbon cluster forms that possibly exist at grain boundaries.

Figure 14.7 shows the TEM and HRTEM of the microstructures of GST and GSTC18% films. The films are annealed at 250 and 400 °C for 5 min. Figure 14.7a,b shows the TEM and selected area electron diffraction (SAED) patterns of the GST and GSTC18% films, respectively. Figure 14.7c,d shows the corresponding HRTEM images. It can be seen that the GST film presents a polycrystalline structure with grains in tens of nanometers, while the GSTC18% film displays smaller crystal grains with amorphous carbon cluster regions, as demonstrated with the dash rings. Smaller grain size of fcc phase is desired to improve the power, speed, and resistance drift performance of the highly scaled PCRAM devices. The lattice distortions over a few atomic distances can be identified in the dash rectangular area of Figure 14.7d, which is probably attributed to the incorporated carbon atoms trapped in the crystal grains. Furthermore, there is little evidence of crystallized carbon in Figure 14.7d, which reinforces the assumption that carbon atoms may partially accumulate in the amorphous matrix environing the reduced grain size. The contrast in microstructure morphology of the two materials is schematically illustrated in the insets of the corresponding figures. For the films heated at 400 °C for 2 min, significant grain growth occurs as shown in Figure 14.7e,f for the GST and GSTC18% films, respectively, and the HEX crystal structure is identified from the SAED rings in the insets of the two figures. One of the most interesting features of the images in Figure 14.7e is the randomly distributed voids among the crystals as compared to the uniform phase morphology in Figure 14.7f with embedded bright and dark areas. The voids development cannot be observed in the GSTC18% film, which is beneficial for enhancing the cell reliability and enabling size scaling down architecture, since there are multichannel high-temperature (about 400 °C) backend processes for PCRAM chip fabrication.

14.2.3.2 Reversible Phase Change Characteristics of C-Doped $Ge_2Sb_2Te_5$

Figure 14.8a shows the $I-V$ characteristics of the PCM cell using GSTC18% material. In the first current sweeping, the phase transition from amorphous to crystalline state is realized since the large snapback of voltage and negative resistance behavior appears as soon as the current is applied to the cell. In the second current sweeping, a crystalline state of the cell is completely achieved since the $I-V$ curves of the second sweeping exactly overlap the crystalline state section of the first sweeping curve. Figure 14.8b shows the reset operations of GSTC18%-based PCRAM cells with various pulse widths. The reversible resistance–voltage ($R-V$) curves of the PCRAM cells based on GSTC18% material are plotted in Figure 14.8c. As seen from Figure 14.8c, the set resistance of the cells decreases with the pulse width, obtaining a lower resistance of approximately 600 kΩ for the 1000-ns-width pulse, which is attributed to the

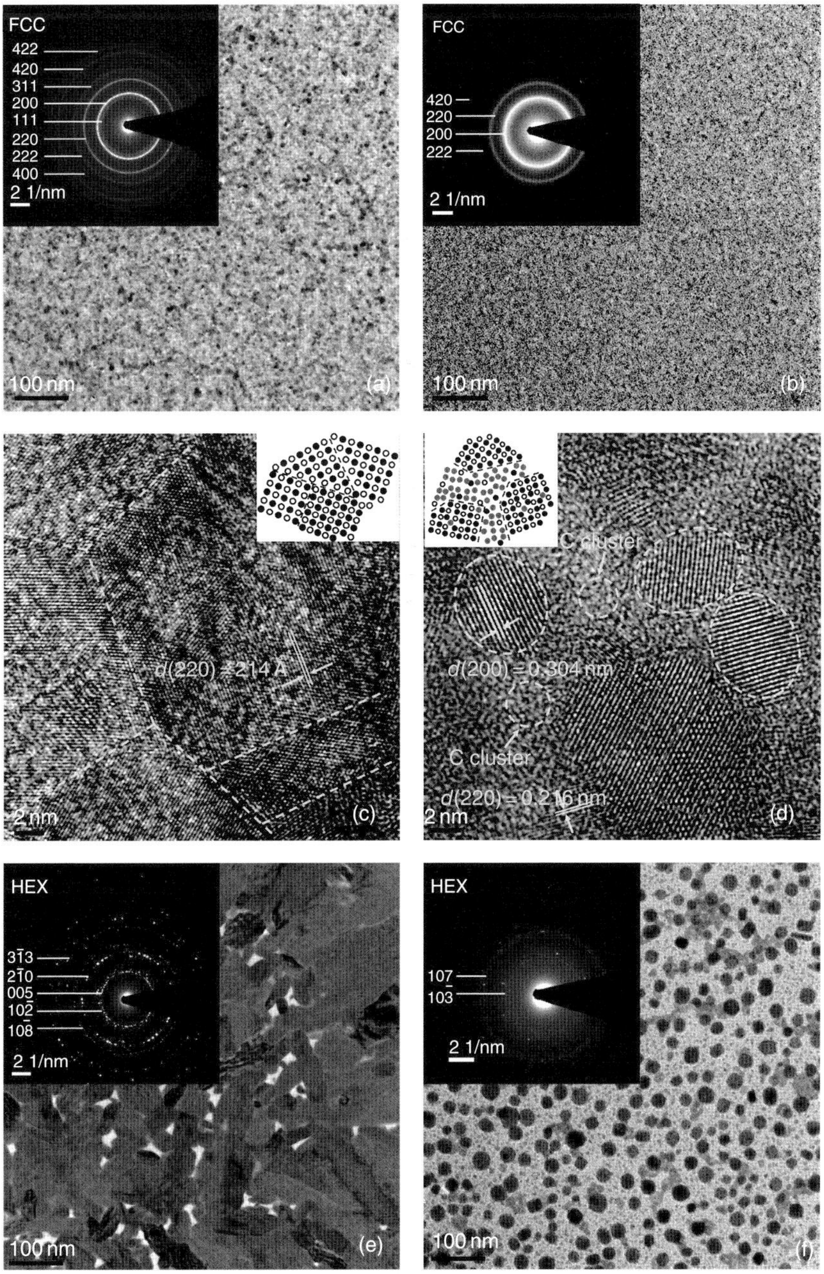

Figure 14.7 TEM and SAED patterns: (a) GST film; (b) GSTC18% film; HRTEM images: (c) GST and (d) GSTC18% film (insets show schematic illustrations of crystalline phases); (e) HEX GST film; and (f) HEX GSTC18% film. (Zhou *et al.* [13]. Reproduced with permission of American Chemical Society.)

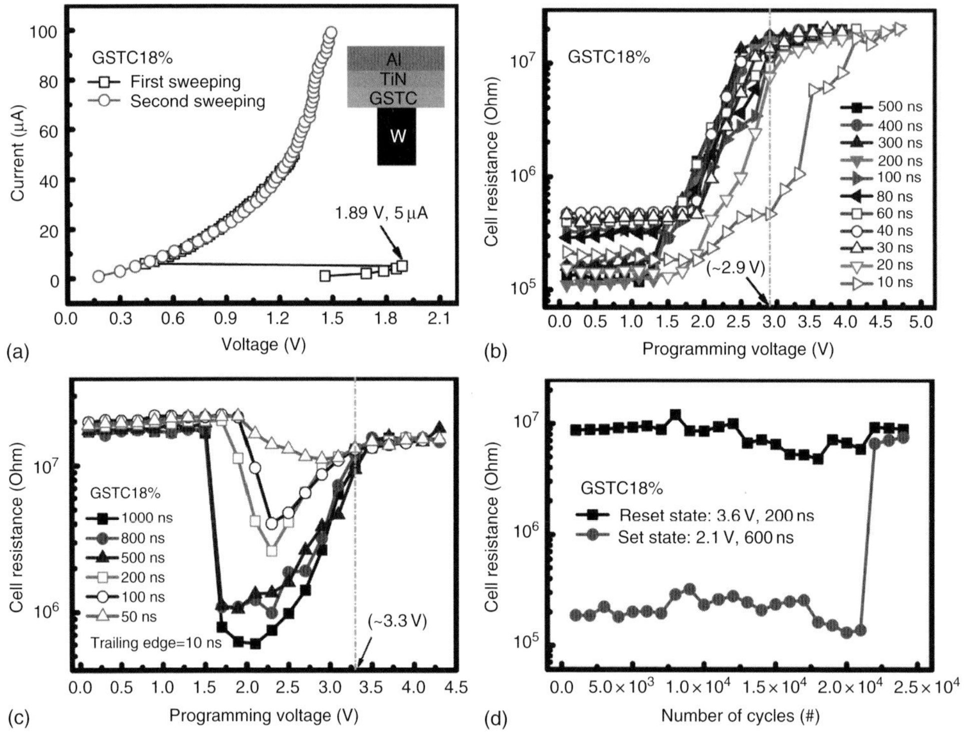

Figure 14.8 (a) *I–V* characteristics of the PCM cell using GSTC18%; (b) reset operations of GSTC18%-based PCRAM cells with various pulse widths; (c) *R–V* curves of the PCRAM cells based on GSTC18%; and (d) endurance performance of a cell using GSTC18%. (Zhou *et al.* [12]. Reproduced with permission of American Institute of Physics.)

larger grains and/or a more complete crystallization state, at the same time the set voltages are correspondingly decreased. As indicated in the figure, the reset voltage of approximately 3.3 V of the cell is clearly lower than that of the GST-based cells (~3.55 V) with the identical cell architecture. The larger set resistance of GSTC will greatly reduce the current magnitude, lowering the voltage since the amorphization behavior is mainly dominated by a Joule heating effect. It is noted that a 50-ns-width pulse fails to set the cell and a pulse width of 200 ns is insufficient for a complete set programming, suggesting that carbon doping indeed leads to a slower crystallization process and thus longer write time for the set operation. It is worthy pointing out that a faster set process and a more conductive set state can be expected with proper programming algorithm, which is currently under investigation. Figure 14.8d displays the endurance performance of a cell using GSTC18% material. The reset/set resistance ratio of approximately 100 is applicable for data resolution, and both the reset and set resistances are relatively stable as the number of cycles increases. The endurance capability keeps about 2.1×10^4 cycles before the presence of resistance disorder with a reset stuck failure mechanism.

14.3 High-Speed and Lower-Power TiSbTe Materials

14.3.1 Film Properties and Microstructure Characteristics

14.3.1.1 Ti-Doped Sb_2Te Materials

Ti-doped Sb_2Te films were deposited by cosputtering of Ti and Sb_2Te targets using magnetron sputtering system at room temperature [14]. Figure 14.9a shows the sheet resistances (R_s) of Sb_2Te and $Ti_{10}Sb_{60}Te_{30}$ films as a function of the annealing temperature ($R-T$). As shown in Figure 14.9a, a sudden R_s drop at about 139 °C is clearly observed for Sb_2Te film. The drop of R_s reflects the phase transition from amorphous state to crystalline state. While for $Ti_{10}Sb_{60}Te_{30}$ film, the crystallization temperature increases to 211 °C, which means that the thermal stability of amorphous state will be remarkably improved. Figure 14.9b shows the data retention characteristics of $Ti_{10}Sb_{60}Te_{30}$ film. As shown in Figure 14.9b, low failure time is obtained with high annealing temperature. By extrapolation from the failure time to 10 years, based on the Arrhenius law, a 10-year lifetime temperature of 137 °C for $Ti_{10}Sb_{60}Te_{30}$ film is provided, with an activation energy of $E_a = 3.5$ eV. These results are much better than those of Sb_2Te and conventional GST.

Figure 14.10a shows the XRD results for Sb_2Te and $Ti_{10}Sb_{60}Te_{30}$ films. Both as-deposited films show no diffraction peak, which indicates that $Ti_{10}Sb_{60}Te_{30}$ film is still in the amorphous state. Diffraction peaks of hexagonal phase are observed in Sb_2Te film annealed at 250 °C. The diffraction peaks of $Ti_{10}Sb_{60}Te_{30}$ film, in comparison to Sb_2Te film, show a decrease in the intensity, some of them even disappearing.

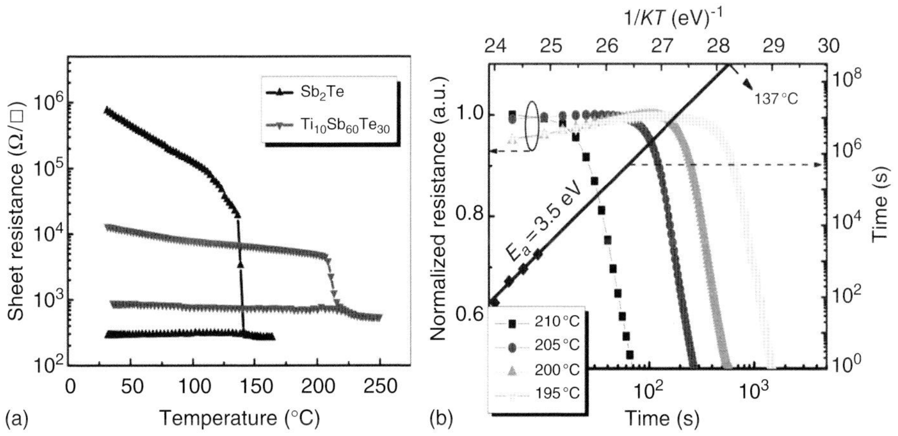

Figure 14.9 (a) Sheet resistances (R_s) of Sb_2Te and $Ti_{10}Sb_{60}Te_{30}$ films as a function of the annealing temperature and (b) data retention characteristics for $Ti_{10}Sb_{60}Te_{30}$ films. (Zhu *et al.* [14]. Reproduced with permission of American Institute of Physics.)

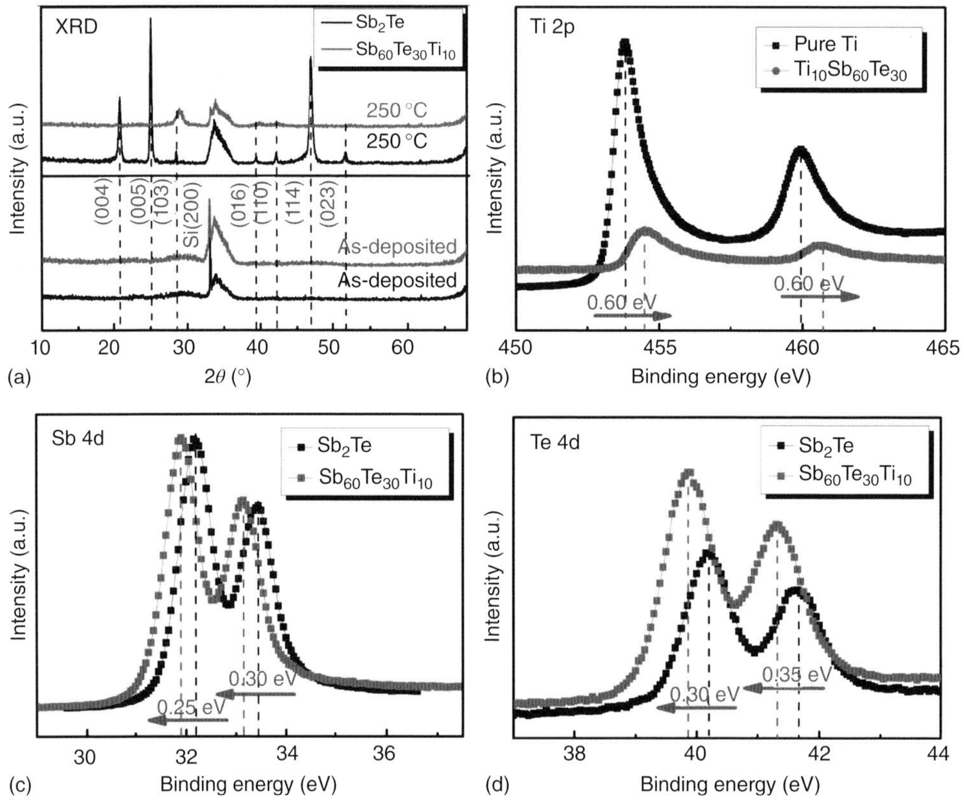

Figure 14.10 (a) XRD results for Sb_2Te and $Ti_{10}Sb_{60}Te_{30}$ films and (b–d) XPS of $Ti_{10}Sb_{60}Te_{30}$ film. (Zhu *et al.* [14]. Reproduced with permission of American Institute of Physics.)

In Figure 14.10b–d, the XPS results for $Ti_{10}Sb_{60}Te_{30}$ film are presented. The binding energies of 454.5 and 460.7 eV are related to Ti $2p_{3/2}$ and $2p_{1/2}$, respectively. These results are 0.6 eV higher than metal Ti–Ti binding energies, which mean that Ti is bonding with other elements. In the case of Sb element, both peaks shift to lower binding energies (~0.3 eV). Similar results appeared in the binding energies of Te element. For smaller electronegativity of Ti (1.54) than that of Sb (2.05) and Te (2.1), the binding energy of Ti increases and the binding energy of Sb and Te decreases when Ti bonds with Sb and Te.

14.3.1.2 Ti-Doped Sb_2Te_3 Materials

Although the thermal stability of $Ti_{10}Sb_{60}Te_{30}$ film is obviously improved and even can satisfy the requirements of automotive systems, the resistance ratio is decreased to about one order of magnitude [15]. In this section, Ti-doped Sb_2Te_3 alloys are discussed. Ti-doped Sb_2Te_3 films were deposited by cosputtering of Ti and Sb_2Te_3 targets using magnetron sputtering system at room temperature. It was found that Ti-doped Sb_2Te_3 alloy has larger resistance contrast and crystallization temperature is greatly increased. Ti-doped Sb_2Te_3 film-based PCRAM cells show fast operation speed, and more than 10^6 cycles were obtained with stable reset and set resistances. Figure 14.11 shows the $R-T$ curves of Ti-doped Sb_2Te_3 film. As can be seen from Figure 14.11, a slow and gradual drop of resistance is observed for pure Sb_2Te_3. The reason is that the as-deposited Sb_2Te_3 film has partly crystallized during the sputtering process due to its low crystallization temperature. While for Ti-doped Sb_2Te_3 film, a sudden resistance drop happens at about 176 °C and the resistance contrast is remarkably increased.

Figure 14.12 shows the XRD patterns of pure Sb_2Te_3 and Ti-doped Sb_2Te_3 films. For as-deposited pure Sb_2Te_3 film, some diffraction peaks appear, indicating that it has crystallized during the sputtering process. The diffraction peaks in 29.67° and 42.30° correspond to (200) and (220) crystal faces of the fcc structure, respectively. With the increasing annealing temperature, the peak intensities of fcc structure are weakened gradually and disappeared finally. Simultaneously, the diffraction peaks of HEX structure become sharp. These imply that the phase transition process has completed. While for Ti-doped Sb_2Te_3 film, the as-deposited film is amorphous. With increased temperature, amorphous Ti-doped Sb_2Te_3 film is crystallized gradually and HEX is the only structure for crystalline phase, indicating that Ti-doped Sb_2Te_3 film has better thermal stability than pure Sb_2Te_3, which is consistent with the aforementioned result of $R-T$ curves. Furthermore, all diffraction peaks for Ti-doped Sb_2Te_3 film become broader distinctly. This signifies that the crystalline grain size in Ti-doped Sb_2Te_3 film is much smaller than that of pure Sb_2Te_3 film.

Figure 14.13 shows the TEM, HRTEM, and SAED results of crystalline Sb_2Te_3 and Ti-doped Sb_2Te_3 films. From Figure 14.13a,b, it can be seen that the grain size of crystalline Sb_2Te_3 film reaches several hundred nanometers. Figure 14.13c

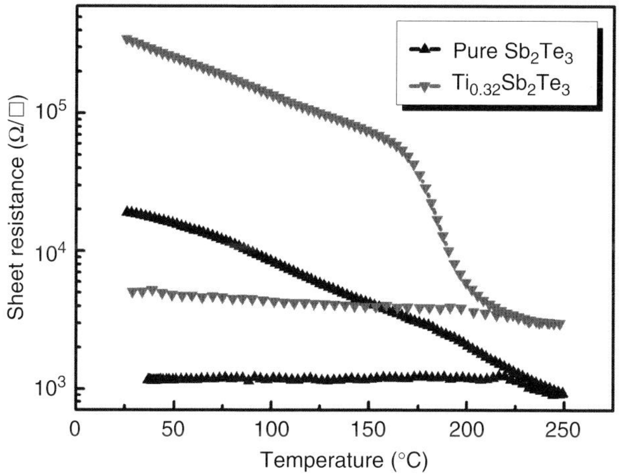

Figure 14.11 $R-T$ curves of Ti-doped Sb_2Te_3 film. (Zhu *et al.* [15]. Reproduced with permission of American Institute of Physics.)

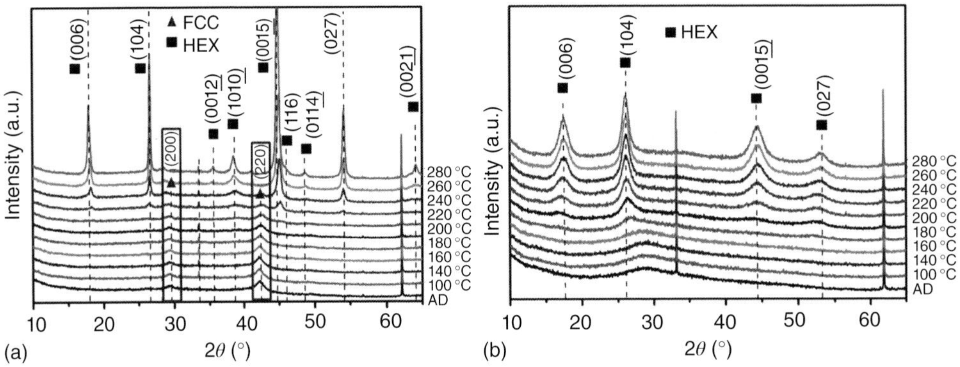

Figure 14.12 XRD patterns of (a) pure Sb_2Te_3 and (b) Ti-doped Sb_2Te_3 films. (Zhu *et al.* [15]. Reproduced with permission of American Institute of Physics.)

shows the HRTEM image of crystalline Sb_2Te_3 film; two lattice fringes with the same spacing of 2.13 Å form an included angle of 60°, which are corresponding to the {110} planes of HEX phase Sb_2Te_3, confirmed by the electron diffraction pattern in Figure 14.13d. Compared to pure Sb_2Te_3, as shown in Figure 14.13e,f, the grain size of Ti-doped Sb_2Te_3 film is much smaller. The HRTEM of the Ti-doped Sb_2Te_3 film in Figure 14.13g displays the similar atomic arrangement of pure Sb_2Te_3 film. The SAED pattern in Figure 14.13h exhibits polycrystalline rings with scattered diffraction spots.

14.3.2
Reversible Phase Change Characteristics of TST Alloy

Figure 14.14a,b shows the $R-V$ characteristics of a phase change memory cell with TST and GST films, respectively. At a pulse length of 100 ns, the set operation for TST test cell is achieved by a pulse amplitude of 1.1 V, which is just half of that for GST PCRAM cell (2.3 V). The set voltage for GST-based device can satisfy the requirement of reset operation for TST-based device. In addition, for TST-based device, except for the set and reset operations, the resistances almost unchanged with increasing pulse amplitude. Noticeably, the cell can readily complete set and reset operations at a pulse length of only 10 ns, with the same R ratio between reset and set states. This proves that TST test cell has the capability of high-speed operation. By contrast, as shown in Figure 14.14b, the resistances of conventional GST-based PCRAM cells are more sensitive to the pulse amplitude and width. Obviously, below 100 ns, the set operation for GST based PCRAM cells is incomplete with abnormal set resistance.

The required time for set operation was measured as a function of applied pulse widths with different amplitudes, as shown in Figure 14.14c. The time for set operation varies regularly with the increase in pulse amplitude: the higher the

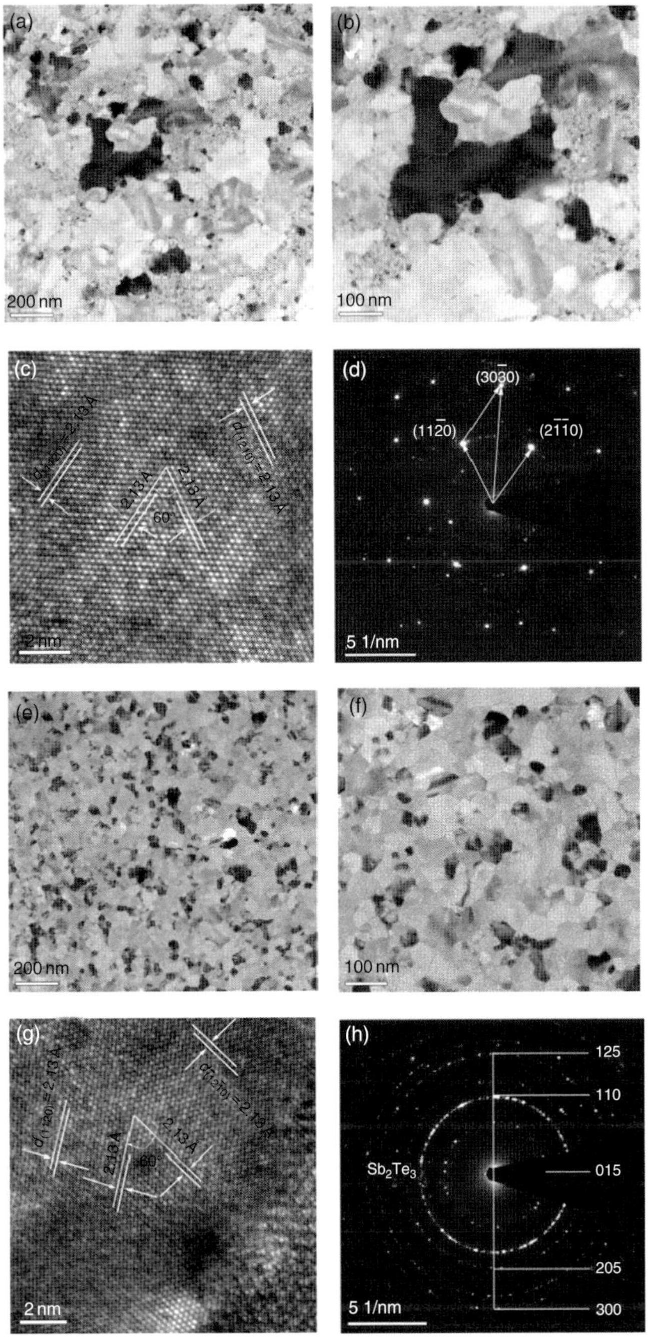

Figure 14.13 (a,b) TEM, (c) HRTEM, and (d) SAED results of crystalline Sb_2Te_3. (e,f) TEM, (g) HRTEM, and (h) SAED results of crystalline Ti-doped Sb_2Te_3 films. (Zhu *et al.* [15]. Reproduced with permission of American Institute of Physics.)

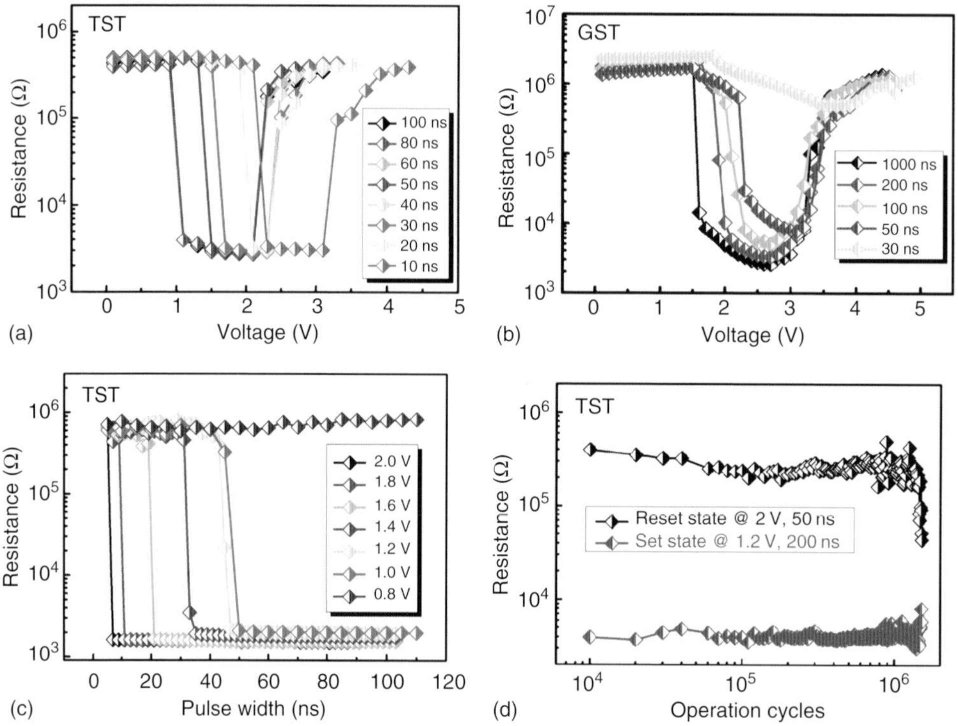

Figure 14.14 *R–V* characteristics of PCRAM cell with (a) TST and (b) GST films; (c) set operation as a function of applied pulse widths at different amplitudes; and (d) endurance of TST-based PCRAM cells.

pulse voltage, the shorter the set time. The set time decreases to only 6 ns as the voltage larger than 2.0 V. The reason for such extremely high speed has arisen from the growth-dominated crystallization mechanism of Sb_2Te_3 matrix and the small grain size for TST phase change materials. The operation speed of PCRAM using TST film is much faster than that of Flash (μs) and even can satisfy the high demand of access times (<10 ns) for DRAM.

Figure 14.14d presents the endurance for TST-based PCRAM cells. Repeated set and reset are maintained more than 1×10^6 cycles. The uniform structure of TST is responsible for the high operation cycles. The operation pulse amplitudes and widths were set to only 1.2 V/200 ns for set operation and 2 V/50 ns for reset operation, respectively. Before failure, the reset and set resistance values during cyclic operation are stable with a fluctuation in only half order of magnitude. Two feasible reasons would be explained as follows: The first point, as mentioned earlier, is that the resistances of TST-based PCRAM cells are less sensitive to the pulse amplitude. The second point is that only one transition happens in TST film from amorphous phase to crystalline phase.

14.4 Summary

PCRAM has drawn much attention as one of the most promising candidates for the next-generation nonvolatile memories. However, there are still several issues with regard to phase change materials that need to be solved for PCRAM applications. For example, the data retention should be improved to meet the needs of higher-temperature applications. The reset current should be lowered to reduce the power consumption, especially for high-density storage. In order to solve these issues, modifying conventional GST phase change material or developing high-performance novel phase change material is one of the important solutions.

In this chapter, GST phase change material, N- and C-doped GST phase change material, and novel TST phase change materials we recently studied are briefly introduced. The C-doped GST phase change material is considered as one of the most feasible schemes to improve the properties of GST. C doping has a beneficial effect on the thermal stability of the amorphous phase. It is indicated that carbon dopants are effective in increasing the stress-rigid character of the amorphous GST state. The set and reset voltages of the PCRAM cells are significantly reduced by carbon incorporation. The endurance property of approximately 2.1×10^4 cycles with sufficient resistance contrast is demonstrated, indicating that the GSTC alloy with moderate C content is a promising candidate for the high-density PCRAM application. Apart from C-doped GST phase change material, Ti-doped Sb-Te alloys are promising new phase change materials. Ti-doped Sb_2Te_3 alloy has larger resistance contrast and crystallization temperature is greatly increased. Ti-doped Sb_2Te_3 film-based PCRAM cells have fast operation speed with low power consumption, and more than 10^6 cycles were obtained with stable reset and set resistances. More importantly, Ti-doped Sb_2Te_3 film has uniform crystalline phase with small grain size, which benefits for operation speed and reducing power consumption.

Acknowledgments

The authors acknowledge the support of National Natural Science Foundation of China (61076121), National Key Basic Research Program of China (Nos 2011CBA00607 and 2013CBA01902), Strategic Priority Research Program of Chinese Academy of Sciences (XDA09020402), and National Integrate Circuit Research Program of China (No. 2009ZX02023-003).

References

1. Ovshinsky, S.R. (1968) *Phys. Rev. Lett.*, **21**, 1450.
2. Yamada, N., Ohno, E., Nishiuchi, K., Akahira, N., and Takao, M. (1991) *J. Appl. Phys.*, **69**, 2849.
3. Lai, S. (2003) Current status of the phase change memory and its future. Technical Digest – International Electron Devices Meeting, pp. 255–258.

4. Wuttig, M. and Yamada, N. (2007) *Nat. Mater.*, **6**, 824.
5. Yamada, N. and Matsunaga, T. (2000) *J. Appl. Phys.*, **88**, 7020.
6. Liu, B., Song, Z.T., Zhang, T., Feng, S.L., and Chen, B. (2005) *Appl. Surf. Sci.*, **242**, 62.
7. Friedrich, I., Weidenhof, V., Njoroge, W., Franz, P., and Wuttig, M. (2000) *J. Appl. Phys.*, **87**, 4130.
8. Jeong, T.H., Seo, H., Lee, K.L. *et al.* (2001) *Jpn. J. Appl. Phys.*, **40**, 1609.
9. Ebina, A., Hirasaka, M., and Nakatani, K. (1999) *J. Vac. Sci. Technol., A*, **17**, 3463.
10. Horii, H., Yi, J.H., and Park, J.H. (2003) A novel cell technology using N-doped GeSbTe films for phase change RAM. Symposium on VLSI Technology Digest of Technical Papers, p. 177.
11. Cheng, L.M., Wu, L.C., Song, Z.T., Rao, F., Peng, C., Yao, D.N., and Liu, B. (2012) *Mater. Lett.*, **71**, 98.
12. Zhou, X., Wu, L., Song, Z., Rao, F., Zhu, M., Peng, C., Song, S., Liu, B., and Feng, S. (2012) *Appl. Phys. Lett.*, **101**, 142104.
13. Zhou, X.L., Xia, M.J., Rao, F., Wu, L.C., Li, X.B., Song, Z.T., Feng, S.L., Sun, H.B., and Zhang, S.B. (2014) *ACS Appl. Mater. Interfaces*, **6**, 14207.
14. Zhu, M., Wu, L., Song, Z., Rao, F., Li, X., Peng, C., Zhou, X., Ren, K., Yao, D., Liu, B., and Feng, S. (2012) *Appl. Phys. Lett.*, **100**, 122101.
15. Zhu, M., Wu, L., Rao, F., Song, Z., Ren, K., Ji, X., Song, S., Yao, D., and Feng, S. (2014) *Appl. Phys. Lett.*, **104**, 053119.

15
Nanomaterials and Devices on Flexible Substrates

Hulin Zhang

15.1
Introduction

Nanomaterials on flexible substrates have attracted plenty of attention due to their wonderful and excellent physical, electrical, and chemical properties. Based on the nanomaterials on flexible substrates, nanodevices have been widely utilized in various applications. Recently, the concept of flexible electronics has been developed in earnest for little more than a decade in order to enable new applications that were impossible using conventional planar technology. In the early stages of development, organic transistors were fabricated on pliable plastic or thin glass substrates, primarily for electronic paper display devices [1, 2], but the inherently poor electrical properties of the polymeric materials, compared with those of high-quality inorganic semiconductor materials, were suggested unsuitable for other applications. In addition, due to the use of elastomeric substrates, the limitation imposed by low-temperature processing has made it difficult to deposit highly controlled active layers on polymer substrates, leading to perishing performance of the constructed electronic devices. However, with the development of novel materials such as conducting polymers [3, 4], carbon-based structures such as metallic and ceramic composites [5–8], and nanomaterials [5–10], together with alternative low-cost, convenient, and facile deposition techniques such as direct synthesis, transfer printing [11, 12], and inkjet printing [13–15], high-quality electronic structures encompassing diverse inorganic, organic, and composite materials can be formed on elastomeric flexible substrates, hence opening up a new electronic field, named as flexible electronics.

In order to realize the applications of flexible electronics, the significant focus is the development of adaptive flexible systems that can support high carrier mobility and good overall electrical performance as well as excellent mechanical and environmental stability. The study of nanomaterials that can be used as active layers or substrates to address the constraints of mechanically inert systems will still be continuous in parallel with the defining mechanics and underlying physics of these intrinsically heterogeneous low-dimensional structures. Certainly, there

Advanced Nano Deposition Methods, First Edition. Edited by Yuan Lin and Xin Chen.

are also several research papers on the nanomaterials and mechanics composed into flexible electronics [16–18]. Not surprisingly, important progress has been achieved both scientifically and industrially, and it is implied that within a few years, the first generation of commercial flexible electronic devices will be put forward. The initial devices will be likely for display applications.

In recent years, the research range has been extended to interdisciplinary areas, including biotechnology and energy technology. The tendency has been pushed toward exploiting and developing a simple and facile method of utilizing flexible systems to realize functions that are difficult or even impossible with rigid systems. By combining rigid materials with elastomeric supports, the invention of novel devices and hybrid structures are possible. The electronics can be deformed into curvilinear, cracked shapes with the electrical performance maintaining well, as well as device reliability, and integration that resemble those of high-quality wafer-based systems.

In this chapter, cutting-edge developments that utilize nanoscale thin films and low-dimensional nanostructures on compliant substrates to exhibit innovative functions were discussed. We try to give a brief understanding on the fundamentals of such flexible systems. Furthermore, insights into future routes and strategies to commercialization, new opportunities, and following challenges for the realization of flexible integrated systems will be proposed.

15.2 Nanomaterials on Flexible Substrates

15.2.1 Nanomaterials Synthesized Directly on Flexible Substrates

TiO_2 nanofilm is an important semiconductor material. Usually, the TiO_2 nanostructures were synthesized at a high temperature. Dr. Miao demonstrated TiO_2 nanofilms that were deposited on polyester fabrics by the reactive magnetron sputtering technology [19]. Prior to deposition, the fabrics were cleaned by ultrasonic and then washed in acetone solution. Following drying treatment in nitrogen, the fabrics were tailored into 6 cm × 6 cm pieces and conditioned in accordance with ASTM D1776-08 before sputtering and measurements. In order to enhance the uniformity of the multilayer films, a sample holder was employed, rotating at a speed of 100 rpm. SEM images of TiO_2-coated polyester fabrics are illustrated in Figure 15.1 [19]. As presented in Figure 15.1, the unique woven structure of blank polyester is exhibited clearly and a smooth surface can be observed. The basic textile structure of the coated fabrics is not changed after deposition. The porous TiO_2 films can be observed on the surface of the fibers and the average size of the holes reduces with the increasing sputtering time. These results indicate that, by adjusting the depositing time, the desired hole size could be obtained feasibly. The amorphous TiO_2 nanofilms might be caused by the lack of heating treatment during the deposition process, as well as

Figure 15.1 SEM images of blank and TiO_2-coated polyester fabric: (a) coating time (90 min) and (b) coating time (180 min). (Miao *et al.* [19]. Reproduced with permission of Elsevier.)

by the discontinuous substrates of the polyester fabrics. The deposition process at a relatively low temperature is suitable for common flexible textile substrates vulnerable to high-temperature destruction. As a result, flexible amorphous TiO_2 thin films can be well coated on the textile substrates by appropriate methods.

Co_3O_4 is an excellent electrode material and widely used in battery and capacitor. The flexibility of electrode materials is of significance in the practical applications. Very recently, a method of Co_3O_4 nanostructures grown on the flexible substrate was reported [20]. Typically, the Co_3O_4 nanoplate arrays were grown on flexible conductive fabric substrates via a simple electrochemical deposition approach, followed by thermal treatment. Prior to the electrochemical deposition, a commercially available Ni-coated polyethylene terephthalate (PET) fiber woven conductive fabric substrate was chosen. Then, the pieces were cleaned by rinsing with acetone, ethanol, and deionized (DI) water, respectively. Meanwhile, the growth solution was prepared by mixing the equimolar concentrations of both $Co(NO_3)_2$ and $C_6H_{12}N_4$ with 800 ml DI water at room temperature. Hence, the growth solution illustrated a pink color in the beaker. After the growth solution was heated at 80 °C, the flexible conductive fabric was carefully immersed into the growth solution with the counter electrode by maintaining a distance of 1 cm. Therefore, a simple two-electrode system was constructed with a working electrode and a platinum (Pt) mesh as the counter electrode [21]. During the electrochemical deposition process, the growth temperature was kept stable and different external voltages were applied from 0.5 to 3.0 V for optimizing growth condition. After 30 min, the sample was carefully picked up from the experimental solution, washed with DI water, and dried by ultrapure nitrogen gas at room temperature. Figure 15.2 [20] shows

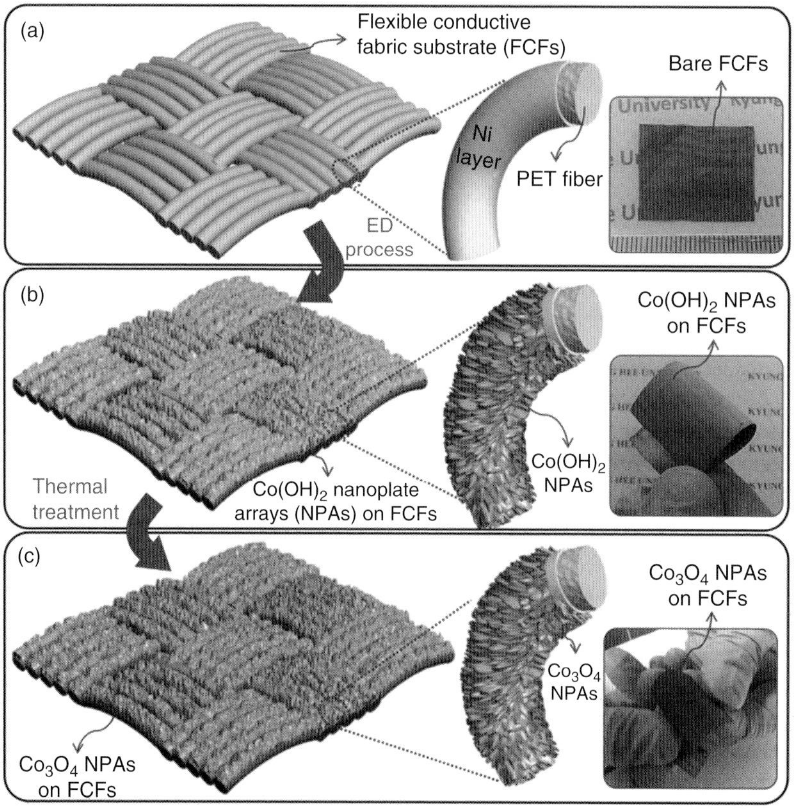

Figure 15.2 Schematic diagram of the Co_3O_4 NPAs on FCFs and the corresponding photographic images. (Nagaraju *et al.* [20]. Reproduced with permission of Elsevier.)

the schematic diagram of the fabrication process of $Co(OH)_2$ nanoplate arrays on flexible conductive fabrics. As shown in Figure 15.2a, the flexible conductive fabrics are electrically conductive and highly flexible because it is composed of woven Ni/PET microfibers. As a result, the flexible conductive fabrics can be suitable for growing the $Co(OH)_2$ nanoplate arrays via the electrochemical deposition process. The $Co(OH)_2$ nanoplate arrays can be crystallized and grown on the overall surface of Ni/PET fibers via the electrochemical deposition routine [22]. When the external voltage was applied between the flexible conductive fabrics and counter electrode in the growth solution, nitrate ions from $Co(NO_3)_2$ were electrochemically reduced on the surface of flexible conductive fabrics to release hydroxyl ions. These induced hydroxyl ions at the flexible conductive fabrics raised the local pH, subsequently forming $Co(OH)_2$ by reacting with Co^{+2} ions [23]. Herein, the morphology of $Co(OH)_2$ nanostructures can be tuned by varying the applied external voltage, because the generation of hydroxyl ions is mainly affected by the induced electrons in the growth solution. The photographic image of the bent sample indicates the excellent flexibility, implying a pink-colored flexible conductive fabric. Through the necessary thermal treatment of the grown $Co(OH)_2$ nanoplate arrays on flexible conductive fabrics at the ambient condition, the $Co(OH)_2$ nanoplate arrays were gradually converted into Co_3O_4 nanoplate arrays by dehydration. As shown in Figure 15.2, the Co_3O_4 nanoplate arrays on flexible conductive fabrics were evenly distributed without any voids or gaps, leading to the uniformly dark color. Moreover, the feasibility characterization of electrochemical supercapacitors, cyclic voltammetry, and galvanic charge discharge measurements were carried out systematically. Herein, the Co_3O_4 nanoplate arrays on flexible conductive fabrics were directly utilized as a binder-free electrode, showing a specific capacitance of 145.6 F/g at a current density of 1 A/g with 89.15% capacitance retention during 1000 cycles at a current density of 3 A/g in 1 M KOH electrolyte. This facile growth of metal oxide nanostructures on conductive fabric substrates provides an alternative approach to fabricate various flexible, wearable, and foldable energy storage units.

Owing to the mutual advantages of wide direct band gap (3.37 eV), high exciton binding energy (60 meV), and strong room temperature emission characteristics, numerous deep studies were focused on synthesizing a variety of ZnO nanostructures. It was demonstrated that a successful fabrication of vertically aligned ZnO nanorods for the first time can be achieved on Teflon sheet as a novel flexible substrate. The growth of ZnO nanorods on ZnO seed layer-coated Teflon (PTFE) substrates contains three procedures. Typically, the PTFE substrates were immersed and washed in isopropyl alcohol solution in ultrasonic bath at a temperature of about 60 °C for several minutes. Then, ZnO seed layer was grown on the cleaned PTFE substrates using radio-frequency magnetron sputtering system at 200 °C with an argon pressure and a sputtering power of 5.5 mTorr and 150 W, respectively, leading to a 100-nm-thick ZnO seed layer after a further post-annealing in a furnace for 1 h. At last, the ZnO nanorods were located on the ZnO seed layer/PTFE substrates using a low-temperature chemical bath deposition method. In the chemical bath deposition growth, 0.05 M zinc nitrate ($Zn(NO_3)_2$)

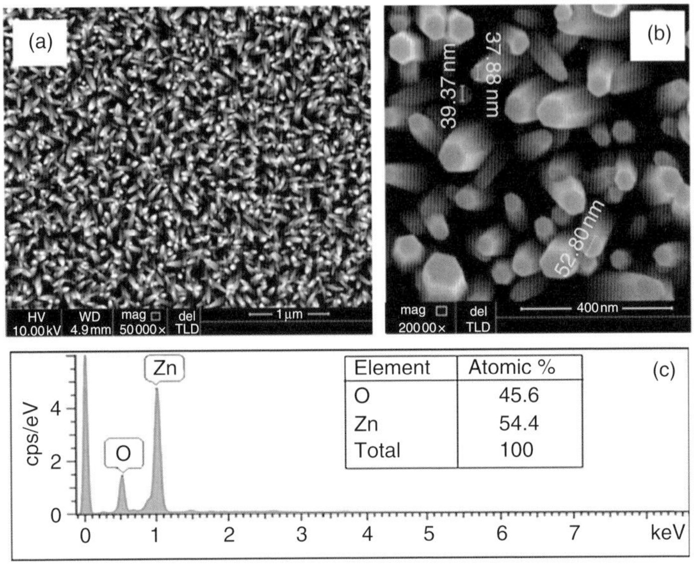

Figure 15.3 (a,b) FESEM images and (c) EDX spectrum of the ZnO nanorods grown on ZnO seed layer/PTFE substrates. (Farhat *et al.* [23]. Reproduced with permission of Springer.)

and an equal molar concentration of hexamethylenetetramine ($C_6H_{12}N_4$) were dissolved separately in DI water at a temperature of 80 °C. Then, the two solutions were mixed in a beaker, and the substrates were placed vertically inside the beaker before being put inside a furnace at 95 °C for 3 h. Finally, the samples were washed with DI water. Figure 15.3a,b [23] shows the typical FESEM images of the ZnO nanorod arrays grown on the ZnO seed layer/PTFE substrates. High-density ZnO nanorods with diameters of about 40 nm were successfully synthesized. No delamination was observed, which highlights the importance of the ZnO seed layer. Figure 15.3c shows the EDX spectrum of the grown ZnO nanorods, suggesting peaks for both Zn and O atoms with approximately similar atomic percentage (Zn/O = 54.4 : 45.6), which may imply the preparation of vacancy-free ZnO nanorods [24]. Meanwhile, it is noted that the obtained ZnO nanorods are nanograined and contain well-developed free surfaces as well as grain boundaries. Some previous reports have shown that the grain boundaries are the intrinsic origin of ferromagnetism in ZnO nanostructures. The small diameter and small grain size of the ZnO nanorods are considered to index to new benchmarks that may be useful for photovoltaic, photocatalytic, and spintronic applications [25–27].

15.2.2 Nanomaterials Transferred on Flexible Substrates

Generally, nanomaterials grown on flexible polymers are relatively inefficient via the traditional methods, such as a low-temperature hydrothermal method,

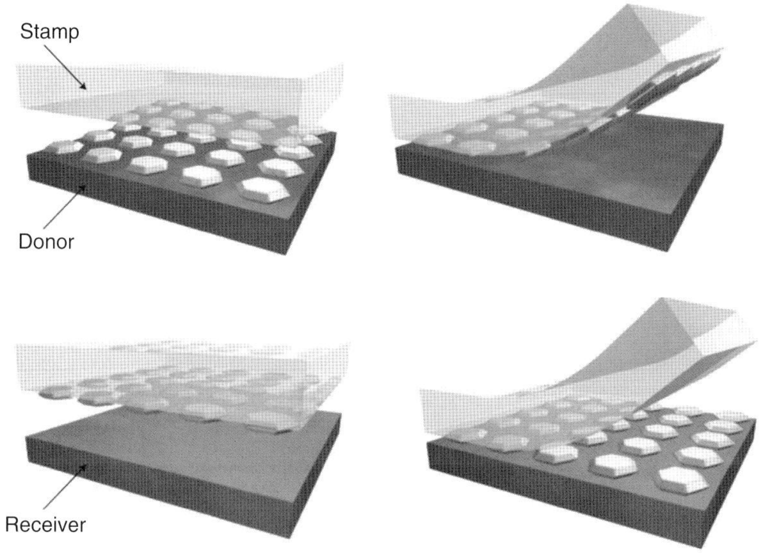

Figure 15.4 Schematic illustration of the generic process flow for transfer printing solid objects. (Meitl *et al.* [28]. Reproduced with permission of Nature Publishing Group.)

physical vapor deposition (PVD) method, and chemical vapor deposition (CVD) method. In recent decades, a novel method of depositing nanomaterials on flexible substrates was introduced by some research groups and their coworkers [16, 28]. It is called as *transfer printing technique.*

Figure 15.4 [28] illustrates schematically the process of transfer printing, beginning with the preparation of a donor substrate that supports fully formed, organized arrays of solid objects (e.g., devices, materials elements, biological entities). The donor substrate can be prepared using top-down fabrication, bottom-up growth, self-assembly, or other means. Contacting a soft elastomeric stamp against these solid objects leads to conformal contact, driven by generalized adhesion forces that are typically dominated by van der Waals interactions. The adhesion between the solids and the stamp is rate sensitive and controllable because of the viscoelastic behavior of the elastomer. Separating the stamp away from the donor substrate with a sufficiently high peel velocity of typically 10 cm/s or faster for the systems presented here leads to conglutination, which is forceful enough to adhere preferentially the solids to the surface of the stamp, resulting in removing them away from the substrate. The stamp, now inked with these objects, is brought into contact with a receiving device substrate. Getting rid of the stamp with sufficiently low peel velocity induces the objects to adhere preferentially to the device substrate and separate from the stamp. The transfer can be performed uniformly with a flat stamp or with a structured element that contacts and transfers only some set of objects from the donor substrate.

It has been verified by rolling experiments to yield quantitative information based on the commercially available polydimethylsiloxane (PDMS) rubber

(Sylgard 184, Dow Corning). In the typical experiments, a steel cylinder rolling down an inclined slab of PDMS rubber reaches a terminal velocity, where gravity works to separate the steel from the PDMS at the trailing edge of the contact region. The loss of gravitational potential is considered as the adhesion energy hysteresis at the measured rolling or separation speed v. The difference between the work requires to separate the steel from the PDMS and the energy evolves at the advancing contact edge. The energy related with the advancing contact area at the front of the rolling cylinder is quite low, and as a result, the adhesion energy hysteresis is taken as the separation energy, or the energy release rate, G.

A similar trend is qualitatively followed by observing pick-up and printing efficiencies in transfer printing experiments. When a PDMS stamp separates slowly from a substrate that supports the microstructures, the separation energy G for the elastomer–microstructure interface is smaller than its counterpart G for the interface between the microstructures and the substrate. Therefore, the elastomer–microstructure interface can be broken more easily than the microstructure–substrate interface. But the separation energy G depends strongly on the delamination speed v (which is due to the viscous behavior of PDMS). It actually corresponds to an increasing function of v, as shown in Figure 15.5 [28]. In contrast, the separation energy for the microstructure–substrate interface $G_{\text{substrate}}$ is totally independent of rate. As the delamination speed increases, G_{PDMS} raises relative to $G_{\text{substrate}}$ until the elastomer–microstructure interface becomes strong enough to laniate the microstructure–substrate interface. Precise criteria for determining which interface fails are beyond the scope of this text, but they depend on the geometry of the microstructures and can be constructed by energy-based arguments similar to those employed to compare competing fracture modes in layered nanomaterials. The core concept of the transfer printing criteria is the strongly rate-dependent separation energy of solid surface from an elastomer. As an example of a type of structure, the transfer elements can retrieve from a donor, Figure 15.5 shows the freestanding silicon beams micromachine from a silicon-on-insulator (SOI) wafer, connected to an unetched part of the wafer that anchors their ends. Slow removal and application of a stamp from these robust freestanding structures leave them intact. However, fast removal often results in fractures of them cleanly at their ends, leaving them adhered to the stamp. The specific stamp, thus "inked" with the silicon beams, can transfer those beams by contact to and slow removal from a receiving substrate.

15.3 Devices on Flexible Substrates

Devices on flexible substrates have drawn much attention since they first appeared in the 1960s. The concept of flexible electronics has been developed in earnest for little more than a decade in order to enable new and wide applications.

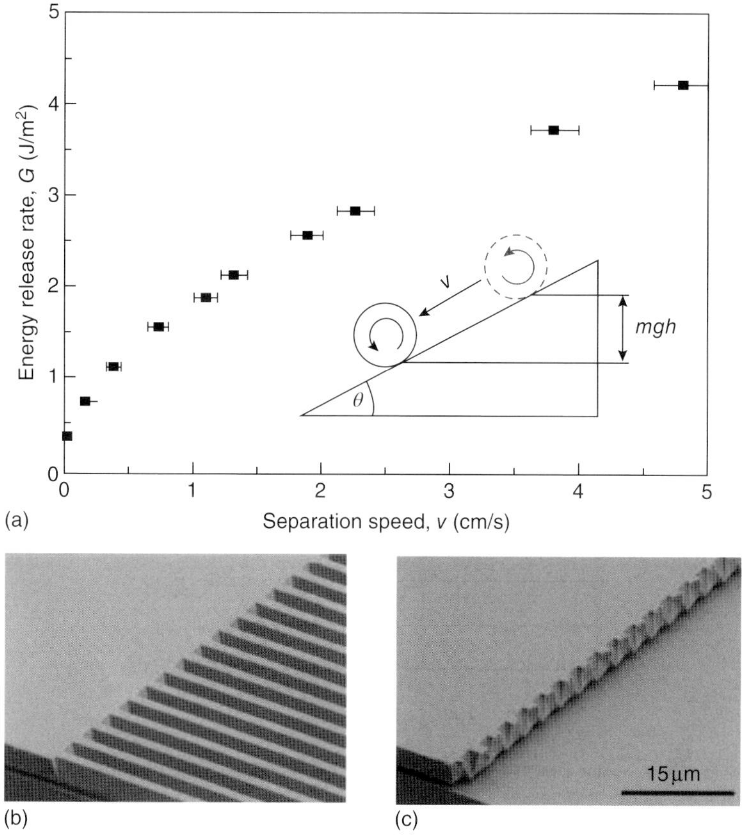

Figure 15.5 Rate dependence of stamp adhesion. (a) The rate-dependent adhesion causes a steel cylinder rolling down an inclined slab of PDMS to assume a constant speed dependent on the angle of inclination, allowing calculation of the energy release rate, *G*, from the loss of gravitational potential energy. (b,c) Slow removal of a PDMS stamp from the silicon beams leaves them intact, but the much higher adhesive force associated with fast stamp removal cleanly breaks the structures from their anchor (c). (Meitl *et al.* [28]. Reproduced with permission of Nature Publishing Group.)

15.3.1
Printing Electronics on Flexible Substrates

Printing electronics on flexible substrates are significantly interesting due to the low-cost fabrication and possibility of obtaining multifunctional electronics on large areas. In recent years, plenty of printing technologies have been introduced to pattern a wide range of electronic materials on diverse substrates. As further expansion of printing technologies is expected in the future for sensors and electronics, it is an opportune to review the common features, the complementarities, and the challenges related with various printing technologies.

Printing technology is the most popular and matured technology for printing electronics because it has been practiced in electronics industries for quite some time to print metallic interconnects on printed circuit boards. It is faster and more versatile in comparison to other techniques, as it has advantages of simplicity, affordability, speed, and adaptability to the fabrication process. The results from screen printing can be reproduced by repeating a few steps, and an optimum operating envelope can be exploited quickly and effectively [29–31].

The accessibility and feasibility of screen printing for flexible electronics have been proved in depth through several printed sensors and other electronic devices and circuits. For example, all screen-printed thin film transistors (TFTs) have been demonstrated in [32–34]. Screen printing was declared to be utilized for the first time to develop OLEDs by the investigation of the process and solution parameters, such as the viscosity of the solution and mesh count of the screen [35]. Multilayer high-density flexible electronic circuits, connected through micro holes with embedded passive and optical devices, have been realized by using advanced screen printing processes [36]. Screen printing is also used for patterning to develop shadow masks for the fabrication of organic TFT. The flexible three-axis tactile sensor shown in Figure 15.6 [37] was fabricated by utilizing carbon nanotube (CNT)–polymer composite. The composite was directly patterned onto a flexible substrate using a screen printing technique to establish a sensor with four sensing cells. All structures of metal plates, interconnects, and sensitive materials were deposited by using screen printing technology [38–40]. Screen printing of cobalt hydroxide has been reported for constructing supercapacitors [41]. The spectral domain optical coherence tomography is used in a simulated process to monitor the structural properties of moving screen-printed interdigitated electrodes [42]. Unlike many other manufacturing techniques, the screen printing does not require high capital investment. Accompanied by some supplemental methods such as inkjet technology, vapor deposition, and laser processing, screen printing is employed in most of the production lines of printed electronics. Using the supplemental technologies often results in cost reduction [43]. Despite these attractions, screen printing also poses a few challenges for developing all layers of a flexible device. These include high wet thickness of the wet film, exposure of the ink to atmosphere, and the dry out of the ink on the mask that deteriorates the mask designs of the screen [44]. Nevertheless, the screen printing techniques have many merits, such as high definition and high precision of multilayer structures, compared to other deposition techniques for large-scale production.

On the other hand, printed sensors and electronics have attracted greater interest as printing enables low-cost fabrication. The strong interest of the researchers in their quest to fulfill the promise of large-area electronics on flexible substrates through cost-effective printing technologies can be found from the raised number of the scientific papers as well as from the experimental demonstrations. Various technologies have been applied so far for the printed devices, such as TFTs, LEDs, sensors, displays, solar cells, RFID tags, printed batteries, energy harvesters, capacitors, and so on.

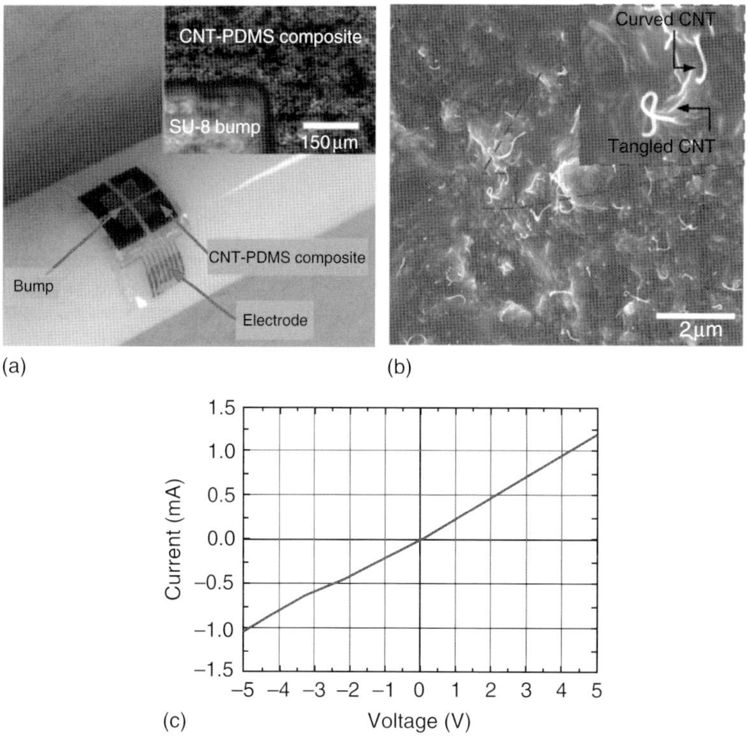

Figure 15.6 (a) Photograph of the fabricated sensor attached to a curved surface and the optical microscope image (inset) of the SU-8 bump on the CNT–PDMS composite. (b) The HRSEM image of CNT–PDMS composite. CNTs are dispersed uniformly within the PDMS matrix, and individual CNTs are tangled or curved. (c) The current–voltage (I–V) characteristics of the CNT–PDMS composite. (Pyo *et al.* [37]. Reproduced with permission of the Institute of Physics Publishing Group.)

Electrospun ultrathin fibers are suitable for constructing flexible and stretchable devices owing to a variety of excellent properties, such as large surface area, high length-to-diameter ratio, flexible surface functionality, tunable surface morphologies, and superior mechanical performance, which have aroused much attention recently. Flexible/stretchable electronic devices are emerging as a new technological achievement, since they can be reversibly bent or stretched while maintaining functionality. To power these devices, rechargeable and stretchable energy storage devices become a necessity. Among supercapacitors that store energy in double layers on the electrode/electrolyte interface with opposite charges for providing a high power density and modest energy density in complementing the battery system [45–47], the flexible/stretchable supercapacitor has aroused broad interest due to its potential applications in portable electronic devices, hybrid electric vehicles, and medical devices [48–50]. Electrospinning can do well in this field due to its ability to fabricate micro-/nanofibers with large interconnected voids between the fibers that create abundant porous structures for the mobile ions in the electrolyte to move freely through the elastomeric separator during dynamic stretching/releasing or bending/releasing [51–53]. Li and coworkers have introduced a dynamically stretchable supercapacitor using elastomeric electrospun polyurethane (PU) membrane as the separator, as well as buckled single-walled CNT macrofilms as the electrodes, and others in propylene carbonate as organic electrolyte, as exhibited in Figure 15.7 [54]. Here, the polyurethane membrane provides low equivalent series resistance, thus allowing a high charge and discharge rate. It can be seen that the specific capacitance of the supercapacitor is enhanced at the maximum strain (31.5% strain) applied over that of the original state at no strain. In addition, cyclic voltammograms (CVs) of the supercapacitors at a scan rate of 100 mV/s at static 0% strain and static 31.5% strain indicate a perfectly symmetrical rectangular shape, and it is found that this device can be reversibly charged and discharged under the dynamic stretching

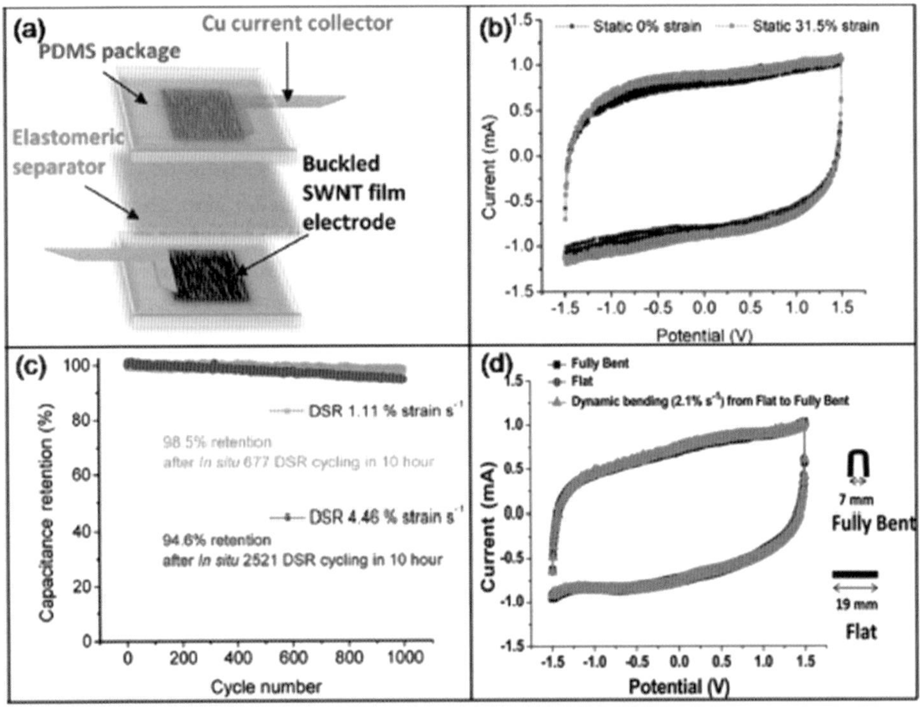

Figure 15.7 (a) Schematic of the main components of the dynamically stretchable supercapacitor. (b) Cyclic voltammogram (CV) curves under static 0% strain and 31.5% strain at a scan rate of 100 mV/s. The device gives approximately 4% larger specific capacitance than that of the cell. (c) With applied dynamic stretching/releasing cycling at a high strain rate of 1.11% strain per second and 4.46% strain per second, the stretchable supercapacitor shows good cycling stability. The current density is 10 A/g. (d) The CV curves of the stretchable supercapacitor at the flat state (light gray), fully bent state, and dynamic bending at a strain rate of 2.1% per second. (Li *et al.* [54]. Reproduced with permission of American Chemical Society.)

and releasing at an extremely high strain rate. At a high current density of 10 A/g, the calculated specific capacitance is 44 F/g, and the total capacitance is 13.2 mF, while the areal capacitance per unit area is in the range between 2.64 (0% strain applied) and 3.47 mF/cm^2 (31.5% strain applied).

In flexible electronic devices, such as supercapacitors and sensors, the electrode is considered as one of the key components with favorable mechanical strength and large capacitance, as well as high energy density. Very recently, Cui's group has invented a facile way to fabricate a transparent electrode based on a metal nanotrough network with the electrospun nanofibers serving as the template [55]. First, continuous polymer nanofibrous mats as random networks or oriented arrays were synthesized by means of electrospinning, as depicted in Figure 15.8 [55]. Then, the polymer nanofibrous networks were covered with a thin layer of material using standard thin film deposition techniques, such as thermal evaporation, e-beam evaporation, or magnetron sputtering, allowing a rich variety of functional materials such as silicon, ITO, or metals such as Au, Ag, Cu, Pt, Al, Cr, and Ni and other alloys to be prepared with a bulk-like electrical conductivity retained stably. Due to the directionality of these deposition processes, the polymer fibers were preferentially coated on one side, thus forming nanotroughs

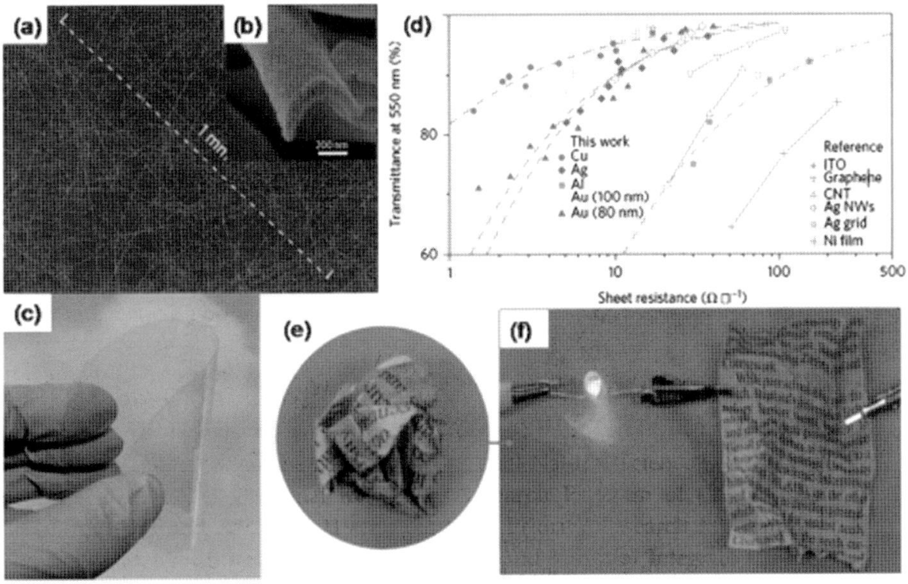

Figure 15.8 (a) Top-view SEM image of gold nanotrough networks. (b) SEM image of the cross section of a single gold nanotrough, revealing its concave shape. (c) Gold nanotrough networks transferred onto PET. (d) Sheet resistance optical transmission for copper, gold, silver, and aluminum nanotrough networks, described by percolation theory. The performances of device-grade ITO, carbon nanotubes, graphene, silver nanowires, silver grid, and nickel thin films are shown for comparison. (e,f) Photographs of "conducting paper," fabricated by transferring Au nanotrough networks onto paper. (Wu *et al.* [55]. Reproduced with permission of Nature Publishing Group.)

with a width of about 400 nm and a length larger than 1 mm with a concave cross section, as shown in Figure 15.8. The coated polymer fibers are mechanically flexible and robust and also can be transferred easily onto various substrates to interface with macroscopic-scale systems. Figure 15.8 illustrates the gold nanotrough networks transferred to a PET plastic. Finally, after transferring the fibers, the polymer templates can be removed by submerging the entire electrode in water or other organic solvents. The as-synthesized transparent electrodes have perfect sheet resistance (R_s) and optical transmittance (T), which are two key parameters that determine the applications of transparent electrodes.

15.3.2 Biointegrated Electronics on Flexible Substrates

Trying to enhance the conformality of such electronic systems on the target tissues is the focus of the recent work toward integrating electronics onto the biological structures by way of reducing the thickness of the substrate or using stretchable, flexible substrates.

Although a practical medical application that enables diagnosing and treating diseases has been achieved, biointegrated circuits can offer important opportunities for modern medical treatment. For example, cardiac arrhythmias can be properly monitored and treated if the irregular heart activities can be mapped in real time by the highly distributed sensors on the surface of the heart. Unfortunately, conventional tools by integrating sparse arrays of rigid electrodes are usually arranged on a cylindrical catheter and manipulated from place to place to record from discrete sites on the heart. Thus, iterative efforts with no capability of real-time monitoring are essential [56]. As such, the major limitations in the integration of electronics to biological surfaces have been the hard, planar surfaces of semiconductors that are invasive in the soft, curvilinear surfaces of biological tissues, along with the instability of the electronic/biological interface that deteriorates both the device and the tissue health. High-performance electronics built on conventional semiconductor wafers are not suitable for intimate, noninvasive integration with the surfaces of biological tissues. In addition, they are vulnerable when in contact with saline solutions and biological fluids, resulting in undesired leakage currents that may damage the contacting tissues. Kim and coworkers [57] overcame the difficulties of constructing an ultrathin electronic device by using a silk support substrate, which can be dissolved after its integration onto a biological tissue (in their research paper, a feline brain) using saline solution, and showed that the device forms an excellent conformal contact with the brain. Figure 15.9 [57] schematically depicts the integration process of electrode arrays designed for passive neural recording on a feline brain. Briefly, ultrathin films of polyimide were spin cast onto silicon wafers and then covered by sacrificial layers of PMMA as a support membrane for neural recording electrode arrays. They were then transfer printed onto a silk substrate, and an anisotropic conductive film (ACF) was bonded to one end of the electrode pads to provide an electrical connection to external data acquisition systems. At last, the

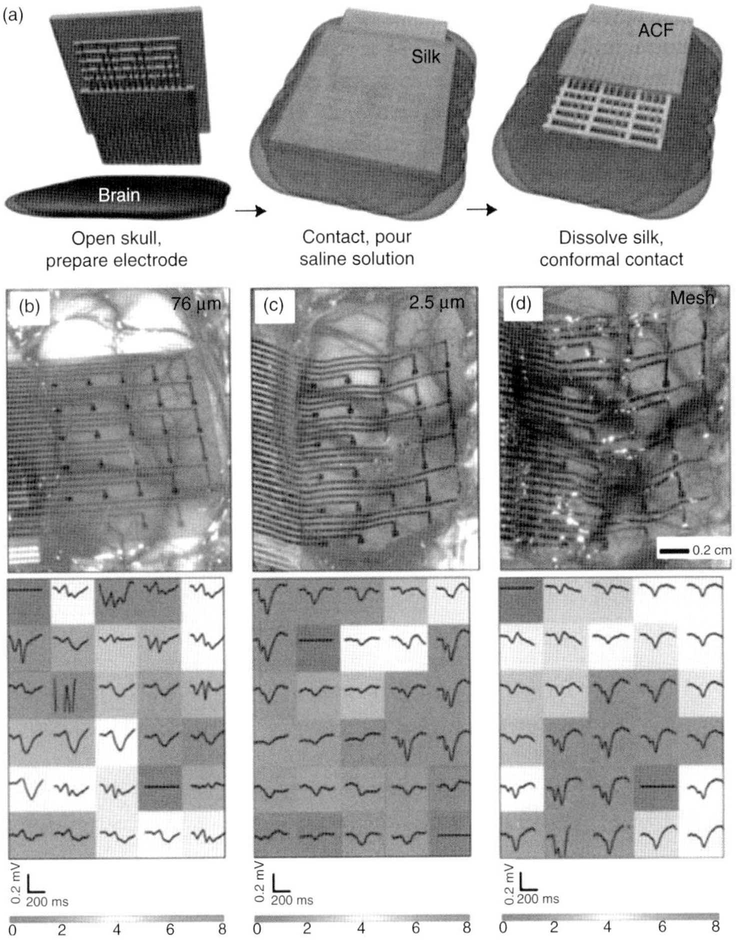

Figure 15.9 (a) Schematic illustration of the steps for implanting PI electrode arrays with a dissolvable silk support. (b–d) Photographic images and the corresponding average evoked response from each electrode implanted on a feline brain with (b) 76-μm, (c) 2.5-μm, and (d) 2.5-μm-mesh electrode arrays. (Kim *et al.* [57]. Reproduced with permission of Nature Publishing Group.)

device was brought into contact with the brain and the silk support dissolved by saline solution, leading to the full integration of the ultrathin device on the brain. The degree of conformality was tested using variable thicknesses of the polyimide membrane. Three different samples were selected and prepared using 76- and 2.5-μm-thick polyimide films and a mesh-structured 2.5-μm-thick polyimide film. It is shown clearly from the photographs that excellent conformality can be achieved when the polyimide support has a 2.5-μm-thick mesh-shaped structure. This is also consistent with the corresponding average evoked electrode response examinations shown in the photographs in Figure 15.9. Various colors show the ratio of the rms amplitude of each average electrode response in the 200 ms window, with the scale of the rms amplitude at the bottom of Figure 15.9. Nevertheless, weak signals with low outputs could be extracted from the electrodes fabricated on 76- and 2.5-μm-thick polyimide films, exhibited by the light gray to dark gray; reliable data were obtained using the ultrahigh conformal electrodes fabricated on mesh-structured 2.5-μm-thick polyimide film, represented by gray in the color map. From the demonstration of the passive electrode systems, it can be concluded that this technology provides a novel route toward intimate electronic integration in which a reliable biotic interface is achieved on living, moving biological structures. This is essential for implanted, high-resolution medical devices based on the flexible electronics.

15.4 Summary

Although the major research on flexible electronics for the past few decades has been focused on the discovery and development of new materials, fabrication techniques, and investigation of the properties of these materials in correlation with compliant substrates, most recent research focuses on methods to deliver novel functions. In this chapter, we reviewed recent efforts and advancements that successfully overcome or circumvent the constraints of conventional technology, covering the materials used, device design, implementation methods, working principles, and performance metrics.

In recent decades, much is understood with respect to the underlying operation mechanisms and capabilities of stretchable and flexible devices through comprehensive experimental and theoretical research. Nevertheless, some challenges in this field have been met and many practical problems have to be solved before the commercialization of fully integrated flexible systems is sustainable. Specifically, the following issues must be settled. First, the new support or substrate must be developed and can endure high levels of strain. In view of that, the sizes of the active elements on such substrates are in the range of nanometers; even a microcrack would severely influence device electronic performance. Second, a method to control changes in the materials and electronic properties in the deformed state is necessary. The desired excellent performance must be guaranteed under bent or strained states, and a steady and reliable output must be achieved with minimal variation. A huge deviation in the device performance suggests that it is vulnerable to stress, and repeated exertion of stress can result in device breakdown. The third is associated with the fabrication of identical devices over large areas. Since a flexible device can often be constructed via a bottom-up approach, the alignment of the nanomaterials must be handled well. This is extremely difficult. The nanomaterials prepared under the same experimental conditions may exhibit dissimilar property. Finally, the problem of residual stress must be addressed. Even if the most robust materials are subject to abrasion after thousands of bending and folding events, due to the accumulated external/internal stresses, the residual stress can affect the device performance, resulting in the reliability issues. Even so, the seemingly unlimited potentials of flexible electronics have drawn global attention. With the continuously increasing demand for smarter and subtler applications, opportunities in the area of flexible electronics using nanomaterials appear almost inexhaustible. Following the trends of flexible electronics, the manufacturing cost of plastic electronics is expected to be lower by the fast speed electronic processing at defined locations. The cost-effectiveness of material preparation technologies and employing them for flexible electronics will enable new various applications and dramatically change the electronics industry landscape.

Acknowledgments

This work is supported by the National Basic Research Program of China (973 Program) under Grant No. 2015CB351905, the National Natural Science Foundation of China (No. 61306015), and the Guangdong Innovative Research Team Program (No. 201001D0104713329).

References

1. Rogers, J.A., Bao, Z., Baldwin, K., Dodabalapur, A., Crone, B., Raju, V.R., Kuck, V., Katz, H., Amundson, K., Ewing, J., and Drzaic, P. (2001) *Proc. Natl. Acad. Sci. U.S.A.*, **98**, 4835.
2. Gelinck, G.H., Huitema, H.E.A., Veenendaal, E.V., Cantatore, E., Schrijnemakers, L., Van Der Putten, J.B.P.H., Geuns, T.C.T., Beenhakkers, M., Giesbers, J.B., Huisman, B.-H., Meijer, E.J., Benito, E.M., Touwslager, F.J., Marsman, A.W., Van Rens, B.J.E., and De Leeuw, D.M. (2004) *Nat. Mater.*, **3**, 106.
3. MacDiarmid, A.G. (2001) *Angew. Chem. Int. Ed.*, **40**, 2581.
4. Heeger, A.J. (2001) *Angew. Chem. Int. Ed.*, **40**, 2591.
5. Avouris, P., Chen, Z., and Perebeinos, V. (2007) *Nat. Nanotechnol.*, **2**, 605.
6. Hussain, F., Hojjati, M., Okamoto, M., and Gorga, R.E. (2006) *J. Compos. Mater.*, **40**, 1511.
7. Cho, J., Boccaccini, A.R., and Shaffer, M.S.P. (2009) *J. Mater. Sci.*, **44**, 1934.
8. Paul, D.R. and Robeson, L.M. (2008) *Polymer*, **49**, 3187.
9. Fan, Z., Ho, J.C., Takahashi, T., Yerushalmi, R., Takei, K., Ford, A.C., Chueh, Y.-L., and Javey, A. (2009) *Adv. Mater.*, **21**, 3730.
10. Cao, Q. and Rogers, J.A. (2009) *Adv. Mater.*, **21**, 29.
11. Quist, A.P., Pavlovic, E., and Oscarsson, S. (2005) *Anal. Bioanal. Chem.*, **381**, 591.
12. Rogers, J.A. and Nuzzo, R.G. (2006) *Mater. Today*, **9**, 32.
13. Eda, G., Fanchini, G., and Chhowalla, M. (2008) *Nat. Nanotechnol.*, **3**, 270.
14. Ko, S.H., Pan, H., Grigoropoulos, C.P., Luscombe, C.K., Frechet, J.M., and Poulikakos, D. (2007) *Nanotechnology*, **18**, 345202.
15. Menard, E., Lee, K.J., Khang, D.-Y., Nuzzo, R.G., and Rogers, J.A. (2004) *Appl. Phys. Lett.*, **84**, 5398.
16. Kim, D.-H. and Rogers, J.A. (2008) *Adv. Mater.*, **20**, 4887.
17. Rogers, J.A. and Huang, Y. (2009) *Proc. Natl. Acad. Sci. U.S.A.*, **106**, 10875.
18. Cordill, M.J. (2010) *J. Miner. Met. Mater. Soc.*, **62**, 9.
19. Miao, D., Hu, H., Li, A., Jiangn, S., and Shang, S. (2015) *Ceram. Int.*, **41**, 9177.
20. Nagaraju, G., Ko, Y.H., and Yu, J.S. (2015) *J. Power Sources*, **283**, 251.

21. Nagaraju, G., Ko, Y.H., and Yu, J.S. (2014) *CrystEngComm*, **16**, 11027.
22. Brownson, J.R.S. and Levy-Clement, C. (2008) *Phys. Status Solidi B*, **245**, 1785.
23. Farhat, O.F., Halim, M.M., Abdullah, M.J., Ali, M.K.M., Ahmed, N.M., and Allam, N.K. (2015) *Appl. Phys. A*, **119**, 1197.
24. Shen, G., Bando, Y., and Lee, C.-J. (2005) *J. Phys. Chem. B*, **109**, 10779.
25. Shabannia, R. and Abu Hassan, H. (2014) *Appl. Phys. A*, **114**, 579.
26. Bedia, E.L., Murakami, S., Kitade, T., and Kohjiya, S. (2001) *Polymer*, **42**, 7299.
27. Xu, S., Adiga, N., Ba, S., Dasgupta, T., Wu, C.F.J., and Wang, Z.L. (2009) *ACS Nano*, **3**, 1803.
28. Meitl, M.A., Zhu, Z.T., Kumar, V., Lee, K.J., Feng, X., Huang, Y.Y., Adesida, I., Nuzzo, R.G., and Rogers, J.A. (2006) *Nat. Mater.*, **5**, 33.
29. Khan, S., Lorenzelli, L., and Dahiya, R.S. (2015) *IEEE Sens. J.*, **15**, 6.
30. Kwack, Y.-J. and Choi, W.-S. (2011) *J. Korean Phys. Soc.*, **59**, 3410.
31. Chang, W.-Y., Fang, T.-H., Yeh, S.-H., and Lin, Y.-C. (2009) *Sensors*, **9**, 1188.
32. Clark, D.A. (2010) Major trends in gravure printed electronics. B.S. thesis. Dept. Graph. Commun., California Polytechnic State University, San Luis Obispo, CA.
33. Noguchi, Y., Sekitani, T., and Someya, T. (2007) *Appl. Phys. Lett.*, **91**, 133502.
34. El Houti El Jazairi, I., Thierry, T., and Moliton, J.-P. (2009) *Micro Nanosyst.*, **1**, 46.
35. Jabbour, G.E., Radspinner, R., and Peyghambarian, N. (2001) *IEEE J. Sel. Top. Quantum Electron.*, **7**, 769.
36. Turunen, R., Numakura, D., Nakayama, M., and Kawasaki, H. (2008) Screen printing process for high density flexible electronics. Proceedings of the IPC Printed Circuit Expo/APEX, 2008.
37. Pyo, S., Lee, J.I., Kim, M.O., Chung, T., Oh, Y., Lim, S.C., Park, J., and Kim, J. (2014) *J. Micromech. Microeng.*, **24**, 075012.
38. Khan, S., Lorenzelli, L., and Dahiya, R.S. (2014) Bendable piezoresistive sensors by screen printing MWCNT/PDMS composites on flexible substrates. Proceedings of the 10th Conference Ph.D. Research in Microelectronics Electronics (PRIME), 2014, pp. 1–4.
39. Khan, S., Tinku, S., Lorenzelli, L., and Dahiya, R.S. (2015) *IEEE Sens. J.*, **15**, 3004.
40. Khan, S., Lorenzelli, L., and Dahiya, R.S. (2014) Screen printed flexible pressure sensors skin. Proceedings of the 25th Annual SEMI Advanced Semiconductor Manufacturing Conference (ASMC), May 2014, pp. 219–224.
41. Ji, X., Hallam, P.M., Houssein, S.M., Kadara, R., Lang, L., and Banks, C.E. (2012) *RSC Adv.*, **2**, 1508.
42. Alarousu, E., AlSaggaf, A., and Jabbour, G.E. (2013) *Sci. Rep.*, **3**, 1562.
43. Cao, X., Chen, H.T., Gu, X.F., Liu, B.L., Wang, W.L., Cao, Y., Wu, F.Q., and Zhu, C.W. (2014) *ACS Nano*, **8** 12769.
44. Krebs, F.C., Jorgensen, M., Norrman, K., Hagemann, O., Alstrup, J., Nielsen, T.D., Fyenbo, J., Larsen, K., and Kristensen, J. (2009) *Sol. Energy Mater. Sol. Cells*, **93**, 422.
45. Conway, B.E. (1999) , Kluwer Academic/Plenum, New York.
46. Liu, J., Jiang, J., Cheng, C., Li, H., Zhang, J., Gong, H., and Fan, H.J. (2011) *Adv. Mater.*, **23**, 2076.
47. Yu, G., Hu, L., Vosgueritchian, M., Wang, H., Xie, X., McDonough, J.R., Cui, X., Cui, Y., and Bao, Z. (2011) *Nano Lett.*, **11**, 2905.
48. Wu, Q., Xu, Y., Yao, Z., Liu, A., and Shi, G. (2010) *ACS Nano*, **4**, 1963.
49. Horng, Y.Y., Lu, Y.C., Hsu, Y.K., Chen, C.C., Chen, L.C., and Chen, K.H. (2010) *J. Power Sources*, **195**, 4418.
50. El-Kady, M.F. and Kaner, R.B. (2013) *Nat. Commun.*, **4**, 1475.
51. Xue, M., Xie, Z., Zhang, L., Ma, X., Wu, X., Guo, Y., Song, W., Li, Z., and Cao, T. (2011) *Nanoscale*, **3**, 2703.
52. Presser, V., Zhang, L., Niu, J.J., McDonough, J., Perez, C., Fong, H., and Gogotsi, Y. (2011) *Adv. Energy Mater.*, **1**, 423.
53. Yan, X., Tai, Z., Chen, J., and Xue, Q. (2011) *Nanoscale*, **3**, 212.
54. Li, X., Gu, T., and Wei, B. (2012) *Nano Lett.*, **12**, 6366.
55. Wu, H., Kong, D., Ruan, Z.C., Hsu, P.C., Wang, S., Yu, Z., Carney, T.J.,

Hu, L., Fan, S., and Cui, Y. (2013) *Nat. Nanotechnol.*, **8**, 421.

56. Aliot, E.M., Stevenson, W.G., Almendral-Garrote, J.M., Bogun, F., Calkins, C.H., Delacretaz, E., Bella, P.D., Hindricks, G., Jaïs, P., Josephson, M.E., Kautzner, J., Kay, G.N., Kuck, K.H., Lerman, B.B., Marchlinski, F., Reddy, V., Schalij, M.J., Schilling, R., Soejima, K., and Wilber, D. (2009) *Europace*, **11**, 771.

57. Kim, D.-H., Viventi, J., Amsden, J.J., Xiao, J., Vigeland, L., Kim, Y.-S., Blanco, J.A., Panilaitis, B., Frechette, E.S., Contreras, D., Kaplan, D.L., Omenetto, F.G., Huang, Y., Hwang, K.-C., Zakin, M.R., Litt, B., and Rogers, J.A. (2010) *Nat. Mater.*, **9**, 511.

Index

Advanced Nano Deposition Methods, First Edition. Edited by Yuan Lin and Xin Chen.